Archaeoastronomy in Pre-Columbian America

Archaeoastronomy in Pre-Columbian America

Edited by Anthony F. Aveni

University of
TEXAS PRESS
Austin and London

For reasons of economy and speed this volume has been printed from camera-ready copy furnished by the Editor, who assumes full responsibility for its contents.

International Standard Book Number 0-292-70310-4
Library of Congress Catalog Card Number 74-17622
Copyright © 1975 by the University of Texas Press
All rights reserved
Printed in the United States of America

Contents

FOREWORD
    Ignacio Bernal.   ix

INTRODUCTION
    Anthony F. Aveni.   xiii

1   NATIVE ASTRONOMY IN MESOAMERICA
    Michael D. Coe.   3

2   THE ASTRONOMICAL RECORD IN CHACO CANYON, NEW MEXICO
    Ray A. Williamson, Howard J. Fisher, Abigail F. Williamson, and Clarion Cochran. . .   33

3   POSSIBLE ROCK ART RECORDS OF THE CRAB NEBULA SUPERNOVA IN THE WESTERN UNITED STATES
    John C. Brandt, Stephen P. Maran, Ray Williamson, Robert S. Harrington, Clarion Cochran, Muriel Kennedy, William J. Kennedy, and Von Del Chamberlain.   45

4   A THOUSAND YEARS OF THE PUEBLO SUN-MOON-STAR CALENDAR
    Florence Hawley Ellis.   59

5   EARLY NAVAJO ASTRONOMICAL PICTOGRAPHS IN CANYON DE CHELLY, NORTHEASTERN ARIZONA, U.S.A.
    Claude Britt, Jr.   89

6   STAR-PATTERNS IN GREAT BASIN PETROGLYPHS
    Dorothy Mayer.   109

7   ASTROARCHAEOLOGY: THE UNWRITTEN EVIDENCE
    Gerald S. Hawkins.   131

8   POSSIBLE ASTRONOMICAL ORIENTATIONS IN
    ANCIENT MESOAMERICA
    Anthony F. Aveni.   163

9   A SCHEME OF PROBABLE ASTRONOMICAL PROJEC-
    TIONS IN MESOAMERICAN ARCHITECTURE
    Horst Hartung.   191

10 THE NATURE AND NURTURE OF ARCHAEOASTRONOM-
    ICAL STUDIES
    Jonathan E. Reyman.   205

11 EFFIGY MOUNDS AND STELLAR REPRESENTATION:
    A COMPARISON OF OLD WORLD AND NEW WORLD
    ALIGNMENT SCHEMES
    Thaddeus M. Cowan.   217

12 THE USE OF ECLIPSE DATA TO DETERMINE THE
    MAYA CORRELATION NUMBER
    Nancy Kelly Owen.   237

13 THE SOLAR ECLIPSE WARNING TABLE IN THE
    DRESDEN CODEX
    Charles H. Smiley.   247

14 PLANETARY DATA ON CARACOL STELA 3
    David H. Kelley.   257

15 A NEW ASTRONOMICAL INTERPRETATION OF THE
    FOUR BALLCOURT PANELS AT TAJIN, MEXICO
    Carmen Cook de Leonard.   263

16 AN ASTRONOMICAL CALENDAR IN A PORTION OF
    THE MADRID CODEX
    Marion Popenoe Hatch.   283

17 OLMEC MOSAIC PENDANT
    Alexander Marshack.   341

Contents   vii

18 MESOAMERICAN ARCHAEOASTRONOMY SO FAR
        Elizabeth Chesley Baity.   379

NOTES ON CONTRIBUTORS.   387

REFERENCES.   395

INDEX.   423

Foreword

Astronomy has preoccupied and interested man since very ancient times. The reasons are obvious if we consider the mystery that must have been represented by the daily course of the sun, the frequent changes of the moon, and the much more complex changes of the planets and stars. For us who at least apparently understand the mystery and therefore have embarked on more profound problems, it constitutes a scientific theme. Astronomy played a part in the curiosity and the doubt that led to Greek civilization.

But the Mesoamerican had different motives for studying the celestial bodies. With our mania for classifying--a mania not in the least indigenous--we can think of two reasons for his interest, one religious and the other practical. The first derives from the mystery of the universe and of the gods, of cosmogony and prophecy; the second is the need to measure and compute time. Whence the appearance of complex calendars that we do not yet fully understand.

These calendars had various functions. The best known are the ritual ones to establish religious holidays and to predict man's future, and the agricultural ones to set times for carrying out farm labors. But I believe that in addition the interest in counting time had at least one other objective. For the Mesoamerican, history was a fundamental element. This was not an antiquarian's curiosity or a scientific attempt to better the future by profiting from past experience. It was much more a political and especially dynastic necessity, and thus it was recorded. In the case of lords, concern with genealogies went much further than simply snobbery, as clearly seen in the Mixtec codices and in Mexican history. I will mention only two examples taken from the latter.

When it was time to choose their first sovereign, the Mexica requested that he be from Culhuacan--a secondary power--and not from one of the more influential towns in the Valley of Mexico. After their triumph over Atzcapotzalco half a century later, the victors distributed royal titles among themselves. Izcoatl, the Mexica, instructed that he be called Culhuatecuhtli, Lord of the Culhuas. Why this insistence on being considered a Culhua? Because, although Culhuacan was a minor group, it had the historical prestige of descending from the Toltecs. Thus the Mexica lord was in reality calling himself Lord of the Toltecs. By this he indicated his descent from Quetzalcoatl, the mythical founder of royalty, and he could consider himself leader by "divine right." Genealogies, authentic or embroidered on, were a political weapon just as history was. We cannot have history without dates, and from this arises the need for a calendar.

Whatever the astral basis of the Mesoamerican calendar was--and there is still considerable doubt about certain parts--it had to be based on astronomical knowledge. This is one of the reasons for the considerable development of that science.

Already in the nineteenth century scholars understood the possibilities offered by eclipses and movements of some of the heavenly bodies not only for reconstructing ancient computation of time but also to establish a synchronization with the Christian calendar. This is part of the unwritten evidence.

Even so, although there are notable exceptions, much of what is said is doubtful and sometimes based on uncertain data or on hypotheses that are hard to accept. From this comes the imperative need to continue these studies and to penetrate even deeper into the theme. Moreover, we need to understand and study in the indigenous manner--so different from our scientific method-- which at times involves profound concepts. For example, Eric Thompson has said that in ancient Mesoamerica time was god. We agree with but do not really understand the implications of this idea and all the confusing ramifications it might have on the mind of a ceremonial society.

Astronomy and the calendar are aspects of mathematics and function on the basis of numbers. Whence the dangerous fascination of playing with figures. By moving them from here to there or giving them different meanings, they apparently can be used to prove any hypothesis. But this does not necessarily cancel out the validity and importance of such studies, as demonstrated by many of the excellent articles published here.

In assembling so many papers on the theme or related to it, this book follows an old study tradition that began in the sixteenth century. It does so of course in modern fashion contributing data of quite diverse types and by specialists in various fields. The text constitutes a valuable contribution to the history of astronomy in ancient Mexico and I am sure its publication will lead to further important work.

Ignacio Bernal

Director General

Instituto Nacional de Antropologia e Historia

Mexico, D.F., Mexico

Introduction

This book is one of the many fruits born of the first joint scientific meeting of Mexico and the United States at which more than 5,000 scientists, students, engineers, government officials, and laymen convened in Mexico City during a two-week period in June 1973 to deliberate a wide range of subjects. The inter-American meeting was organized jointly by Mexico's Consejo Nacional de Ciencia y Tecnologia (CONACYT) and the American Association for the Advancement of Science (AAAS) in the belief that the cause of science transcends national boundaries.

The circumstances (including the summer solstice date) provided an ideal backdrop for the first large-scale meeting of scholars interested in Pre-Columbian Archaeoastronomy, here defined as the study of the extent of the astronomical knowledge and practice of the ancient people of Mesoamerica, a region spanning the desert southwestern United States through Central America. For more than a thousand years up to the time of the conquest, there flourished on this soil a civilization long known to have displayed a keen awareness of celestial phenomena. But what knowledge did these people actually possess, how did they acquire it, and how did their understanding of astronomical events interact with other aspects of their culture?

From the publication of Sir Norman Lockyer's Dawn of Astronomy in 1894 until the Stonehenge controversy of the sixties, archaeoastronomy has been a highly controversial field having no solid roots in either the sciences or the social sciences. Properly placed as a cooperative interdisciplinary venture among interested anthropologist, astronomer, and historian of science, it might have revealed much more about ancient man's awareness of his large-scale environment than is already known, but until this decade little effort has been made to unite interested parties from the various established disciplines. As a result much of the literature on the subject, particularly that relating to the advanced civilizations of Mesoamerica, is written from widely varying viewpoints and usually is available only to scholars within the context of the narrow discipline in which it is written. Worse still, much of the extant literature abounds in error, false generalizations, and far more speculation than documented fact.

The editor of this volume and Arquitecto Horst Hartung of the University of Guadalajara co-arranged the archaeoastronomy section at the Mexico City meeting in order to cross-pollinate the ideas about the subject emanating from various traditional disciplines. The program, when it was finally completed and presented, consisted of 26 papers by scholars from a number of American and foreign countries, who exhibited a wide range of professional backgrounds; presenters included seven anthropologists, six astronomers, four historians, an architect, an archaeologist, and four laymen. We were particularly pleased to have included in the program review papers by Michael D. Coe of the Department of Anthropology of Yale University and astronomer Gerald S. Hawkins of the Center for Astrophysics at Cambridge, Massachusetts.

As they were presented, most of the papers seemed to fall naturally into one of three subject areas: (a) ceremonial and rock art relating to the practice of astronomy in the southwestern United States, (b) the astronomical orientation of buildings, and (c) the native American calendar and the correlation question. The selected papers included in this volume are grouped accordingly.

The central question posed at the conclusion of the meeting was does Mesoamerican archaeoastronomy exist as a bona fide interdisciplinary field composed of more than just various segments of traditionally established fields? It is my opinion that from the interest and cooperation among all conference participants both during and outside the conference, as well as from the quality of the papers presented, the field has established itself. Whether it can and should be linked in a cross-cultural approach with old world archaeoastronomy has yet to be determined. At the outset it would seem that such a study of ancient man's oldest and most widely practiced science would bear great relevance to our understanding

Introduction

of the development of civilization and culture. The number of questions and problems proposed by contributors to this volume is staggering, for we know so little about the scientific-cultural heritage left us by the people of Pre-Columbian America as a consequence of the nearly total destruction of the historical record in conquest times. The conferees have made a beginning by employing the Mexico City meeting and this proceedings volume as a platform on which to define some of the problems and by open-mindedly sharing viewpoints seen through different eyes for the first time.

I am grateful for the assistance of Sharon L. Gibbs, my Research Associate at Colgate University on a Sloan Foundation Post Doctoral Fellowship, who translated from Spanish to English; Barbara A. Toner, Watson Fellow in the History of Astronomy at Colgate University; Susan Forster and Yvonne Taylor of Colgate, who assisted in assembling and proof reading the manuscript; Lorraine Aveni and Helen Payne who typed the final manuscript; and Arqto. Horst Hartung, who assisted admirably in the organization of the archaeoastronomy section. Thanks also are due to the members of the AAAS/CONACYT Executive Committee, particularly to Dr. Walter G. Berl, for their superlative efforts in organizing the general meeting. Finally, I wish to acknowledge the Colgate Research Council for financial support in preparation of the manuscript.

Anthony F. Aveni

Department of Physics & Astronomy

Colgate University

Hamilton, New York   13346

Archaeoastronomy in Pre-Columbian America

1

Native Astronomy in Mesoamerica

Michael D. Coe

Department of Anthropology

Yale University

New Haven, Connecticut

Introduction

If any one trait can be said to be distinctive of the native cultures of prehispanic Mesoamerica, it is a deep concern with the heavenly bodies and the passage of time as marked by the apparent movements of these objects. We know of the Mesoamerican obsession with the sun, the moon, and the night sky from the testimony of carved stone monuments, of the surviving native books, and from statements of the native intelligentsia set down after the Spanish conquest. For instance, we are told by the historian Torquemada of the meritorious qualities of Nezahualpilli, king of Texcoco, in the following terms:

It is said that he was a great astrologer; that he was much concerned with understanding the movement of the celestial bodies. Inclined to the study of these things, he would seek in his kingdoms for those who knew of these things, and he would bring them to his court. He would communicate to them all that he knew. And at night he would study the stars, and he would go on the roof of his palace, and from there he would watch the stars, and he would discuss problems with them. (Leon-Portilla 1963, p. 142)

There is abundant evidence for much of Mesoamerica that the study of the heavens was the province of specialists, generally the priests as among the Aztecs, and we even have a Nahuatl word for "astrologer," <u>ilhuica tlamatilizmatini</u>,"the wise man who studies heaven."

It comes, then, as a surprise to learn how poor, scanty, and misleading our information is on native astronomy in Mesoamerica. In Spanish writings and compilations of data on central Mexico and the Maya area on the eve of the Conquest, the subject is hardly mentioned at all, and then in the most equivocal terms. Even the great Sahagun devotes only a few pages to it. And yet these men were contemporaries of Copernicus, and lived during a time when all of Europe was astounded and perplexed by the revolution in our knowledge of the universe that was then taking place. I am inclined to think that this revolution largely bypassed Spain and the soldiers and missionaries that she sent to the New World. These were men who, if they thought about astronomy at all, thought about it in terms of the judicial astrology then in vogue, and were inclined to dismiss with contempt native concepts of the heavens. I will later point out how utterly confusing the Spanish accounts really are. The failure of the Spaniards to properly record or to understand resulted in one of the greatest intellectual losses in all history.

Our present knowledge of Mesoamerican astronomy thus comes largely from the study of codices and monuments, particularly those from the Maya area. Since the nineteenth century, scholars such as Forstemann, Seler, Nuttall, Spinden, Thompson, and Caso have devoted themselves to this pursuit, and have revealed a world of knowledge unsurmised or ignored by the Spaniards.

But there are obviously entire areas of native astronomical concepts and practices not clearly visible in the reliefs and surviving books. As an example, large numbers of deities in the central Mexican and Maya manuscripts have stellar attributes. What does this mean? In the present state of our knowledge, we are not at all certain. What instruments did the specialists use to observe and measure the heavens (Nuttall 1906)? We can only guess at this stage. What constellations and asterisms were important to them, and why? Did they have a zodiac, and if so, was it solar or lunar?

There is a final source of information that has usually been overlooked: the American Indians of Mesoamerica and adjacent regions whose cultures have largely survived the continuing onslaught of European civilization. This area has hardly been touched, and the reasons are not hard to find. In the first place, there is scarcely an ethnologist or social anthropologist who can identify anything other than the moon and the Big Dipper in the night sky; the so-called natives are a great deal wiser. As a Huichol told Carl Lumholtz (1900, p.59), who was recording information about a native planisphere, "People think we Indians don't know anything, but we know more than the whites." Secondly, the local subculture of the social anthropologists who have worked in Mesoamerica has generally ignored the problem of survivals of nature culture in favor of acculturation and community studies, which are for the most part of little or no interest to archaeologists.

I am convinced that there is still much to be learned from modern ethnoastronomical research on these peoples, particularly on those who are still relatively isolated from the processes of ladinoization. The fragmentary data suggest that many groups retain native constellations and names for the bright stars, and this is an area which sorely needs further research. It is true that the great specialists in astronomy, those who had deep scientific and esoteric knowledge of the heavens, were effectively eliminated by the Spanish overlords after the Conquest. But specialists (on what might be called the "folk" level of organization) probably still survive in remote areas. One must not expect "the man in the street" to have much of this knowledge, any more than one could find out from the modern New Yorker how a television set works. In what is probably the most complete study ever made of the starlore of any American Indian group, Father Berard Haile (1947) found that the average Navajo is unacquainted with this body of knowledge; not even most singers know anything about it. Singers who wish to reach a high degree of proficiency in the star-gazing art must lie out under the stars night after night, and through the seasons, with older practitioners to memorize the heavens, and then this person must be able to reproduce what he has seen in

colored sands within the hogan. Navajo specialists of
high degree are able to prescribe the proper ceremonial
for sick clients by viewing the refracted colors of
first-magnitude stars through crystals or glass.

In this brief survey, I am going to leave the extremely important subject of archaeoastronomy aside,
and concentrate only upon native astronomy as revealed
to us in the documents, whether pre- or posthispanic.
It is my hope to be able to single out those celestial
phenomena that seem to have been of most significance to
the native Mesoamerican mind, so that those investigating the possible orientation of ancient buildings and
cities, and perhaps their complete layout, might know
which correlations are more likely than others, and
which ones do or do not conform to the Mesoamerican
mental set. As Burland (1952, p.26) has so aptly put
it, "Each correction to past work is a step nearer the
truth, and we must, if we wish our researches to progress, give up our preconceptions and try to understand
native American cultures by 'thinking Indian' and seeing nature as they saw it--simply and with a respect for
its mystery."

The Mesoamerican Universe

To understand Mesoamerican astronomy, one must study
their conception of the cosmos in which the heavenly
bodies acted out their role. For this, we have abundant evidence from both the central Mexicans and the
Maya, well analyzed by Thompson (1950, 1970), Soustelle
(1940), Caso (1954), Leon-Portilla (1963), and Nicholson (1971). The picture that has been built up is almost Ptolemaic in its scope. In place of the concentric
spheres of Ptolemy, in which astral objects fulfilled
their geometric roles, the Mesoamericans conceived a
layered universe, well illustrated by two pages of the
Codex Vaticanus A (Figure 1). The earth itself was conceived of at times as a large wheel or disk, at other
times as a figure resembling our four-leaved clover; apparently this formed the back of an enormous saurian
lying in water, surrounded by water-lilies and other
aquatic vegetation.

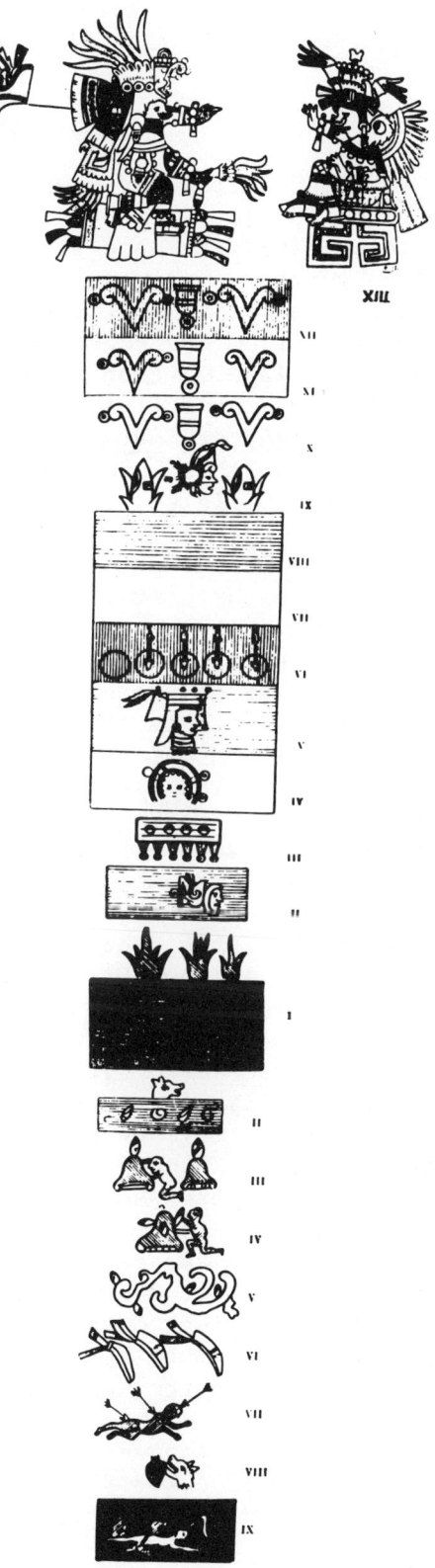

1 *The heavens, the earth and the levels of the underworld as depicted on page 2 of the Codex Vaticanus A (Rios) (after Nicholson 1971, Fig. 7).*

The water surrounding the earth was called by Nahuatl speakers <u>teoatl</u> or "divine water," and <u>ilhuica</u> <u>atl</u>, "celestial water," since the seas which bordered the land extended up to the heavens on the horizon. These heavens were conceived of as thirteen in number, although it is obvious from our codex that the earth was counted as the first layer. Each heaven contained celestial objects as well as certain gods. Through the second and lowest, <u>Ilhuicatl</u> <u>metzli</u>, traveled the moon on its course, and from this layer were suspended the clouds. Above it was <u>Citlalco</u>, the place of the fixed stars and the abode of <u>Citlallicue</u>, She of the Starry Skirt, an astral goddess who seems to have been a female aspect of the dual Creator Deity. The sun traveled its diurnal road in the fourth layer, <u>Ilhuicatl</u> <u>Tonatiuh</u>; while Venus, usually given a name meaning "great star," inhabited the next layer up. The sixth layer was <u>Ilhuicatl</u> <u>Mamalhuazocan</u>, the Heaven of the Fire Drill, a constellation the identity of which is debatable; in it were seen the comets, called "stars that smoke." Also in this heaven were the Fire Serpents, reptiles whose duty was to conduct the sun each day from the eastern horizon to the zenith. The seventh layer was the blackish or green heaven in which were winds and storms, while dust filled the eighth or blue heaven. Nicholson (1971, Table 2) thinks it possible that thunder came from the ninth heaven, called <u>Itztapal</u> <u>Nanatzcayan</u>, "where the stone slabs crash together." Layers ten through twelve were associated with the colors white, yellow, and red. Finally, in the thirteenth layer, <u>Omeyocan</u>, resided the dual, male-female, creator god whose all-embracing role as the progenitor of time, space, the gods, and all things has been so well worked out by Leon-Portilla.

Of course, this scheme has been largely derived from central Mexican sources, but there is strong reason to believe that it also applies to the Maya. Furthermore, both Maya and Mexicans seem to have had a nine-layered underworld, beginning with the earth again as the first layer.

Another extremely important concept in the Mesoamerican world-view is that of color directions. Each cardinal point was associated with a color (to the Aztecs these were east-red, north-black, west-white, south-blue), a tree, a bird, and (according to Vaticanus A), a part of the human body. Most importantly, to each direction was assigned a day in the 260-day count and a division of the same. Thus, space and time became inextricably intertwined in a kind of all-embracing mechanism.

Perhaps most important of all in their cosmological thinking was the calendar itself. At its heart was the sacred 260-day count, the origin of which remains obscure. This was based upon the permutation of thirteen numerical coefficients with a sequence of twenty days, the names of which are largely of animal origin, although plants and natural forces also enter into the scheme. At the time of the Conquest, almost all prognostications were based upon this calendar, a fact which caused Sahagun to utter the testy comment:

> For the art of judiciary astrology, common among us, is founded upon natural astrology, which is in the signs and planets of the heavens and in their courses and aspects. But this art of soothsaying followeth, or is founded upon, some characters and numbers in which no natural foundation existeth, but are only an artifice made by the devil himself. Nor is it possible that any man could have made or invented this art. For it hath no foundation in any science nor in any natural order. Rather it appeareth to be a thing of fraud and deceit than rational or ingenious. (Sahagun 1957:145)

A contrary view has been advanced by several students of the subject (summarized in Thompson 1950, pp.98-9), namely that there *is* a rational basis for the count, if at one time, as among some modern Maya, the count was fixed rather than cyclical. It so happens that at about 15° North, which is the latitude of Copan and many Maya sites, the two passages of the sun through the zenith take place at an interval of 105 days; this is also the interval between the two planting dates in this area. If the 260-day count began as an immovable segment of the agricultural year, but at some later point in the distant past was set rolling as a perpetual cyclical count, then this *might* explain its adoption in the first place. I consider these arguments extremely tenuous. At any rate, since it was associated with the color-direction concept, with the gods, and with the affairs of men, this ritual count was the most significant mental construct in Mesoamerica.

Detailed information from central Mexico and the Maya area shows that most or all Mesoamericans conceived of a dynamic universe, one in a constant state of change. The cosmos had been initiated with an old male-female couple, situated in the navel of the world and in the center of the thirteenth heaven; this dual divinity was at times manifest as the Old Fire God, who acted as the lord of the household hearth, of time, and of the solar year. To start the universe on its space-time course, the male-female divinity produced four offspring, each assigned to the four directions and appropriate colors; among the Mexicans, these were the four Tezcatlipocas, among the Maya the quadripartite deities known as the Bacabs. The struggle between these offspring, particularly between the Black Tezcatlipoca, the ruler of the north, and the White Tezcatlipoca of the west, called Quetzalcoatl, resulted in a series of cyclical creations and destructions known as "suns." Each of these creations was, however, imperfect, and it was only in the fifth sun, our own, that the Sun, Moon, and men were created as we know them. Interestingly enough, the stars had been formed in earlier creations and thus pre-existed the Sun.

In summary, the Mesoamerican cosmos was one in constant flux, in which space and time were co-terminous, in which the heavenly bodies moved in fixed layers, and which was in constant peril of cataclysm. It was this world-view which guided native astronomy.

The Sun and the Year

Among the central Mexicans, the sun was deified as a young, red-visaged personage with the name Tonatiuh (Figure 2), often symbolized in eagle form. To the Maya, he was an old god with large eye and Roman nose (Figure 3), and his avian counterpart was the scarlet macaw. According to Aztec mythology, once the perfect, fifth sun had been created in Teotihuacan, the ancient capital city of highland Mexico, it had to be fed with hearts of brave

2   *Tonatiuh, the Aztec Sun God, on page 12 of the Codex Cospi.*

captives, and mankind was specifically created for this sanguinary end.

The so-called solar calendar among both Mexicans and Maya numbered exactly 365 days, divided into eighteen 20-day periods (or <u>veintenas</u>, in Spanish) totalling 360 days, with five "days without name" at the end. This has been aptly termed the Vague Year, since, in spite of assertions in some of the ethnohistoric records, there is no evidence that the Mesoamericans ever intercalated days or leap days. This Vague Year permutated with the 260-day count to produce a Calendar Round of 51 Vague Years of 18,960 days, a time span which to them contained some remarkable numerical properties. Among the Maya, each Vague Year was named from the 260-day count position on which its first day fell, and among the Aztec (according to Caso 1971, p.34b), from the 260-day count position of the final day of the eighteenth month. These four days were the so-called Year Bearers.

The fact that the length of the solar year is actually 365.2422 days means that the Vague Year calendar was constantly gaining on the seasons by a factor of thirteen days every 52 Vague Years. It was once advanced by the astronomer John Teeple (1930, pp.70-85) that the Maya had a system of "determinants" by which the Maya expressed the accumulated error since the inception of their Long Count Calendar, which was based upon a 360-day count. It has been conclusively shown by

3 *The Maya Sun God, from page 11a of the Dresden Codex.*

Proskouriakoff that these alleged "determinant" dates are actually historical events occurring at irregular time intervals. Thus results a fundamental problem in Mesoamerican astronomy: the Vague Year was constantly running ahead of the sun and the stars. On the other hand, their calendrical system was an eminently rational one which, by avoiding lunations for the most part, was similar to that of the Egyptians and was a prototype of the system of Julian days by which modern astronomers record time intervals. And furthermore, we know from their lunar calculations that the Maya, at any rate, had a remarkably accurate knowledge of the true length of the tropical year.

The beginning of the Vague Year therefore changed by about a quarter from year to year. There has been some dispute, based on conflicting evidence, as to the <u>veintena</u> which actually began the Vague Year among Nahua speakers. According to Caso (1971, p.341), this was Izcalli, the first day of which fell on 24 January in the year 1521; Nicholson (1971, Table 4) would make it Atlcahualo, which commenced on 13 February in that year. Landa, writing of the Yucatec Maya, correctly begins their Vague Year with the first of Pop, and has it fall on 16 July, 1533. There is, therefore, not the slightest justification for the following statement in the Histoire du Mechique:

They counted the year from the spring equinox, when the sun makes a straight shadow, and as soon as it was felt that the sun was rising, they counted the first day, and [thence] the days by twenties. (Garibay 1965, p.69)

In fact, I have not been able to discover any other references to the solstices and equinoxes in the early data, whether archaeological or ethnohistorical, although these would have been easy to calculate using instruments as simple as a gnomon, sighting sticks, and horizon landmarks. I have no doubt, however, that they were important. Rafael Girard (1966) has discovered some remarkable information from the contemporary Chorti Maya of the Guatemala-Honduras border. The Chorti priests calculate the beginning of the Vague Year on 8 February, after the sun has completed an extensive "rest" in the winter solstice. The night of 30 April-1 May they take to signal the first passage of the sun across the zenith, based upon the position in the night sky of Orion's Belt, the Southern Cross, and the Pleiades; according to his informants, it is the position of at least some of these following sunset that is important. The summer solstice is marked on 21 June, 52 days following zenith passage. After another 52 days occurs the second passage of the sun through the zenith, heading south toward the equinox. All of these points are marked by important festivals, including a winter solstice celebration which takes place from 19 to 27 December, at which time the native priests say that the Pleiades and Orion's Belt rise at sundown and vanish at dawn.

Girard's discovery of the importance of these solar observations (which unfortunately he does not describe in enough detail), and their correlation with the celestial sphere, is so far unique for Mesoamerica. But he goes one step further, by showing that for the Chorti Maya, the four directions are not the cardinal points but the solstices, a finding that has been confirmed among the highland Maya of Chiapas. How widespread this concept was in ancient Mesoamerica remains unknown.

A further problem relating to the solar calendar is the point at which the Mesoamericans began their day, and the division of it into so-called "hours." Caso (1971, p.345) suggested that the central Mexicans began the day at noontime, this point in time being easy to observe (presumably with a gnomon). But for the Classic Maya, Thompson (1950, pp.174-4) feels that it started at sunrise, although the modern Jacaltec and Ixil Maya begin it with sunset. At any rate the data are far from conclusive.

Seler claimed that the well-known sequence of Nine Lords of the Night and Thirteen Birds of the Day, which accompany in unvarying succession the 260 days of the ritual calendar, proved that the day had thirteen hours, and the night nine. No other scholar has followed him in this. There is evidence that the Aztec priests divided our 24-hour day into nine ritualistic periods, four in the day and five by night (Thompson 1950, p.177); one of the principal duties of the priests was said to be the observation of the stars so as to tell the nocturnal divisions, and to sound trumpets and drums for the observance of the appropriate rites. The Zapotec supposedly had nine divisions for the day, and nine for the night.

Now, one can devise simple means to tell the time of day. The question is, how did they divide the night accurately into equal parts without timepieces? There is absolutely no evidence that the Mesoamericans had water clocks or any other kind of non-solar mechanism. One thus is led to doubt statements that they could determine midnight or any other night hour except in the vaguest way; even then the divisions would have been of different lengths as the seasons advanced.

The Moon and the Lunisolar Calendar

The moon was almost generally a female deity, although in the central Mexican accounts of the creation of the Fifth Sun at Teotihuacan, the god seems to have been a male. As a female, the lunar orb was for the Mesoamericans the very embodiment of the fair sex. The young, waxing moon was seen as a beautiful woman, forming part of a complex of youthful goddesses associated with sexual love; her image can be seen in the central Mexican codices and, above all, on many pages of the Maya Dresden Codex (Figure 4). As the moon waned and gradually slipped back towards the eastern horizon, she became an old and somewhat malevolent deity, with snakes

4 *The Young and the old Moon Goddess, on pages 16c and 43b of the Dresden Codex.*

in her hair or on her skirt, or with spindles placed in her headdress as an indication of her role as patroness of weaving. Another domain over which she ruled was that of childbirth. Again, she apparently formed part of a larger complex of aged goddesses and merged in many ways with some of these, particularly with the female half of the dual Creator God.

The association between the Moon Goddess and water was a close one. The heaven in which she traveled was that of the Rain God, and her symbol among the central Mexicans was a kind of cross-sectioned vessel--or womb-- in which water can be seen. This concept persists today among the Chorti Maya, who explain the waxing and waning moon as a pot being filled and then gradually emptied of water. Another strong association is the rabbit (Figure 5), a creature which all Mesoamerican people see on the face of the moon; the rabbit is the symbol of drink and drunkenness, and it is likely that the moon's domain extended to the complex of pulque deities.

It is only for the Maya that we have information on the role that the moon played in the calendar, and on lunar observations. Landa (Tozzer 1941, pp.133-4) states of the Yucatec Maya, "They divide [the year] into two kinds of months, the one kind of thirty days and called U, which means 'moon,' and they counted it from the time at which the new moon appeared until it no longer appears;" his other kind of "month" was the twenty-day veintena, called uinal in Maya. Notwithstanding

5 *The goddess Tlazolteotl as Moon Goddess, standing before the rabbit in the moon. From the Codex Borgia, page 55.*

Landa's claim, neither the Maya nor any of the Mesoamericans attempted to construct a grand lunisolar calendar for civil and religious purposes. In failing to do this, they are probably unique among the early civilizations, but they were lucky. The moon has the unfortunate property of following a path similar to the sun's and of being the same apparent size as the sun; these facts, and religious considerations, led ancient astronomers in the Old World to try to correlate the two main heavenly bodies into one grand scheme, an attempt which led to centuries or even millenia of confusion.

The problem is that the moon's orbit is elliptical, as is the earth's; this means that it travels at different speeds at different times of the year. Lacking gravitational theory, the ancients could never satisfactorily account for the variation in the length of the synodic month, which can be as short as 29.26 days, or as long as 29.80 days.

Nonetheless, the Maya kept a very close account of synodic lunations over a very long period of time, beginning at least as early as A.D. 300. In spite of claims to the contrary, the only astronomical calculations which surely are present on Maya monuments of the Classic Period are lunar; these are given following the initial Long Count date at the beginning of an inscription. Since the majority of dates in Classic times are now recognized as historical, or as Period Ending dates which occurred in the life of a ruler and were celebrated by him, it is clear that the moon was felt to exert a powerful influence on terrestrial events. The lunar data are presented in a passage of up to six glyphs, and include the following information: 1) the age of the current moon, 2) the number of moons already completed in a lunar half-year, and 3) the length of the present lunation, either 29 or 30 days. Parenthetically, neither on the monuments nor in the codices is there any indication that the Maya reckoned the sidereal period of the moon, which averages 27.32 days.

Of greatest interest to modern scholars are the attempts by the Maya to correlate synodic months with the solar calendar. At first, each center had its own formula for correlating the two by groupings of moons. As Teeple was able to show, by A.D. 682, Copan began using the formula 149 moons equals 4,400 days, which means that the average length of a lunation was given the remarkably accurate value of 29.53020 days. This system was rapidly adopted all over the Maya area, but lack of uniformity again appeared after A.D. 756.

Among the lunisolar formulae adopted by the Maya was one which appears to have been in use at Palenque, in which 405 lunations equal 11,960 days (Thompson 1950, p. 246). This apparently foreshadows the famous eclipse tables on pages 51-58 of the Dresden Codex, believed by Thompson to have been compiled in the 12th century. Among the properties of the number 11,960 which must have seemed remarkable to the Maya was the fact that, in addition to correlating the solar round with the moon, it also contains exactly forty-six 260-day counts. This coincidence is taken advantage of in the eclipse tables. Thompson (1950, pp.245-6) has summarized the workings of these tables as follows:

> The arrangement of the groups of moons within the table in Dresden is such that there is no doubt whatever that the cycle of 11,960 days had been divided in such a way as to give a series of days, at intervals of 177 (occasionally 178) and 148 days, on which eclipses might,

but not necessarily would, occur. After each occurrence of a five-lunation group of 148 days there is a picture. Most of these carry symbols indicative of an eclipse or at least of conjunction... It has been suggested that these pictures may indicate lunar eclipses between two partial eclipses of the sun one lunation apart.

The nodes, when the path of the moon and that of the sun cross, occur every 173.31 days, the eclipse half-year, as the Maya astronomers well knew, for eclipses could only take place within about 18 days from the node. It so happens that three eclipse half-years are exceedingly close to twice the 260-day count, another coincidence that the Maya exploited in this table.

Some remarkably acute astronomical records must have been kept over a period of time to work out such a table, although Alexander Pogo (1937) had the feeling that a knowledgeable native astronomer could have worked it out successfully from observations of lunar eclipses stretched out only a third of a century. The question still remains whether they were aware of the regression of the nodes, but Thompson feels that even if they had no knowledge of this event, the Maya astronomers at least had observed their effect and devised means of periodically correcting the eclipse table.

We are not exactly sure how the Maya reckoned the age of the moon. Spinden once advanced the idea that they counted the days of a lunation from full moon, but Thompson (1950, pp.236-7) has effectively disposed of this idea. Although Landa's testimony states that the count was from first appearance after conjunction, Thompson feels that linguistic evidence indicates that disappearance or conjunction (astronomical "new moon") were the more likely starting points. Sahagun's description, on the other hand, favors the Landa hypothesis:

When the moon is born anew, it seems like a delicate little arch of wire, no radiance does it emanate; little by little, it begins to grow. (Leon-Portilla 1963, p.49)

Finally, interesting though the seasonal movements north and south of the risings and settings of the moon might be to investigators of ancient orientations, all of our sources remain silent on the subject.

Native Astronomy in Mesoamerica

The Celestial Wanderers

If the vagaries of the moon must have been puzzling to the ancient Mesoamericans, those of the planets must have been equally so. The apparent loops or retrograde motions in the orbits of the brightest planets must have been apparent to careful observers like the Maya astronomers, and they must have begun compiling records of their motions at an early date. Curiously, vocabularies of native languages give no indication that the planets were viewed as different in any way from the fixed stars. This can be seen in the Maya Books of Chilam Balam, compilations of pre- and post-Spanish materials, in which the Spanish term planetob (pluralized in Maya fashion) is employed in the astrological sections.

While a profound concern with the moon seems to have been effectively confined to the Maya, the Venus cult was pan-Mesoamerican. All of the peoples of our area realized that with the Morning and Evening Star, they were dealing with the same heavenly body. All of them seem to have calculated its synodic period as 584 days, the nearest whole number to its actual average value, 583.92 days. It so happens that 5 X 584 = 8 X 365, so that five synodic periods of Venus exactly correspond to eight Vague Years. It is an even more remarkable coincidence that in 104 Vague Years, or two times the Calendar Round which coordinates the 260-day count and the Vague Year, there are exactly 146 260-day counts and 65 Venus periods. Among the heavenly bodies, only the moon could not be coordinated into this grand system. Small wonder that the Mesoamericans considered their calendar to be divine.

In almost all lexicons, Venus as Morning Star is glossed by a compound word which can be translated as "Great Star," although the Maya Motul Dictionary (1929) also gives the term xux ek, meaning "wasp star." In the central Mexican and Maya codices, the Venus period began with its heliacal rising in the east as Morning Star. There is abundant evidence that this event boded ill for the inhabitants of the earth. At each appearance with the dawn sun at 584-day intervals, the Venus regent threw his spear at a victim symbolizing an aspect of Mesoamerican daily life: at a water goddess, signifying impending drought (Figure 6); at a jaguar throne, symbol of the rulers; at various deities; at the jaguar warriors, i.e. the soldiery; and at the Maize God, indicating starvation.

In the central Mexican codices, the table is laid out in five sections to present 65 Venus periods or 104 Vague Years, and there are no subdivisions of the full synodic period. In the Maya Dresden Codex, and in the newly discovered Grolier Codex which can be ascribed to the Maya-Toltec (Coe 1973), the 584 days are separated into four subperiods: 1) Morning Star (236 days), 2) Superior Conjunction (90 days), 3) Evening Star (250 days), and 4) Inferior Conjunction (8 days). The days in the 260-day count which initiated these subperiods are outlined in tables which also cover 104 Vague Years.

The problem that arises with these Venus tables is the small error that accumulates over the centuries, gradually displacing the true heliacal rising of Venus at the start of the 104 Vague Year count from its official position, the day 1 Flower in the 260-day count. Teeple, followed by Thompson (1950, pp.226-7), has shown how this correction was made, by subtractions of small numbers of days at the end of 57 and 61 Venus periods.

Venus was enormously important in Mesoamerican religion and mythology. A large body of myth relates to the apotheosis of Quetzalcoatl-Kukulcan, the Feathered Serpent, as the Morning Star, and he and the Evening Star were conceived of as a pair of Hero Twins. At other times and for other purposes, at least in central Mexico, the Morning Star was coterminous with Mixcoatl-Camaxtli (Nicholson 1971, pp.426-7), a god complex associated with the northern hunters known as the "Chichimecs," and especially with the sacrifice of captives, emphasizing the basically malevolent character of this great heavenly body.

There is still debate about other members of our solar system that might have been consistently observed by Mesoamerican astronomers. It has long been recognized that certain bands appearing in Maya reliefs and codices represent the sky, or at least a segment of it. Recognizable signs or glyphs in these celestial bands include the sun, the moon, and Venus, but there are

Native Astronomy in Mesoamerica

6 *The Venus God spearing the Water Goddess, from the Codex Borgia, page 53.*

several other signs yet untranslated. From this alone one would draw the conclusion that the band represents the ecliptic, and that the remaining signs are planets other than Venus.

On pages 22-23, 43-45, 58 and 59 of the Dresden Codex are tables giving multiples of 78 and 780 days; the accompanying pictures show a strange monster with upturned snout and cloven hoofs descending from a celestial band. The synodic period of Mars being 779.936 days, Ernst Forstemann (1906, pp.215-6) and the astronomer R. W. Willson (1924, p.30) drew the conclusion that these tables deal with Mars, and that the creatures depicted were the "4 Mars beasts." Teeple and Thompson (1950, pp.257-8) have doubted this assertion, but the coincidence seems to me to be convincing, especially since the synodic period of Mars would have been easy to observe for Maya specialists experienced in the recording of celestial events. More problematical is a table on pages 30-33 which has a base of 117 days, very close to the average synodic period of Mercury of about 116 days, for that planet is not easy to observe. I continue to be puzzled at the absence of any plausible reference to the synodic period of Jupiter, 398.88 days. At any rate, I think that the data do suggest the Maya were deeply interested in the planets, above all, in Venus.

The Fixed Stars

There are said to be some 3,000 stars which the average observer can pick out with the naked eye at any one moment on a clear night (Moore 1965, p. 1). It could certainly be expected that at least all of the first and second magnitude stars visible in the latitudes of Mesoamerica as well as the Milky Way would be of considerable interest to their astronomers. Furthermore, there are certain asterisms and perhaps even constellations which would form significant groupings no matter who the observer, such as the Pleiades, Orion's Belt, the Hyades, Castor and Pollux, the Southern Cross, possibly the Northern Cross in Cygnus, the Big Dipper in Ursa Major, and Cassiopeia. At any rate, native ways of classifying and grouping the fixed stars comprise one of the most interesting areas of the young study called ethnoscience.

I have mentioned the poverty-stricken nature of the Spanish or Spanish-influenced sources on native astronomy, particularly as these apply to Mesoamerican asterisms and constellations. Just about all the information which Sahagun wishes to impart on this subject is presented on two pages of the unedited Madrid manuscript, accompanied by drawings of the sun, the moon, Venus, comets, and apparent Aztec constellations. I will return to these drawings later. But to point up the pitfalls involved in using these sources, let me quote from the account given us by Tezozomoc (1944, p. 396) of the admonitions given to Moctezuma Xocoyotzin on his election as emperor. He was especially to make it his duty to rise at midnight (and to look at the stars):

...at yohualitqui mamalhuaztli, as they call "the keys of Saint Peter" among the stars in the firmament; at the citlaltlachtli, the north and its wheel; at the tianquiztli, the Pleiades (Spanish, "las cabrillas"); and the colotlixayac, the constellation of the Scorpion, these mark the four parts of the world, governed by heaven. Toward morning he must also carefully observe the constellation xonecuilli, which is the Cross of St. James (Spanish, "la encomienda de Santiago"), which appears in the southern sky in the direction of India and China; and he must carefully observe the morning star, which appears at dawn and is called tlahuizcalpan tecutli. (translation adapted from Seler 1904, p. 355, with emendations).

Now, just what does this source mean by such terms as "the keys of Saint Peter" and "the Cross of Saint James?" Ever since the Venerable Bede, Christian scholars had been attempting to substitute personages drawn from the Bible and Christian hagiography for the pagan deities of Classical mythology as designations for the European constellations; this effort reached its culmination in 1627, with the star atlas of Julius Schiller (Allen 1963, p. 28). Apparently, such piecemeal changes had crept into popular parlance in Spain during the 15th and 16th centuries, and it is unfortunate that our scanty sources usually use this terminology rather than the Classical names of serious astronomers, astrologers, and navigators.

Sahagun's figure of the Pleiades (Figure 7a) is surely that, since it closely conforms to the appearance of that asterism in the sky. The Nahuatl name given by Tezozomoc, tianquiztli or "marketplace," is matched by words in Maya languages meaning "a multitude" or "something heaped up". Perhaps more significant is the Yucatec and Lacandon Maya term, tzab or "rattlesnake rattle," for there is a clearcut association between the asterism, the native term, and the supreme Maya creator god, Itzamna. There are strong reasons for believing that Itzamna was the counterpart of the dual creator god of the central Mexicans, and of his avatar, the Old Fire God, the lord of time. Throughout much of the native New World, the Pleiades seem to have played a role closely connected with the creation of the night sky, with the hearth and its fire, and with the agricultural cycle (for its role in the determination of the seasons, see Levi-Strauss 1964, pp. 222-45). I suspect that it, and not Polaris, was thought of as the center of the firmament. Among the well-documented Navajo, the Fire God or Black God created all of the stars in an orderly fashion, only to have some of them strewn about by a trickster Coyote; his own symbol was the Pleiades, which hopped from his foot up to his face, where it remained permanently lodged.

Among the Aztecs, the most celebrated feast was the Toxiuhmolpilia, the Binding of the Years, a ceremony which took place every 52 Vague Years. According to Sahagun (1957, pp. 143-4), all fires were put out at the end of this period, which equalled a Calendar Round. The priests and temple servants of the great temple proceeded to the Hill of the Star which they reached before "midnight."

Having reached there, they looked at the Pleiades to see if they were at the Zenith, and if they were not, they waited until they were. And when they saw that they had now passed the Zenith, they knew that the movements of the heavens had not ceased and that the end of the world was not then, but that they would have another fifty-two years, assured that the world would not come to an end.

This ceremony took place at 52-year intervals in the year 2 Reed and in the month Panquetzaliztli, which in the time of the conquest lasted from 21 November to 10 December (Nicholson 1971, Table 4).

As Burland (1952, pp. 23-6) has noted, there are several problems with this account. The first is, how did they know when midnight occurred? Secondly, if no intercalations were made, by the end of 52 years the midnight culmination of the Peiades would have been 13 days off. And finally, because of the effect of precession, the Pleiades would cross the zenith at midnight about one day earlier every 69 years. Informed Spanish observers could have questioned the native priests on these matters, but they failed to do so.

It is obvious that the Vague Year never could have effectively functioned as an agricultural calendar, and

Native Astronomy in Mesoamerica

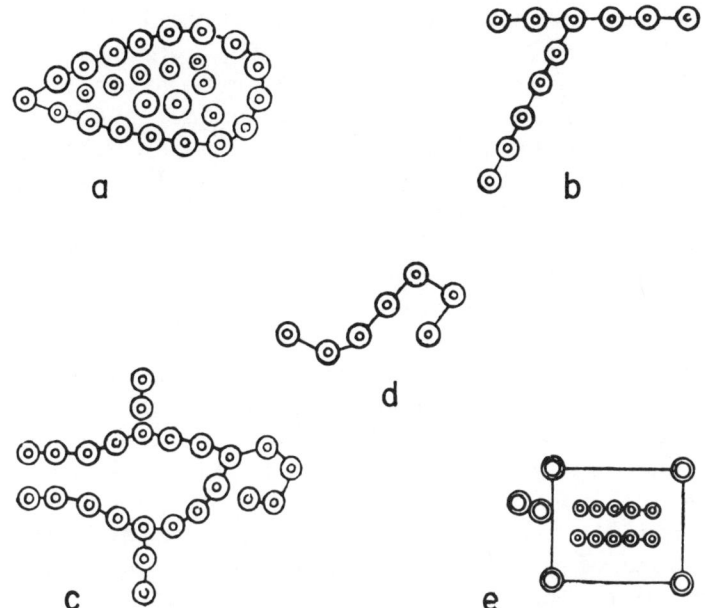

7 *Aztec constellations as given by Sahagun. a, Pleiades. b, Mamalhuaztli. c, Citlalcolotl. d, Xonecuilli. e, unnamed constellation.*

it could be expected that Mesoamerican farmers would have used striking asterisms like the Pleiades to mark points in time for their agricultural round; this seems to be the case for the Chorti, and also for the Lacandon, for whom Baer and Baer (n.d., p. 136) state, when the Pleiades have reached the tops of trees by the dawn, the milpa must be burned and the corn planted.

Another Sahaguntine constellation is called <u>Mamalhuaztli</u> (Figure 7b), or the Fire Drill, and interpretation of this group has given rise to great confusion. Sahagun says:

> Hazia esta gente particular reverencia y particular sacrificios a los mastelexos del cielo que andan cerca de las cabrillas que es en el signo del toro. (Sahagun 1953, p. 60)

It will be remembered that Tezozomoc calls the Fire Drill "the keys of Saint Peter." In their translation of

Sahagun, Dibble and Anderson have uncritically identified it with Castor and Pollux in Gemini, which is unlikely on the face of it since these stars are nowhere near the "sign of Taurus". Their reasoning is probably based on an alternate term, <u>astillejos</u>, which Sahagun gives for the group; this is a word which, like <u>mastelexos</u>, might be translated as "little sticks," not an unreasonable description of the group as it is shown, or the fire-making apparatus itself. However, the Diccionario de la Real Academia, 1970, derives <u>astillejos</u> from the Latin <u>aster</u> or star, and following earlier editions, glosses the word as Castor and Pollux. Suspecting this identification, I looked in the Nebrija Dictionary, published in 1493 or 1495, and probably available to Sahagun; there we find <u>astilejos</u> defined as the constellation Orion.

I am convinced that the Fire Drill is in fact, the Belt and Sword of Orion, which form a figure closely conforming to the picture, and which certainly do "march near the Pleiades"; the latter, of course, are part of Taurus. Just to show, however, that there is always room for doubt, Seler (1904, p. 357), following Jose Ma. Vigil and the Julius Schiller atlas, saw "the Keys of St. Peter," or the Fire Drill, as the Hyades in Taurus. We may never track this down, which is unfortunate since to the central Mexicans the Fire Drill was of great importance, being addressed as "Lord of the Night" in nocturnal offerings and sacrifices. Sahagun (1953, p. 60) informs us that the figure of the constellation was even burned onto the wrists, lest, after one went to the land of the dead, fire might be kindled on those parts.

Less problematical is <u>Citlalcolotl</u>, the Scorpion (Figure 7c). Both ethnohistoric and ethnological evidence (i.e. from the Huichol and Maya) leads me to believe that this is our constellation, Scorpius. It will be seen that as Sahagun's figure is inverted, it is remarkably close to Scorpius as we know it, with the curved tail prominently displayed. Either this group naturally looks so much like a scorpion that it has received this name independently in both hemispheres, or it was in some way diffused to the New World.

The identity of Sahagun's <u>Xonecuilli</u> (Figure 7d) or "twisted" is insoluble at present. He calls these the

"S-chaped stars in the mouth of the trumpet (Spanish, bocina)," which would presumably indicate that the group formed our Little Dipper in Ursa Minor, an identification accepted by Zelia Nuttall (1901, p. 33). However, Tezozomoc glosses Xonecuilli as "the Cross of St. James, which appears in the southern sky in the direction of India and China," from which Seler (1904, p. 358) concluded that this is the Southern Cross, a constellation recognized by many peoples in Mesoamerica. Later on, however, in his commentary on the Borgia Codex, Seler (1963, 1, pp. 193-4) suggests that Xonecuilli might be certain stars in Hercules and Draco, an identification for which he has little supporting evidence.

One further constellation is shown by Sahagun (Figure 7e) but not described; Seler thinks this is Tezozomoc's Citlaltlachtli, the Star Ballcourt, and ascribes it to stars which circle the Pole, but this is extremely tenuous.

I cannot go into individual stars which were, and in some places still are, given special names in Mesoamerica, such as Polaris. Information from the Tzeltal (gathered by my former student Allen Turner) and from the Lacandon is very suggestive. The data given by Bruce et al (1971, p. 15) for the Lacandon are extremely rich: certain asterisms like "turtle" (also given in the Motul) are not identified, but these people have names for Jupiter, Venus, Orion's Belt ("Peccary"), Rigel ("Woodpecker"), Betelgeuse ("Red Dragonfly"), Ursa Minor ("Alligator"), and Sirius (large species of woodpecker). Turner's brief research with a Lacandon informant suggests that further work will throw light upon such questions as what the Maya had in mind when they refer to a "turtle" constellation, which the Motul Dictionary places in Gemini.

The Milky Way was and is of universal significance in Mesoamerica. The Nahuatl word for the Galaxy was Citlallicue, (Starry Skirt), which is in fact, the name of the goddess who created the stars. The Yucatec Maya term was Tamacaz, which is also the name of the fer-de-lance, and associations with snakes seem to be universal in our area. In a modern dictionary of the Quiche Maya language (Leon 1954), there are two terms for the Milky Way, one (sac bey) for summer, and one (xibal bey) for the winter, when it is bifurcated; the bifurcation is identified with the Underworld, and it is quite probable that the Maya, like many other American Indians, thought of the Milky Way as the road of the souls journeying to that region.

Fragmentary allusions suggest that many of the Mesoamerican gods had their abode in the heavens. The Historia de los Mexicanos por sus Pinturas says that Tezcatlipoca and Quetzalcoatl, after creating the Milky Way, now live in the Galaxy, and Tezcatlipoca is specifically associated with Ursa Major by a myth having to do with his expulsion from heaven by his rival Quetzalcoatl (Garibay 1965, p. 30). Other deities are said to have once lived in the sky, but later descended to earth, such as <u>Yacatecuhtli</u>, the Merchant God, and <u>Mixcoatl</u>, the god of hunting. Probably all of the gods had stellar associations.

The Question of a Zodiac

Stansbury Hagar (1912), a confirmed diffusionist, was the first to advance the claim for a solar zodiac in the Maya area, based on evidence which has later been critically examined by Spinden (1916). In the Maya Paris Codex, on pages 23 and 24, are two horizontal celestial bands, beneath which are pendant thirteen animals each connected with a sun symbol (Figure 8). Hagar says that this is a zodiac. Because at least some of the figures might be star groups (i.e. rattlesnake - Pleiades, turtle - Gemini?, jaguar - Tezcatlipoca), Hagar wanted to see this as a zodiac. Ever since Forstemann (1903) tried to work out the meaning of the numbers and day names appearing in this table, there has been much discussion of what this is all about. Recently, my colleague Floyd Lounsbury has reached the conclusion that these pages comprise an eclipse table; the animals are in the process of eating the solar eclipse symbols themselves, recalling the Yucatec Maya word for eclipse, <u>chi'ibal kin</u>, "to eat the sun."

Just as dubious as a zodiac is a stellar band on the east wing of the Monjas at Chichen Itza (Figure 9), in which animal and other figures connected with star symbols are placed at intervals between sky signs. By an

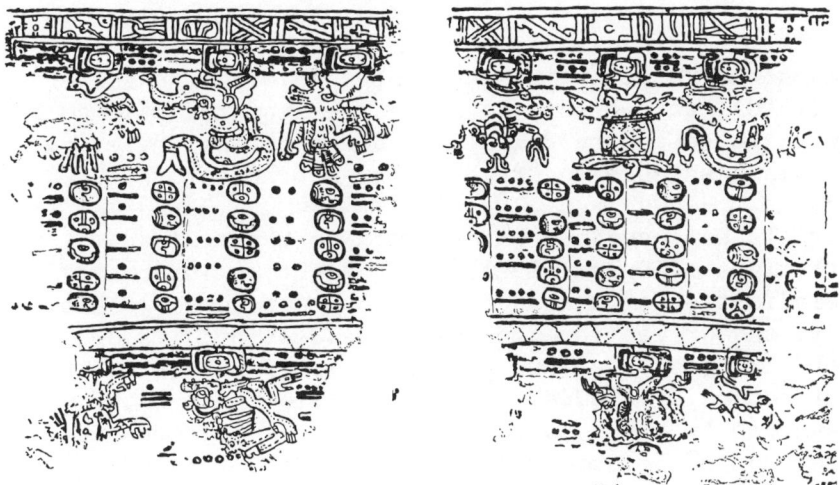

8 *Probable asterisms in the Paris Codex, pp. 23-4.*

9 *Sequence of probable asterisms in the stellar band on the East Wing of the Monjas, Chichen Itza, Yucatan.*

ingenious rearrangement of the stones forming the relief, Seler (1910), was able to match up the sequence of figures with those in the Paris. Needless to say, however, nobody has been able to prove either of these to be zodiacs.

There is some curious information in the text and illustrations of Duran (1971) which <u>might</u> be interpreted as an indication of a kind of zodiac. According to him, each "month" or <u>veintena</u> in the Vague Year had its own "planet" or constellation, some of which he shows in the sky surrounded by clouds. These are also known as <u>veintena</u> symbols in other sources, such as the pierced bird for the "month" Tozoztontli.

This is clearly a subject which needs more research before it is dismissed. The native zodiac (if it existed at all) may not have been like ours, i.e. constellations in a band extending on either side of the ecliptic, but a system of lunar mansions like the <u>hsiu</u> of China, which extend from pole to pole and which are designated by star groups which can occur anywhere within them.

Conclusions

Certain generalizations can be made about native astronomy in Mesoamerica. Firstly, they seem to have dealt exclusively with synodic periods rather than sidereal. Secondly, all of the calendars which we have, including the Venus table of the Maya, are <u>official</u> calendars, codified long ago in the past and only seldom offering the native priests or astronomers the opportunity to build in necessary corrections. As Spinden (1916, p. 66) has said, "It has long been recognized that interpolated corrections would vitiate the elaborate calculations of the Maya where solar, lunar, and Venus periods are correlated over vast stretches of time." Numerology ruled supreme in Mesoamerica, allying their astronomy much more closely with that of Mesopotamia than with the Greeks, whose obsession was geometry.

It is certain that Mesoamerican astronomy was far more complex and advanced than our fragmentary data indicate. Some aspects of it may have been the result of diffusion. Kelley (Moran and Kelley 1969) has made the

Native Astronomy in Mesoamerica

ingenious suggestion that the sequence of twenty named days may be a reduction of an oriental system of 28 lunar mansions, and Needham (1959, p. 407) has commented upon the unusual coincidence that the Maya astronomers and those of the Han Chinese worked with an eclipse calendar of 11,960 days. Finally I might point out that the practice of indicating constellations by circles connected with straight lines, to be found not only in the Primeros Memoriales (the first version of the Florentine Codex; see Sahagun 1905) but also on the borders of the Aztec Calendar Stone (Figure 10), has no antiquity whatever in Europe, but goes back as far as the Han Dynasty in China. In fact, it does not appear in the western part of the Old World until 1785, as a conscious effort of French astronomers to reconcile their star maps with those of the Chinese (Deborah Jean Warner, personal communication).

I will end with a plea to everyone concerned to collect ethnoastronomical data from the surviving native peoples of Mesoamerica; there is not much time left to salvage this body of information, which must surely help answer many of the problems raised here.

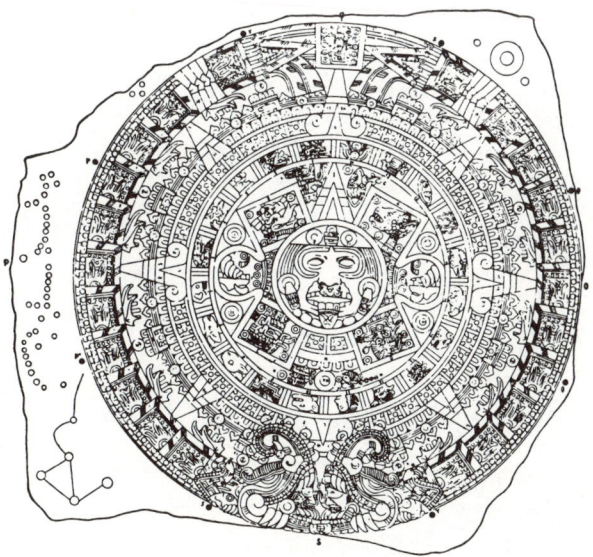

10 *Probable constellations on the border of the Aztec Calendar Stone. (after Nuttall 1901, Fig. 56).*

2

The Astronomical Record in Chaco Canyon, New Mexico

Ray A. Williamson

Howard J. Fisher

Abigail F. Williamson

St. John's College

Annapolis, Maryland

Clarion Cochran

Chaco Canyon National Monument

New Mexico

Introduction

It is well known that present day Pueblo Indians observe the motions of celestial bodies in order to time and regulate their ceremonial life (Ellis and Hammack 1968). Of particular importance to most groups is the yearly cycle of the sun from winter to summer solstice and back (Stephen 1936; Parsons 1939). Stellar groupings, such as the Pleiades and the Belt of Orion and various bright stars are considered important as well. For example, because they make their first yearly appearance just before summer solstice sunrise, the Pleiades are called the summer stars by the Tewa (Parsons 1939, p. 182). The presence of Orion in the winter sky signals the approach of the winter solstice.

Although the actual timing of the winter solstice ceremony is highly variable, depending on the U.S. calendar, phases of the moon, etc., many pueblos still appoint a "sun watcher" to observe the change of rise point of the sun along the horizon. In most pueblos the solstice horizon rise-set directions are known in advance by the alignment of horizon markers, either natural (Stephen 1936, p. 29) or artificial (Lange and Riley 1970, p. 70; Cushing 1941, p. 128) or a combination of the two. As reported by Stevenson (1904, p. 149), the determination of the location of these markers is aided by the fact that near the solstices, the sunrise azimuth appears to "stand still" for several days (Table 1).

Table 1

Sun Azimuth for Latitude 36°03'N

1973 date	Apparent Declination for 0^h Ephemeris Time	Azimuth*
June 18	+23° 24' 00"	119° 00'
June 19	23° 25' 15"	119° 02'
June 20	23° 26' 06"	119° 03'
June 21	23° 26' 32"	119° 03!5
June 22	23° 26' 33"	119° 03!5
June 23	23° 26' 09"	119° 03'
June 24	23° 25' 20"	119° 02'
June 25	23° 24' 07"	119° 00'

*Corrected for refraction to nearest 0.5'.

There is increasing suspicion that these celestial interests of historic Pueblo Indians are the remnants of more sophisticated, pre-Columbian astronomical observations made throughout the American Southwest and Mexico. For instance, Molloy et al. (1973) have shown that the upper room of "Big House" at Casa Grande Ruins National Monument was probably used as a lunar and solar observatory. Reyman (1971), in his study of the possible astronomical alignments of buildings and ceremonial structures in the major southwestern pueblos, found that there were a significant number of possible alignments to the solstices and to the rising and setting azimuths of various bright stars.

Our Work

Our interest in pueblo astronomy began in 1972 when Dr. Thomas Simpson reported some of Reyman's work to us. We were particularly interested in the problems raised by astroarchaeological measurement and spoke with Dr. Reyman about it. Subsequent conversations with him led us to make an independent study of the astronomical record of Chaco Canyon. We chose Chaco Canyon to try out our measurement methods because contained within the National Monument boundaries are several large pueblos and many small ones. Also some of the large pueblos have been well studied and detailed reports on them are readily available. Because of the limited time we had to spend in the Canyon, the nature of our report is necessarily preliminary. More extensive measurements are now being made and will be reported at a later date.

We began our work with the Great Kivas because they were the most accessible structures with which to test our methods of measurement, and they had been previously studied by Reyman (1971). We concentrated our attention on Casa Rinconada, Kiva A in Pueblo Bonito, and Great Kiva Chetro Ketl I. We chose these because the architectural features of each suggest that they are in some sense a homogeneous group (Vivian and Reiter 1960) and they were readily accessible for study. One disadvantage in studying these Great Kivas is that they have all been reconstructed and it is impossible to depend on the edges of doorways or edges of floor features for sight lines. The Great Kivas are remarkably symmetrical, however, and it seemed a reasonable working hypothesis to assume that the axis of symmetry of each Great Kiva is the determinant of orientation (Table 2).

Measurements

Because measuring azimuth orientations of man-made structures is essentially a problem of surveying, most investigators have used a surveyor's transit for the task. This instrument has the advantage that angular orientations can be determined immediately and accurately. It has two major disadvantages, however. (1) The magnetic variation of the site being measured must be known or determined by solar observations. (2) Because the transit is normally most easily used in daylight the archaeoastronomer tends to be removed from the object of his investigation, the relationship of celestial objects to buildings or alignment markers. We have attempted in this study to develop a method of determining alignments which will avoid these two disadvantages and retain an acceptable accuracy.

Table 2

Axes of Symmetry (from South-North)

CASA RINCONADA	Inner east edge of south doorway - center firebox - center north doorway
PUEBLO BONITO A	Center Niche 17 - center firebox - center north doorway
CHETRO KETL I	Center of firebox - center doorway

We used two independent methods to determine the orientation of the kivas, a tripod-held camera, and a hand-bearing compass. The photographic measurements were made as follows: The axis of symmetry was determined by consulting the drawings in Vivian and Reiter (1960) or Judd (1964) and by on-site inspection. For each kiva, a 35-mm camera was carefully positioned at night on the line of symmetry to the south of the central firepit. The northern doorway was centered in the camera viewing screen by artificially lighting the kiva, and the camera was leveled. Two independent 10 min. black and white exposures were made for each kiva.

Two pieces of information are immediately available from the photographs. (1) The declination of a star which would rise or set at the intersection of the kiva axis and the local horizon can be directly determined from the photograph. (2) The azimuth of the axis of symmetry can be calculated using the known declination of stars whose paths are tangent to the axis when it is extended vertically. The standard formulas for converting between altitude-azimuth and right ascension-declination coordinates can be employed for this purpose.

The advantages of this method are that it does not depend on knowing the magnetic compass variation in a given location and it preserves a record of the appearance of the sky at the epoch of measurement. (It should be noted here that because of precession, the coordinates of specific stars will not be equal to their values during construction of the site. Thus one cannot determine directly from the photograph to which star the structure may have been aligned when constructed.)

Once the declinations are measured from the photographs, one can solve the spherical triangle in the standard way (Hawkins 1968, 1973) to determine whether any bright stars rose or set at the computed declinations at plausible dates of occupation. Another welcome advantage of this method is that in cases where the local horizon is considered to be the orientation referent, no additional correction need be made for horizon elevation or atmospheric refraction.

As a simple check on our method, handbearing compass measurements of the line of symmetry were made independently by two of us (R.W. and H.F.) and the results averaged. A comparison of the photographic and compass measurements and their estimated errors is presented in Table 3. The dates in the second column are from Vivian and Reiter (1960) and a magnetic variation correction of $13°.5$ has been employed in deducing the (compass) azimuth.

Solstice Markers

We discovered two possible solstice markers at Chaco Canyon. While hiking along the mesa top near the ancient stairway north of Pueblo Bonito one of us (A.W.) found a petroglyph in the form of two crossed perpendicular lines 15 cm long on the horizontal surface of the mesa. One line was oriented to an azimuth of 301° and the other to 30°. At Chaco Canyon the summer solstice sun sets at an average azimuth of 300°04'. A complete tabulation of solstice sunrise and sunset values for Chaco Canyon circa A.D. 1000 is presented in Table 4. These were interpolated from Aveni's tables (1972).

We found a second solstice marker across the Canyon at the top of the ancient stairway east of Casa Rinconada. It consists of a series of holes cut in the sandstone. Referring to Fig. 1, holes A and B lie along an azimuth line of about 301°. Holes A and C, D, E lie along an azimuth line of 30°. Line BF is parallel to AE.

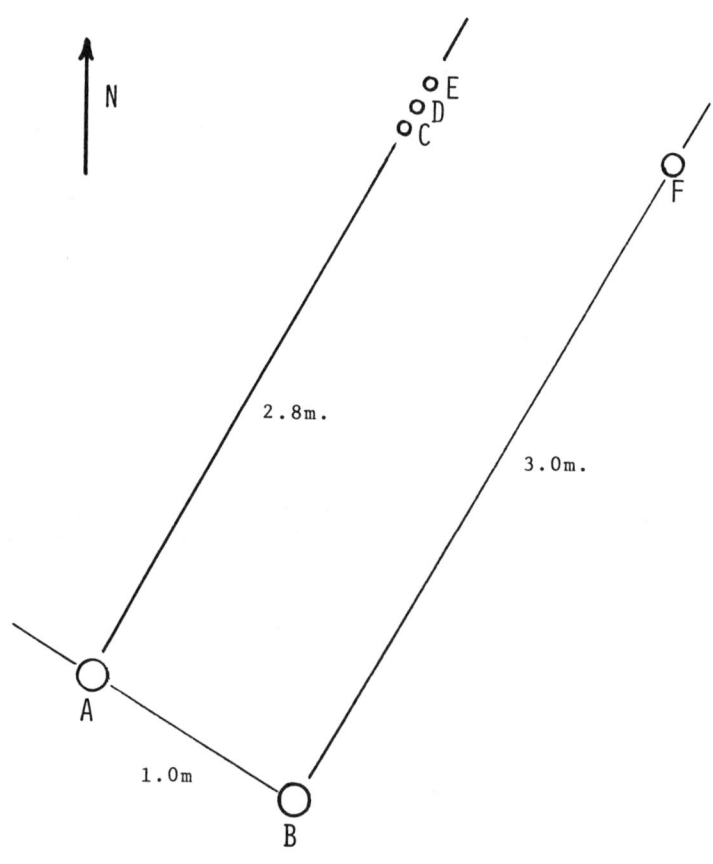

1 *Solstice marker east of Casa Rinconada*

Table 3

Comparison of Photographic and Compass Measurements

Kiva	Dates A.D.	Latitude of Site	Azimuth (Photograph)	Azimuth (Compass)
Pueblo Bonito A	919–1130	36°04'N	359°30'±15'	359°30'±30'
Casa Rinconada	1100–1116	36°03'N	359°25'±15'	359°45'±40'
Chetro Ketl I	1062–1090 110–1116 (remodeled)	36°04'	344°17'±15'	344°00'±30'

Table 4

Solstice Azimuths for Chaco Canyon

Summer Solstice	Rise	Set	Winter Solstice	Rise	Set
First gleam	59°46'	300°16'	First gleam	119°00'	241°00'
Center sun	59°58'	300°04'	Center sun	119°12'	240°48'
Last gleam	60°12'	299°52'	Last gleam	119°25'	240°35'

Average latitude = 36°3'N

Obliquity of Ecliptic (A.D. 1000) = 23°24'

The measurements of these possible solstice markers were made with a magnetic compass. Our reported azimuth values include an average magnetic variation correction of 13.5°. Since this value may be in error by as much as a degree, accurate transit measurements of these markers will soon be made. Since the horizon is obscured by higher mesas in the direction of the winter solstice sunrise, it is likely that these observation sites were used to observe the summer solstice sunset.

Reyman (1971, p. 255) had previously found a line of burned spots on the mesa top at an azimuth of 240° from Casa Rinconada which he interprets as marking a winter solstice sunset line. These three sites constitute strong evidence that observations of the solstices were made by the inhabitants of Chaco Canyon.

It is an interesting fact that the Canyon is oriented very roughly to an azimuth of 300° (120°). A perpendicular to the summer solstice sunset line (or winter solstice sunrise line) therefore, extends directly across the canyon. Transferred to the floor of the Canyon, this line (ACDE in Fig. 1) could have been used to construct the winter solstice sunrise or summer solstice sunset line in the Canyon, since any line perpendicular to the 30° azimuth line would have an azimuth of 300°. Dow (1967) reports that such orientation markers may have been used at Teotihuacan as surveying tools. As yet we have found no archaeological support for this possibility in Chaco Canyon, but a search for such evidence is under way.

Rock Art

We also consider the pictograph near Penasco Blanco, which was reported on in two other papers at this meeting, to be a part of the astronomical record of Chaco Canyon. Whether it is indeed a record of the A.D. 1054 supernova (Brandt et al. 1973) or whether it marks a sacred shrine (Ellis 1973), it is further proof that the ancient inhabitants of Chaco Canyon had a profound interest in the sky.

One of us (C.C.) is conducting a systematic study of the rock art of Chaco Canyon. Except for this pictograph

Astronomical Record in Chaco Canyon

and numerous pecked sun symbols consisting of concentric rings, we have found no other examples of rock art which can be directly related to astronomical observations or observances in Chaco Canyon.

Discussion

As is clear from Tables 4 and 5, we have found no certain evidence for stellar alignments to stars brighter than magnitude 3.0 for any of the three Great Kivas we measured. The north-south orientation of Casa Rinconada and Kiva A in Pueblo Bonito could have been derived from solar and/or stellar observations. The alignment azimuth we obtain from the axis of symmetry of Chetro Ketl I corresponds to no bright star setting either on the local horizon or the plane of the horizon at the suggested dates of construction. Ellis (1973, private communication) has emphasized that the construction dates for Chetro Ketl are very poorly known and the published data should be accepted with great caution. By going to an earlier construction date for the great kiva in Chetro Ketl, we find two possible orientations to setting stars (Table 5).

Table 5

Computed Alignments for Chetro Ketl

Azimuth = $344° \, 17' \pm 15'$

Declination (Local Horizon) = $+61° \, 54' \pm 15'$

Declination (Horizon Plane) = $+50° \, 43' \pm 20'$

Date A.D.	Star (Local Horizon)	Star (Horizon Plane)
1062-1116	None	None
960	None	α Cassiopeia
925	ε Ursa Major	None

In considering these results it is necessary to raise the question of how important the local horizon may be for the astronomical orientation of Chaco Canyon structures. The solstice markers discovered so far indicate that the important direction was the position of the sun on the plane of the celestial horizon. For orientation to the stars, however, there is no clear evidence whether the local horizon or the horizon plane was the orientation determinant. In the absence of clear evidence for one or the other hypothesis, both must be considered. Because of this we have included both possibilities in Table 5.

It should be emphasized here that the choice of an axis of symmetry is inherently subjective. Reyman's measurement (Reyman 1971, p. 306) for the azimuth alignment of the Great Kiva Chetro Ketl I, $A = 341° 30'$, is substantially different from ours. If the Chetro Ketl Great Kiva is indeed aligned to the setting of a specific star, we would expect the edges of kiva components to be more important than for a N-S alignment. For this reason, a range of possible alignments should be considered. Also, because the Great Kiva in Chetro Ketl has been rebuilt at least twice, an effort should be made to determine the possible orientation of earlier construction.

It seems clear from other archaeoastronomical studies (Dow 1967; Fuson 1969; Molloy 1973) that an important distinction should be made between archaeological sites which can be considered to have served as observatories and those which align to important celestial directions, but which, because of an obscured horizon, could not have been used to observe the event. Although these distinctions often overlap, both possibilities may exist at Chaco Canyon.

For example, the mesa top east of Casa Rinconada (above site BC-53) seems to be a kind of observatory of the sun and may have served as a religious shrine as well. Neither Casa Rinconada, nor Kiva A in Pueblo Bonito, because of their position within the Canyon, could have been used as observatories, but their alignment along cardinal directions may have served an important ceremonial function.

It will be important in future explorations of the astronomical record of Chaco Canyon to measure the orientation of the major walls of Pueblo Bonito, Chetro Ketl, and old Pueblo Alto. The available site maps indicate that several walls of these pueblos are astronomically aligned.

We wish to express our thanks to Tom Simpson for arousing our interest in Chaco Canyon and to Henry Harris and Debbie Krikorian for their help in obtaining the measurements. We extend special thanks to Richard Hardin, former Superintendent of Chaco Canyon National Monument, for his cooperation in making our studies possible and our stay at Chaco Canyon a pleasant one. One of us (R.W.) is indebted to the National Science Foundation for partial support of this study: NSF Academic Year Extension Grant (NSF-Y008700).

3

Possible Rock Art Records of the Crab Nebula Supernova in the Western United States

John C. Brandt and Stephen P. Maran
Laboratory for Solar Physics
NASA-Goddard Space Flight Center
Greenbelt, Maryland 20771

Ray Williamson
University of Maryland
College Park, Maryland 20742

Robert S. Harrington
U.S. Naval Observatory
Washington, D.C. 20390

Clarion Cochran
Chaco Canyon National Monument
New Mexico 87313

Muriel Kennedy and William J. Kennedy
1012 Del Monte Boulevard
Pacific Grove, California 93950

Von Del Chamberlain
Abrahms Planetarium
Michigan State University
East Lansing. Michigan 48823

Introduction

Dynastic records of the Sung Dynasty, the Sung-Shih, describe a bright object in the constellation Taurus which appeared on 4 July, A.D. 1054 (Julian calendar), was visible in daylight for 23 days, and faded from the night time sky after 653 days (Duyvendak 1942, Needham 1959). Association of the Crab Nebula with the object described in the Sung-Shih was apparently first suggested by Hubble (1928) when he noted that about 900 years would be required for the Crab Nebula to reach its present size at the measured (Duncan 1921) rate of expansion. Elaborate determination and discussions of the expansion time of the Crab Nebula have been carried out (Duncan 1939; Baade 1942; Trimble 1968) and the currently accepted date of outburst based on the assumption of unaccelerated motion is A.D. 1140 ± 10. The difference of 86 years between the oriental records and the purely astronomical extrapolation is a problem to which we will address ourselves later.

Acceptance of the A.D. 1054 event as the supernova cause of the Crab Nebula came in the 1940's (Duyvendak 1942; Mayall and Oort 1942) and this view prevails today (e.g., Mayall 1962; Scargle 1967; Davies and Smith 1971).

Many authors had noted the absence of written records of the Crab supernova in Europe and the Middle East (e.g., Mayall 1962; Shklovsky 1968, p. 52). Hence there was considerable interest in Miller's (1955a,b) announcement of two possible rock art records (Figure 1) of the event in Northern Arizona. Each consists of two design elements, interpreted by Miller as representing the crescent moon and the supernova. Mayall and Oort (1942) estimated the visual magnitude at approximately -5 or a brightness about 6 times that of Venus (but see Minkowski 1971 where the assumed type of supernova is questioned). This is brighter than either Tycho's supernova of A.D. 1572 (-4) or Kepler's supernova of A.D. 1604 (-2). Chinese historical records concerning the color of the A.D. 1054 supernova near maximum are probably not reliable because of the Imperial yellow color (Duyvendak 1942). The astronomical circumstances that made Miller's suggested interpretation quite plausible was the close conjunction of the crescent moon and the position of the Crab Nebula just before dawn on the morning of 5 July, 1054. This was pointed out by Miller, and our independent calculations based on the U. S. Naval Observatory's positional data for the "ancient moon" are shown in Figure 2. Analysis of pottery fragments by R. C. Euler,

Possible Rock Art Records

 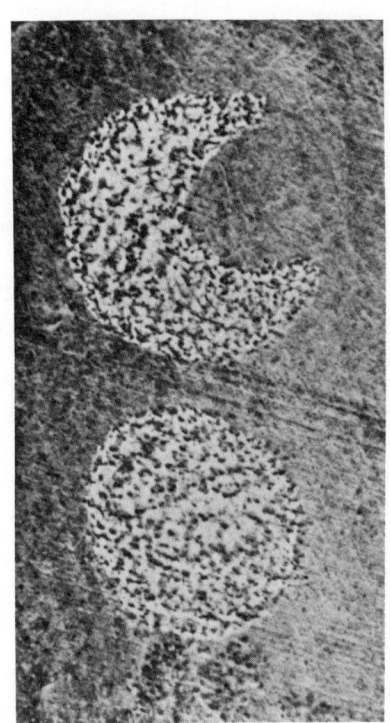

1 *The northern Arizona sites described by Miller (1955a,b).*
Left: The White Mesa pictograph (Cave designated NA 5561). The drawing appears to have been made with a lump of red hematite. The diameter of the circle is approximately 10 cm.
Right: The Navaho Canyon petroglyph (site NA 5653). The circle is approximately six inches in diameter. (Photographs courtesy of Wm. C. Miller.)

Museum of Northern Arizona, indicate that the region of the rock art cited by Miller was inhabited for several centuries including the year of the supernova (Miller 1955 a,b). Because the moon moves approximately its own diameter with respect to the celestial background in one hour, the close conjunction of the moon and the supernova was basically only observable in western North America. Clearly "close" is a subjective judgment, but the moon would have moved at least eight times its own diameter to the east of the point of closest conjunction before it would have been observable from China. Positional astronomy was well developed in eleventh-century China (Needham 1959; Needham, Wang Ling, and Price 1960), and the bright object described in the Sung-Shih was referred not to the moon, but to a nearby 3rd magnitude star, ζ Tauri. (Note that a discrepancy occurs in the position with respect to ζ Tauri. We discuss the situation below.)

In this paper, we describe several other sites in California and New Mexico where rock art exists that resembles the two cases reported by Miller, and which, therefore, may also represent the Crab Nebula supernova event. Preliminary results have already been reported (Brandt et al. 1972; Brandt et al. 1973). In addition, Miller (1972) has informed us of the existence of another candidate in northern Arizona. In each case, the relationship of the star to the crescent is reasonable, the eastern horizon is visible from the vicinity of the rock art, and the available archaeological evidence is compatible with habitation of the site circa the middle of the eleventh century. We believe that this evidence is consistent with the observation and recording of the conjunction of the Crab Nebula supernova and the crescent moon by several groups of American Indians.

It should be noted that the interpretation of ancient rock art is a very uncertain subject, beset with pitfalls and inherently subjective. We have approached this investigation in the spirit of attempting to see what evidence might exist for rock art records of a spectacular astronomical phenomenon. We recognize, however, that it is not possible to prove the conclusions suggested below.

Our Work

Our involvement in this work began with an effort to find archaeological evidence for a much brighter supernova in the southern hemisphere (see Brandt et al. 1971b; Alexander et al. 1971; Maran et al. 1973). An appeal was made for archaeological help in locating a dateable record of the Vela supernova (Brandt, Maran, and Stecher 1971a) and a description of this appeal

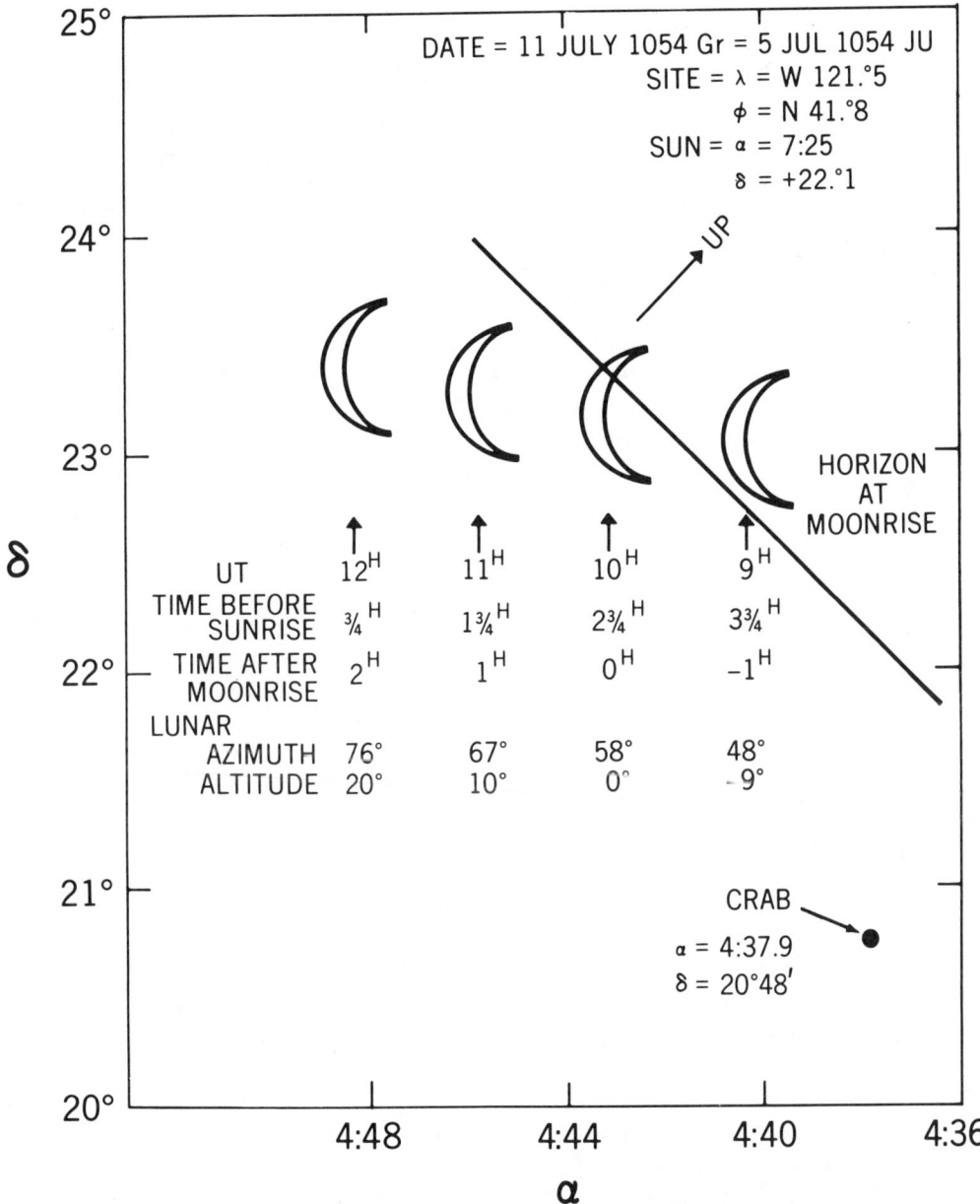

POSITIONS OF MOON AND CRAB 5 JULY 1054 AS SEEN FROM FERN CAVE

2 *Astronomical circumstances of the conjunction on the morning of July 5, 1054 as seen from northern California. The calculations are based on the U. S. Naval Observatory's data for the "ancient moon".*

appeared in the 27 March, 1972 issue of Time. One of the pictures discussed by Miller was used to illustrate the Time article and a response from Muriel Kennedy at Lava Beds National Monument initiated the present study.

The search for and tentative identification of possible rock art records of the supernova is enhanced by the extreme rarity of crescents in rock art as noted by Miller (1955a,b) for Northern Arizona rock art, and by Schaafsma (1973) and Bain (1973) for New Mexico rock art. We have searched the rock art literature and confirm Miller's observation, as discussed below. Details relevant to a particular crescent motif site are given below when they are available.

The Fern Cave Site

This possible record of the supernova is on the west wall of Fern Cave, in the northeast part of Lava Beds National Monument in Northern California. The cave is listed as archaeological site Tlk-2, and it contains a large number of rock paintings. In 1935, test trenches were dug in the floor of the cave, and charcoal, stone awls, mat or basket fragments, arrowheads, and other evidence of past occupancy were found.

Fern Cave is one of many archaeological sites in the Monument area described by Swartz (1964). He found evidence for occupancy at various times dating back to before 1500 B.C. Although he noted the radiocarbon date of one timber in a housepit as A.D. 803 (± 160 years), he described the relevant period of occupation, called "Component III", as having been probably of short duration sometime between A.D. 500 and 1869.

Fern Cave was formed in a lava flow in the late Pleistocene or Recent Epoch and is classified geologically as a lava tube. Entrance is possible at only one point, a hole in the ceiling above a mound of rocks. The most impressive array of pictographs is on the east wall of the cave (Grant 1967, p. 102, figure inverted). These paintings have received a fair amount of attention and were photographed by archaeologists more than 30 years ago. At that time, chalk was applied to increase the contrast in making the photographs. This practice,

Possible Rock Art Records

3 *The Fern Cave pictograph as described in the text.*

once common, is now recognized as undesirable. However, on the opposite (west) wall of the cave, many less spectacular pictographs were spared this treatment. Among them is the rock painting that includes a crescent (Figure 3) done in a dark pigment that may be charcoal.

The crescent is about 140 centimeters above the cave floor and its diameter is 14 cm. The distance from the left-hand cusp to the circular figure directly below is 17 cm, while the distance from this cusp to the circle located above and to the left is about 21 cm. The crescent and the lower circle are located on a rock panel that slopes down and away from the viewer at an angle of about 35° to the vertical. The upper circle is drawn on a rock surface with a different slope. This photograph was taken with a camera held near the floor of the cave, with its optical axis essentially orthogonal to the surface containing the crescent and the lower circle (which may represent the supernova). This pictograph is recorded in Swartz (1963).

The correct orientation of the moon and the supernova is shown in Figure 2. These computations were made at the U. S. Naval Observatory to determine the phase and location of the moon as seen from the Lava Beds region with respect to the precessed location of the Crab, and with respect to the horizon, on the morning of 11 July 1065 (Gregorian) or 5 July 1054 (Julian). Note that the crescent, although resembling the cave drawing in phase, is concave towards the right and upwards. The moon reached an altitude of 10° at $1^h 45^m$ before sunrise, and an altitude of 20° at 45^m before sunrise.

We must now deal with the incorrect orientation of the crescent (Figure 3) and its reversal (Figure 1). Miller (1955a,b) has suggested that the reversal of one of the Arizona crescents was due to what may be a common error in recording one's recollection of the appearance of the moon. Although there is clear danger in assuming that people of very different cultures will make the same kind of error in recalling the orientation of a celestial body, we tried an experiment. At the University of Maryland, John Carlson arranged for 29 liberal arts undergraduates who were taking the elementary astronomy course to view the moon and Jupiter on 18 July 1972 when the moon was at first quarter. The class assumed that this was an ordinary observing session. Two days later, however, they were asked to draw the moon and Jupiter as they had appeared in the sky. Twenty-three students did a fairly good job of it, but six others reversed the bright and dark halves of the moon, and one of the six even drew Jupiter on the wrong side of the moon. The results are consistent with Miller's suggestion that errors in recording the orientation of the crescent moon are common.

We now return to Miller's point that crescents are rare in Arizona rock art. The rock art of the Lava Beds region is not thought to be related to that found in

Arizona. Instead, it is classified with the "Great Basin" region pictographs and petroglyphs of Nevada and eastern California. These have been studied in great detail by Heizer and Baumhoff (1962). Although they identified 58 basic shapes or "design elements", crescents were not included in their list. We examined the sketches of several thousand Great Basin petroglyphs and pictographs in their 1962 book and found only a few crescent-like figures. We also examined several thousand rock art figures from the Klamath Basin (which is a smaller region that includes the Lava Beds) as depicted by Swartz (1963) and again found only a handful of crescent-like figures, one of which is the Fern Cave painting discussed here. Finally, we looked at the hundreds of pictographs and petroglyphs from all over North America that appear in Grant's (1967) book. The only crescent among them is one of the two reported by Miller in northern Arizona.

The Symbol Bridge Site

This site, which may be related to the Fern Cave site, is approximately 5 miles southwest of Fern Cave. Symbol Bridge is on the southeast slope of Schonchin Butte in Lava Beds National Monument and is marked on the National Park Service maps of the Monument. The pictographs are shown in Figure 4; they have also been recorded by Steward (1927, Figure 4, p. 61, inverted).

Our interest in this pictograph is in the three crescents shown adjacent to round objects. Refer to the arguments concerning the rarity of crescents in rock art given in the discussion of the Fern Cave site. Here, we have three crescents in a pictograph on the same rock and the overall result is suggestive of the supernova-moon conjunction.

The Chaco Canyon Site

This pictograph (Figure 5) was discovered in June 1972 by an archaeological survey team from the University of

New Mexico. It is located 6 meters above the canyon floor on the ceiling of a very shallow sandstone cave in the west wall of Chaco Canyon (site no. 29 SJ 427), about 500 meters northeast of Penasco Blanco ruin. Penasco Blanco is one of six large pueblos in the Canyon. Built on a mesa on the southwestern edge of the Canyon, it commands an impressive view of most of the Canyon. The vertical sandstone walls east of Penasco Blanco contain many petroglyphs typical of Chaco Canyon designs; except for a sun symbol on the vertical wall directly below the suspected supernova drawing, no other pictographs are known in the area.

Although habitation of the site at some time in the past is indicated by pottery sherds and viga holes adjacent to the pictograph, no dating is yet available for the sherds. However, tree ring dating of the nearby Penasco Blanco ruin (Bannister 1965, p. 200, Figure 43) establishes that it was inhabited between A.D. 900 and 1100.

The reddish-brown pigment of the pictograph is probably hematite, a substance commonly found in the canyon. The diameter of the crescent is about 18 cm, and the distance between the center of the crescent and the center of the star is about 30 cm. The image of a hand just above the crescent is about life size. Because the cliff wall is oriented north-south at this point in the canyon and the pictograph is painted on a surface perpendicular to the cliff, the star image appears south of the crescent, and the horns of the moon point westward in Figure 2. The hand is therefore east of the crescent.

Just what connection the hand bears to the star-moon image is problematic. Handprints painted on rock are common in places sacred to present day pueblos (Ellis and Hammack 1968), and the presence of a sun symbol on the vertical wall may indicate that the spot was a sacred one. Cushing (1941, p. 128) in his popular account of his first stay at Zuni mentions "a pillar sculptured with the face of the sun, the sacred hand, the morning star, and the new moon", but exactly where this pillar is or what the relative orientation of the design elements is we do not know. If the petroglyph seen by Cushing at Zuni is an ancient one it might be a record of the A.D. 1054 supernova and could be considered by modern Indians to be a record of Venus in conjunction with the crescent moon.

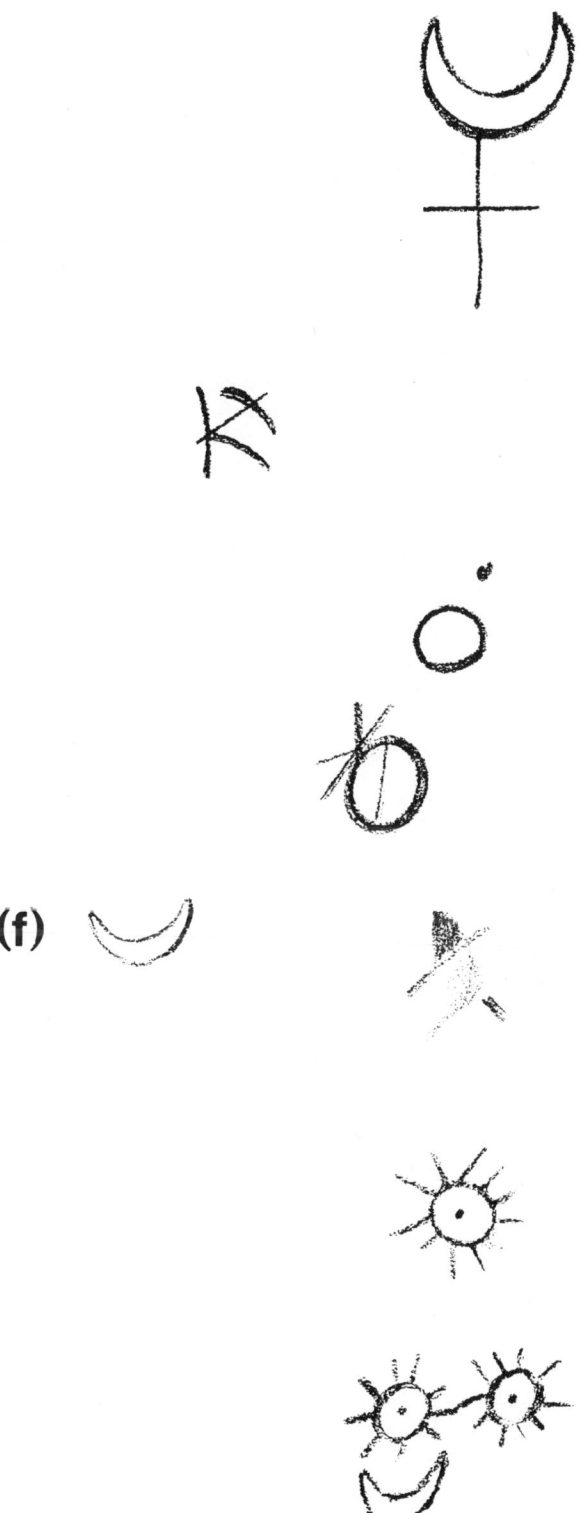

4 Charcoal sketch of the Symbol Bridge pictographs based on photographs and direct inspection of the site. Part of the pictograph is weathered and the feature marked (f) is faint but distinct. (Sketch by Dorothy Brandt.)

5 *The Chaco Canyon pictograph as described in the text.*

Other Sites

Col. James G. Bain of Albuquerque, New Mexico has kindly supplied us with photographs of two other possible Crab Nebula petroglyph records in New Mexico, one in the Village of the Great Kivas, and the other near Scholle. Both of these examples are associated with other rock art, but no other information is yet available. These sites are presently under study.

Discussion

We can ask whether the ancient native Americans were in fact likely or inclined to record astronomical phenomena in their rock art. We know of no written record nor oral tradition going back to the eleventh century that can answer this question. Perhaps the best we can do is to look at Indian customs in historical times.

Navajo gourd rattles and sand paintings were often decorated with symbols of the Pleiades and Orion (Haile 1947), and there are likewise Navajo cave paintings of the eighteenth century and perhaps earlier that have been interpreted as representing constellations (Schaafsma 1966, 1973; Britt 1973). Certainly recognizable constellations are common in modern Navajo sand paintings. Stephen (1936) gives many examples of Hopi designs of celestial objects, including unmistakable representations of the Milky Way, the Pleiades, Orion, the crescent moon, the sun and the Morning Star. The kiva murals of Kuaua (Dutton 1963) also contain celestial motifs.

There are pictographs and petroglyphs from various parts of the United States and Mexico that LaPaz (1948) described as records of meteors; most impressive of all are unmistakable paintings of the Leonid shower of 1833, made on a buffalo skin by the Dakota Sioux (Mallery 1893, p. 723). The Dakota Sioux also recorded the total eclipse of the sun which passed through their territory on August 7, 1869 (Mallery 1893, p. 286). Finally, Grant (1966, plate 25) has noted two pictographic representations of comets in the rock art of the Chumash; one of the authors (J.C.B.) has inspected a striking petroglyph of a comet near Los Alamos, New Mexico. Thus astronomical phenomena do appear in Indian art.

We have also investigated the possibility that the rock art in question might be representations of another supernova - the oriental records (e.g., Xi and Po 1966; Shklovsky 1968) indicate a supernovae in May of A.D. 1006. However, during the months of May, June, and July in 1006 the moon and the alternate supernova candidate were never closer than $21°$ in the sky and the moon was always gibbous at the time of closest approach. As a test of the Sung-Shih record, we have also checked the positions of the bright planets for 5 July, 1054 and none of them were near the crescent moon or ζ Tauri.

Ho, Paar, and Parsons (1972) have questioned the identification of the Crab Nebula supernova with the Guest Star discovered by the Chinese on 4 July, 1054 as described in the Sung-Shih. The principal evidence for this contention is (1) the reversal of the directions of the supernova from the star ζ Tauri; the Crab is actually northwest of ζ Tauri while the Sung-Shih states southeast. Also (2) the difference in the extrapolated date (A.D. 1140 ± 10) from A.D. 1054 is regarded as significant. We do not regard these objections as convincing. The difference in directions is in fact just a <u>reversal</u> of directions - the kind of error one might expect in a report transcribed many times. The difference in dates is probably not serious because the quoted error represents only the internal consistency of the astronomical measurements and <u>not</u> the uncertainty in the interpretation (Trimble 1968, 1971).

These considerations notwithstanding, if the identification of the Crab Nebula supernova with the 4 July, 1054 Guest Star is correct, then a spectacular conjunction of the crescent moon and the supernova was visible from western North America on the morning of 5 July, 1054. We have listed several cases of rock art that appear to resemble such an event, and which are found in Arizona, California and New Mexico. In each case the archaeological evidence is consistent with habitation in the eleventh century A.D. Thus, these records are consistent with and supportive of the <u>Sung-Shih</u> interpretation criticized by Ho, Paar, and Parsons (1972).

As we noted at the outset, we cannot prove that the rock art records discussed here refer, in fact, to the Crab Nebula supernova. All that we can do is to build a circumstantial case. Since the close conjunction would have been visible from Mexico, we would appreciate the assistance of Mexican archaeologists and astronomers in locating additional possible records of this fascinating phenomenon.

Ray A. Williamson, ordinarily at St. John's College, Annapolis, Maryland, acknowledges support from the NASA-Goddard Space Flight Center (Grant NGS 21002-033).

4

A Thousand Years of the Pueblo Sun-Moon-Star Calendar

Florence Hawley Ellis

Professor Emeritus of Anthropology

and

Collaborator, Chaco Center

National Park Service

In the fall of 1972 two photographs were sent me showing the petroglyph of a handprint, a horizontal crescent moon upside down, the Great or Morning Star, and the Pueblo sun symbol (two concentric circles with a center dot) in a combination (29SJ427) running down a protected ledge facing east just below Penasco Blanco, the westernmost of the big ruins in Chaco Canyon in northwestern New Mexico (Fig. 1). One photograph came from Mr. Tom Windes of the Chaco Center, for which I act as consultant, and which had discovered the petroglyphs during their survey. The other came from Mr. Clarion A. Cochran who had been working as a summer ranger in the Chaco Canyon National Monument and was struck by the parallel between the petroglyphs and a group described by Cushing near Zuni, with which I also was familiar. The find was, indeed, something over which to become excited, not because it was spectacular but because it permits an unusual glimpse into the agricultural-ceremonial-calendrical system used by the Pueblo Indians of Chaco Canyon a thousand years ago.

Tree ring dates prove that the great Chaco sites were occupied from the 10th into the 12th century, and the suggestion that this combination of symbols recorded the advent of a supernova in the sky in A.D. 1054 has been made by astronomers. This concept, however, might be

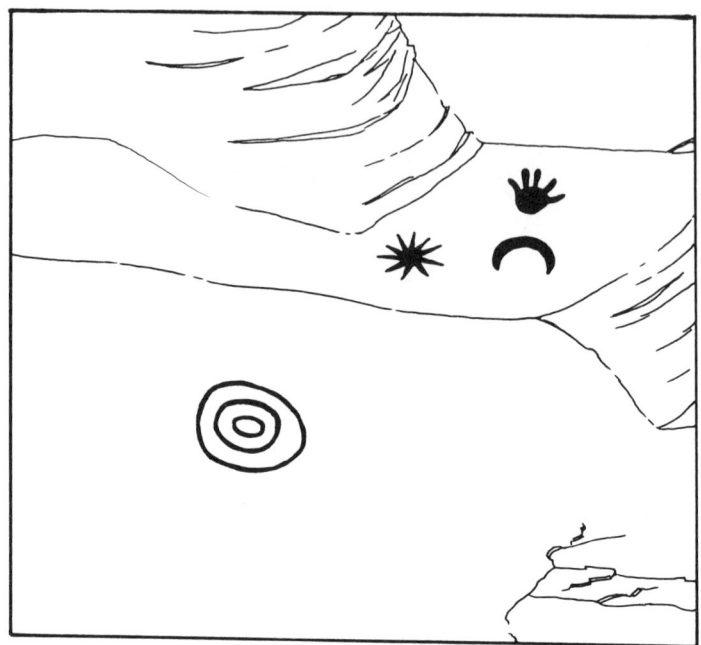

1 *The group of petroglyphs comprising what we believe to be an 11th century sun-watcher's station and shrine below Penasco Blanco, Chaco Canyon National Monument, New Mexico.*

questioned by anyone who knew Southwestern ethnology in some detail. If the supernova was visible in the Southwest, as is said to have been the case, the prehistoric Pueblo people no doubt saw it and were impressed. This is no reason to suppose that they recorded it. In his article on the Crab Nebula which is believed to represent the debris of the 1054 explosion, Oort (1957) notes that "Strangely, although the event must have been witnessed by practically everybody in Europe, not a single mention of it has been found in any European chronicle," though in China and Japan the advent of this new star was duly set down. Although at first it could be seen even in the daytime, after two years it no longer was visible to the naked eye. The literature on Southwestern Indians of the historic period and my own questioning in various villages indicates that the Pueblo people characteristically did not record exciting events. For example, although we are positive that the volcanic eruption of Sunset Crater in the San Francisco Mountains of northern Arizona in A.D. 1067 must have been visible to the Hopi on their mesas only 80 miles away and that peoples living near the crater were forced to abandon their ash-filled houses, their ash-buried fields, and even the region as a whole for some years, no petroglyphs recording this event are known. The Hopi themselves state that they do not even have legends dealing with this traumatic experience in which some part of their ancestors quite certainly were observers if not participants. We are equally without legends pertaining to some of the painful and exotic punishment of Pueblos in the Spanish conquest of New Mexico. The Pimas and Papagos of southern Arizona had calendar sticks, the closest thing to an actual record of events in the Southwest, but although one or more men of a village might choose to keep such a stick, it could be "read" only by the owner because his symbols were no more than mnemonic devices serving to recall selected events to his own mind. He used what symbols he liked to remind himself of what events he thought worth remembering just as we tie a string around one finger to remind a man to buy a bottle of whiskey, his wife to go to the hairdresser, and his son to put gas into the car. Only one symbol was consistently employed by all the Pima calendar keepers and always for the same event: a dot which specifically referred to their major ceremony. This "Prayer stick" ceremony paralleled the Pueblo solstice celebration in some respects, though given earlier in the fall and only every four years north of the present international border and every two years in northern Sonora (at least in the historic period), and one may wonder whether the dot could have been an abbreviation for the sun symbol.

Other standardized symbols were used in the historic Southwest, but for a different type of purpose. Sun, moon, four pointed stars, stepped elements, and triangles representing clouds which may have a fringe of falling rain, katcinas, the flute player, the horned serpent, mountain sheep, corn, snakes, the four or six lines crossed within a circle which Parsons (1939, p. 359) refers to as glyphs of the directions, the friendship marks of interlocking half circles, and the squares within squares which symbolize a sipapu (all identified by reliable Pueblo informants involved in their use) have been found on cliffs, in caves, on kiva walls, on boundary stones, as well as on masks, pottery, and other items. We know that when the Hopi went on one of their ritually-embellished salt gathering trips to the Grand Canyon, each man pecked his clan symbol into a certain great boulder as his signature, and when Eagle clansmen went out together for some duty they left their clan mark on a rock (Parsons 1939, pp. 358-360). The people of New Mexico similarly claim to have used clan symbols for signatures at places of prayer and repeated group activities; clanless Pueblos such as those of the Tewa presumably employed symbols to represent the names of their bilateral extended families. The most widespread sun symbol of the Pueblos (other than their sun mask used for decorative events) is the Maya glyph for sun (Ellis and Hammack 1968), obviously an import from the south as apparently was true of the water symbol and probably of some other glyphs. The main question of concern to us is why they were placed upon the rocks or elsewhere. Except for occasional scenes of hunting or warfare, the symbols could be said to be nouns with no implied verbs other than "was here" or "occurred" or "pertains to this spot". As discussed by Pueblo Indians, they relate only to the usual, the repeated. A major katcina associated with the Hopi Winter solstice ceremony, for instance, carries a crescent moon and the Big or Morning Star on his mask, designs associated with the solstice. Cushing (1967) wrote about Zuni sun-watching in 1893 done at the nearby ancestral pueblo of Matsakya; the pillar marking the spot carried exactly the complex of symbols found on the Chaco ledge. I have in my notes the account of a Rio Grande sun and new moon-watch from a conical white stone outcrop marked with a big star symbol, this watch coming in connection with the winter solstice, as we shall later describe. With Alexander Stephen's considerable body of data on the process of dating ceremonies and work periods and the symbols involved, collected largely between 1891 and 1894 when he was living at Keams Canyon and on the First Hopi Mesa, we need not guess about the Chaco group of symbols.

One way of knowing that the Chaco find of the four associated symbols does not pertain to the supernova is having the data to prove that this complex does pertain to observations pertinent to dating the solstice and the agricultural round. But basic to our understanding of this is some knowledge of the old native Pueblo ceremonial and agricultural calendar.

The Pueblo Sun, Moon, and Star Calendar

At the peak of the Pueblo was a Sky deity, sometimes said to have been sexless but commonly equated with the Christian God by the Rio Grande Indians today and possibly related to the Old Fire God. Of more immediate concern to the Pueblo people was Father Sun and Mother Earth who provided fertility, food, and warmth. Sun's apparent movement from the north to the south was understood to create the seasons. Since it appeared that Sun hesitated a few days in his southern "house", the Pueblos deemed it wise to put on a ceremony which would spur his decision to again take up his march northward. As part of the ritual at that time, the ceremonialists sought omens relating to the length of the growing season and hence to its success in the months ahead. The succession of horizon points over which the sun rose or set became calendar markers. The periodic reappearance of a new moon provided another reference for measuring time, and the consideration of moon and sun together was thought to strengthen chances for good omens because Moon could aid in influencing Sun. Some say that Morning Star, symbol of the beloved Twin War gods and also of the Star of Sky god already mentioned, was watched because it followed Moon so closely, but this is not all of the story. Prayers were made to Sun, Moon, the Morning Star, Orion, and the Pleiades at this time and on some other occasions. It is interesting to note that these heavenly bodies are pricked into the surface of gourd rattles used by Navajo ceremonialists, most of whose lore was borrowed from the Pueblos.

It is unfortunate that in generalizing, some students of native Southwestern calendar systems have rather overlooked the lunar and certainly the Venus features, though in other places the same persons may have described these matters. We read, for instance, that the

Hopi of northern Arizona establish dates for all their winter ceremonies by the position of the sun setting over the western horizon and the dates of their summer ceremonies by its position in rising above the eastern horizon (Fewkes 1897; Nequatewa 1931, p. 386; Parsons 1939, pp. 496-497). The statement also has been made that the Eastern Pueblos of the northern Rio Grande put less emphasis on sun-watching than the Western Pueblos, but we believe that this should be understood only in the present tense. The use of our own calendar was forced onto the Rio Grande peoples centuries ago by Spaniards who made dates to be met and imposed saints' days to be observed.

The lunar calendar which nature has neatly divided could be used alone even though lunar periods vary somewhat, but that would mean taking no account of the seasons. Moreover, lunar periods can be rather easily lost. Each so-called Pueblo "moon" begins with the appearance in the sky of the new crescent moon. The two preceding days in which no moon is visible are not counted; they fall between lunar periods. Named intervals in a lunar period at Hopi are spoken of as "new moon" or "crescent first seen", "moon vertical", "crescent moon horizontal", "moon half gone", and "last of moon" (Parsons 1939, p. 1040). The lunar calendar is not purely ceremonial, for both Stephen and Titiev report the Hopi planting according to the lunar as well as the solar sequence. Major ceremonies such as those which lead up to the winter solstice and those others which comprise the very important spring pre-planting complex are dated by the positions of both sun and moon, as if one were reading time by the two hands of a clock. In Zuni, prayer stick offerings still are made in each lunar period by the curing societies, the katcina groups of the Shalako complex, and the impersonators of Sun Youth. On some years we impatiently wait for a late announcement of the big Zuni Shalako celebration and finally hear that the Pekwin has found difficulties in reconciling the position of the Sun, which should be near his southern "house", and Moon, which should be full, a type of problem familiar to the priests of Mexico. Acoma and Laguna explain that Moon, to all the Keres except Zia, a female (as to the Zuni and the Maya), though male to the other Pueblos (as to the prehistoric people of the Valley of Mexico), must be in correct position so that she can bring her influence to bear on Sun for the good of mankind at the winter solstice.

The Pueblo calendar of thirteen lunar months (paralleling the old Maya and Mexican religious calendar), like that of the solar divisions, is best preserved today at Hopi, where every ceremony is associated with a moon and every moon with a ceremony. When a sequence of retreats (the private portion of a ceremony) is to be held by the religious societies, the first is dated by solar and/or lunar observation and the others by the count of days elapsed since that observation or since the preceding society's retreat. Tally cords with knots to be untied, sticks into which notches are cut, and marks on a floor or wall, which could be erased as the days passed, customarily were used in cutting the calendar into shorter intervals when necessary. If an official lost a lunar month, as Titiev (1972, p. 166-167) noted for Hotevilla, the mistake was covered as long as possible by patient and loyal followers of the watcher but eventually must be rectified by a public announcement that a later month must be skipped.

The data collected by Titiev (1938, fn. 2; 1944, pp. 174-176) in the 20th century and by Stephen in the late 19th (Parsons 1939, pp. 1036-1037) give us the names of successive moons, the month of our calendar in which each supposedly should fall, and the major activities reserved for it. To make the scheme more graphic we have added a translation of the names, the date on which the new moon initiating each lunar period actually appeared in 1892 and 1893, and also the calendar date for those moons 80 years later in 1972 and 1973.

Winter

Kel-muya --- November moon. Crescent on 20 Oct., 1892; 6 Nov., 1972. Known as "the initiate or sparrow-hawk moon" when youths receive their second or tribal initiation and thereafter (in the past) became warriors, of which various hawks are symbols. The making of new fire and the first appearance of katcinas also mark this ceremony, which leads up the Soyal or Winter solstice ceremony as the Shalako sequence does in Zuni. Wuwuchim is dated by the chief of one of the three important societies involved in watching the setting sun. At the right point he informs the Chief of Wuwuchim who calls his men that evening to smoke over prayer plumes and decide that the ceremony must be announced next day at dawn by the Crier Chief, but the date, 16 days hence, is that of the closing of the ceremony when offerings are deposited on shrines, a customary type of announcement and confusing to whites.

Kya-muya --- December moon. Crescent on 19 Nov., 1892; 5 Dec., 1972. Known as "the sacred but dangerous moon" because it includes the 5 days of imminent disaster from witchcraft, falling through the now fragile covering of the earth, and other frightening possibilities, a complex borrowed from prehistoric Mexico's feared 5 days at the end of their Vague year. The Winter Solstice ceremony, dated by observation of Sun's reaching his "house", the most southern sunrise point on the horizon, is elaborate and lengthy with all religious societies participating. It opens about 12 Dec., the date depending on being in certain relative positions, as elsewhere in the Pueblo area.

Pa-muya --- January moon. 18 Dec., 1893; 3 Jan., 1973. The "play moon"; when gaming and some minor dances comprise the entertainment because weather is too cold for outside activities. During full moon, beans and some corn are secretly planted in containers carried into Chief kiva which is kept heated to produce early sprouting and growth.

Powa-muya --- February moon. 17 Jan., 1893; 3 Feb., 1973. "The purification moon". This is the second lunar period after the winter solstice. The Mayan people of Yucatan held a festival at this time to open their agricultural season. The Hopi term for the period refers to the first initiation of young children in which the katcinas, who have come for the big Powamu ceremony to insure fertility and precipitation for the fields, administer a purificatory whipping to initiate and disclose the secret of masking to represent gods. The Powamu pre-planting

Pueblo Sun-Moon-Star Calendar

ceremony clearly is based on the elaborate masked month-long preplanting ceremony of post-Classic Mexico (Spence 1912, pp. 48, 56-58), which began early in May at the opening of the rainy season. At "horizontal crescent moon," the Hopi pull up the bean and corn sprouts to be distributed by Sun Youth katcina to women during his visit to springs, clan houses, and katcina clan kivas. More corn and bean seed is planted in containers in kivas, the height reached being interpreted as an omen foretelling success of the coming agricultural season. Uninitiated children receive gifts of katcina doll representations, rattles, etc. from the masses of katcinas who appear. At the Pamurti ceremony held at First Mesa in late January, Sun katcina arrives accompanied by the family of Chakwena, whose mask is glossy black with a horizontal crescent moon on one cheek and the many spiked Great Star on the other (duplicates of the Chaco petroglyph symbols). The Chakwenas are involved with warfare but also with childbirth.

Isu-muya --- March moon. 16 Feb., 1893; 4 Mar., 1973. The "moon of moderating winds". A second term, "cactus moon," refers to the considerable dependence on cactus for food at this time when stores from the previous summer's harvest were running short. During this moon the sun is watched, for when it reaches "the first point" a little early corn is planted in protected spots so that the katcinas will have it to bring in during their final Niman or Home-going ceremony, which falls before the main harvest.

Kwya-muya --- April moon. 17 Mar., 1893; 2 Apr., 1973. "The moon of setting up windbreaks for small plants". When the second solar point is reached, people get their seeds ready for planting.

Hakiton-muya --- May moon. 16 Apr., 1893; 2 May, 1973. "The waiting moon", when major planting cannot yet be done. At the 3rd solar point (mid-April), the leading families are told to plant in succession and others as they please, the first crop put in being corn. At the 4th solar point the first string beans, melons, and squash are planted and at the 5th the main planting of these is done. Any planted after the 6th point would not mature. At the 7th, lima beans are planted. At the 8th the general bean crop is put in.

Summer

Kel-muya --- June moon. 15 May, 1893; 31 May, 1973.
"The planting moon". The 7th and 8th points mentioned for the preceding lunar period may fall into this period if we take the actual dates given. Men are called for communal work in planting the fields of the Town Chief.

Kya-muya --- July moon. 13 June, 1893; 30 June, 1973.
At the 10th solar point, fields to be irrigated with run-off water are planted and at the 11th, sandy areas are planted, for winds have died down. The summer solstice rites at the 12th point, Sun's "summer house," mark the end of all planting for the plants could not mature before frost.

Pa-muya --- August moon. 13 July, 1893; 29 July, 1973.
"The moisture moon" when rains should begin. The big Niman katcina ceremony which opens 25 days after summer solstice ends the growing season, and katcinas go home to rest in the San Francisco Mountains until November. Chief of the Snake society watches the sun for 16 days after this moon opens and then starts the Snake-Antelope or, alternately, the Flute ceremony.

Powa-muya --- September moon. 11 Aug., 1893; 27 Aug., 1973. "The big feast moon". Four days after the Snake ceremony comes the smoke meeting of one of the three Women's societies, usually followed by a dance.

Angok-muya --- October moon. 9 Sept., 1893; 26 Sept., 1973. "The harvest moon". One of the four names which Stephen gives for this month is that which Titiev gives for the 13th month. The Oa'qul harvest basket dance by one of the Women's societies is followed by unimportant dances such as the buffalo and the butterfly.

Isu-muya --- October moon. 25 Oct., 1973. Not given by Stephen, but listed by Fewkes and Titiev.

The problem in length of growing season and in actual dating of moons is well illustrated in the difference of sixteen days between the record of the same new moon in Stephen's time and our own, but even in using Stephen's dates alone, one sees that the moon expected to fall in a certain one of our months often actually appears within the preceding month.

The Hopi think of the moons of their calendar as being in a continuous succession which we can best picture as a moving chain encircling a wheel, the upper half of which represents our world and the lower half the underworld. With a strong feeling for duality they sometimes speak of two divisions of the year because they believe, theoretically at least, that when we are experiencing any given winter moon, the underworld moon is at an equal position in a summer season. Major ceremonies given in one season characteristically have a short counterpart performed in the opposite season when the dead who reside in the underworld are believed to be performing their ritual, for everything pertaining to the dead is opposite to that which pertains to the living. This explains why the summer solstice ceremonies are minimal in comparison to the big celebrations of the winter solstice. Stephen could obtain no names for the last five months of his year, the group being lumped as moons without names. In the 1930's Titiev found five of the names of the first seven successive moons, all referring to seasons, agricultural conditions, or ceremony, repeated in the names of the next six lunar periods, but even some of the chiefs had difficulty remembering the names of the lunar periods in order between September and the first of November. The shift in dates of successive new moons and of the same moon in successive years is one cause for confusion of the natives and even of some anthropologists. If we figure our 13 moons as averaging 29 days from crescent to crescent, we have 377 days, some 17 more than those of the solar year. By accurate reckoning, the named moons thus would move farther and farther from the seasonal breakdown of the solar year for which the ceremonies were primarily intended as an aid to growth. Moreover, crops would suffer as lunar dates for planting and other agricultural activities shifted in relation to the solar calendar. Something obviously would have had to give, possibly by the priest-chiefs occasionally skipping one of the nameless moons to bring the calendar back to reality.

Stephen found that in his day (late 19th century) the two portions of the native year, taken together, could be referred to either as one year or as two years

(Parsons 1939, p. 1039). Apparently the Hopi, and presumably the other pueblos, were thinking in terms of two chains encircling their wheel at once, one chain made up of alternate groups of seven and six successive moons and the other of two divisions marked off by the solstices in the 365 day solar year. In one of Stephen's notebooks (1883-1884) from ten years earlier, however, he remarked that the katcina or ceremonial year was reckoned in 13 moons but the work year was computed in two successive periods of eight moons, each of these 240 day periods being known as a "lesser year". The first lesser year began with the Initiates' moon in November which opened the winter solstice sequence. It ended in June. The second opened in July and ended in February. The third began in March and ended in October. The fourth, like the first, began in November. Each bundle of four lesser years were known as "the Year of the Great Moon". Stephen does not refer to this eight month work division again and may have been in error, though this seems unlikely if we think of the detail in this scheme. Parsons thought it quite possible that this was the old Hopi system, already dying out in the 1880's, and being replaced, with attendant confusion, by the seven and six month divisions of the 13 moon katcina calendar, which bought the native system closer to that of the whites.

The old Zuni system duplicated that of the Hopi, with a two part year of seven named moons followed by six moons sometimes said to be nameless, though Cushing states that they took the color names of the six directions. Stevenson (1904, p. 108) speaks of the two six month divisions of the year at Zuni, her informants evidently modernizing their system at that time. Each of these months then was broken into three parts, a division not mentioned elsewhere. Like the Hopi, the 20th century Hopi-Tewa recognize a 5 moon summer and a 7 moon winter season, 5 of the moon-period names being repeated in each season (Parsons 1925, p. 1041), a capitulation to the total of 12 in our calendar.

Official Watchers of the Celestial Bodies

Who does or did the sun and moon-watching and made adjustments when necessary? Today some peublos such as Santa Ana are without sun-watchers and merely note when the gold of turning aspens creeps down the western slopes of a near-by mountain. Other pueblos first observe the date on one of our wall calendars and then send out a representative to observe the sun and/or moon shortly before a specified ceremony, but in the past and even today in some of the really conservative villages several men have been involved at once in sun and moon-watching and in discussion and interpretation of the results.

Answering the question of sun watchers necessitates taking a moment to meet the array of Pueblo officials in the old native system who represent supernatural beings of the pantheon from whom they take their authority. The secular officers of the New Mexico pueblos (governor, lieutenant governor, etc.) were imposed by Spain so that dealings with the religious hierarchy which controlled each village could be avoided. For present purposes we can forget the secular slate. Although the native political systems and officials differed somewhat between villages, the two most important religious officers in each were the Town Chief, now generally known as the cacique in Rio Grande villages, and the War priest or Outside Chief. The Town Chief (two in each Tewa town of the northern or Upper Rio Grande who alternated between the long summer and the short winter seasons but only one in each of the other pueblos) was chief priest and generally said to represent Mother Earth. He spent much time in prayer for rain and fertility but he also was a major sun-watcher and after the observations were made he scheduled the principal ceremonies of the year for his own village. Pertinent data from the New Mexico pueblos are meager except for Zuni and one village of the Middle Rio Grande, to be discussed later. At Hopi in northern Arizona where the Town Chief is said to be the earthly representative of Eototo, leader of the katcinas (Titiev, p. 114), the system differs somewhat. On First

Mesa a Sun chief, as such, watches the sun after the
summer solstice and through the winter solstice. The
leader of the Flute Society then takes up the duty, to
run through the summer solstice. Oraibi, the Third Mesa
Hopi village which was so disrupted in the split of 1906,
shows an approximation of this system (Parsons 1939, pp.
502-504 and fn.; Titiev 1944, p. 17). The Hopi have one
site for summer sun-watching and another for winter observations, which seems to have been common in the other
pueblos. The division of sun-watching periods does not
correspond to the two sections of the Hopi ceremonial
year in which ceremonies of the agricultural period, from
February (Powamu) to late July (Niman), which heavily
feature the katcina rain-fertility spirits, are controlled
by the Katcina clan chief and the shorter period from
Niman ceremony (July) to Powamu (February) by the head
of the Badger Clan, claimed to have been first to emerge
from the underworld and hence a man of power. Here we
see again the ceremonial division into 13 parts and the
agricultural division into two, though the attempt to
tie plant growth, heavenly observations, and the katcina
rain-fertility spirits into a threesome is obvious.

Among supernatural beings Sun, as the embodiment of
male qualities, is a warrior and patron of warriors,
though his two sons, the Twin War gods, especially the
older and leader of the two, were even more specifically
the patrons of warriors. In the pueblos of New Mexico
these sacred twins are thought of as youths but the Hopi
tales depict the twins as children and give the War Star
God, the Sky God already mentioned, the qualities of the
older of the Twin gods as known in New Mexico. Both personages are symbolized by the Morning Star. In the
Southwest, this was the most important of stars, even in
Mexico where it symbolized Quetzalcoatl, who was more or
less the prototype of our Sky or Star god. As far as we
can make out, in early times each pueblo had a War Priest
or Outside Chief who served for life as representative
of Sun. Working with him were two men who represented
the Twin War gods. As the War Priest was required to
have taken a scalp, that office has largely disappeared,
though it still is found, without scalp requirement, in
Hopi, Zuni, Acoma, and Jemez. In Zuni we still have the
Pekwin or War Priest and the two Priests of the Bow represent the War gods. The representatives of these two

War Gods in the Rio Grande once held lifetime offices, it is said, but since Spanish days, and by Spanish decree, they have been re-appointed or re-elected annually and are known as captains, a position the Spaniards could understand. They have taken over some of the duties of the old War Priest and the Town Chief has had to accept others. The Star God is represented in Hopi by a Star Chief who emphatically twirls the sun symbol in one of the winter solstice rituals as a hint that Sun should hurry his start northward once more.

The association between stars and war was universal throughout the pueblos. A five pointed star (the Evening or Big Star could have four or more points) was drawn on the wall of the War Chief's chamber at Hopi and on interior walls of a number of other religious chambers and effigies of Orion and the Pleiades were used on his altar (Parsons 1939, pp. 9, 84, 87, fn. 1; Figs. 2, 5, 64a). Bourke reported that for Hopi in 1883 "upon the walls of the estufas (kivas) are rudely etched and painted the symbols of sun, moon, and Morning and Evening Star, and Pleiades" (Bourke 1884). In Hano, also known as Hopi-Tewa (the First Mesa village established in the early 1700's by Tewas from the northern Rio Grande who chose to avoid the Spaniards by moving to Hopi) the sun-watcher for the winter solstice was keeper of the sacred images of the War gods, whose symbols were the Morning and the Evening stars. The Town Chief is the watcher for the summer solstice, and the six stars of Orion, which make a backwards figure seven, are watched in May to be interpreted as an omen relating to a long and hence good summer. These stars, which are painted on the wall of the Chief Kiva on the Third Hopi Mesa, are shown with four points, and in this case Morning Star also has but four.

In the Rio Grande, the eastern edge of Pueblo country, the oldest religious society of which we have record is the Koshare, whose members are known as "sons of the sun". A bowl decorated with a Koshare figure found in Sapawe, one of the ancestral Tewa pueblos, proves that this society is at least 600 years old. The chief of the Koshare, as might be expected, is one of the sun-watchers, as well as being of utmost importance in village councils. Both the caciques or Town Chiefs do sun-watching in the northern Tewa pueblos, the place of the sun's setting being watched for the winter solstice and the place of its rising for the summer solstice and other dates. The duty of watching the Evening and Morning Stars and the constellations fell to heads of the religious societies. The movements of the Great Dipper as well as of the six stars of Orion and of the Pleiades were used for timing during the night and the shadow of a small stick set upright into soft soil or the position of the sun in its arc served for timing during daylight. In Nambe, solstice dating was handled by the caciques noting where sunlight coming through an east window struck a buckskin hanging on the west wall. The rising stars of Orion's belt are watched in relation to both solstices; the statement is that if these appear on 4 and 6 May there will be neither early nor late freezes (Parsons 1939, pp. 175-176, fns.). The Morning Star and the constellations of Orion and the Pleiades are clearly depicted on one of the ritual vessels which held corn meal for morning offerings to the sun at Sapawe, the Tewa ruin mentioned above (excavated under the direction of F. H. Ellis for the University of New Mexico, report in preparation).

We can summarize by saying that a religious officer associated with Mother Earth customarily functioned as sun and moon-watcher and possibly also as star-watcher during the long growing or summer period, which included the summer solstice, and that an officer associated with Father Sun and winter handled this watching from that time through the very important winter solstice, at least while such an officer was available. Working with them were representatives of the Sky or Star God and/or the older of the Twin War gods.

Symbols and Ceremonies Pertaining to the Most Important Celestial Events

Let us now come back to the problem of the group of associated petroglyphs found in the Chaco, our approach to be through known ethnographic data. The simplest form of sun symbol throughout the Pueblo area consists of two concentric circles with a dot in the center, the outer symbol symbolizing his rays, the second circle symbolizing his body, and the dot being his umbilicus from which all good things are supposed to come for mankind (Ellis, Field notes pertaining to Rio Grande and Hopi). Rays sometimes were added all around the outside circle or in projecting groups, usually four in number. A very fine example of a large sun symbol with rays and lesser representations of several in smaller scale, with and without rays, are to be seen on boulders just below the village of Shabik'eschee toward the eastern end of Chaco Canyon seven miles up from Pueblo Bonito where they unquestionably mark a sun-watcher's shrine. On the date of the fall equinox, 21 Sept. 1973, the sun rose over a very distinctive projection on the eastern horizon. Its first rays settled on the big Shabik'eshchee sun symbol (C. M. McLeod, observer). On a direct line between the rising sun and the stone, a $94°$ azimuth, at a point 2/3 mile out from the stone on a small point below the cliff top is a collapsed cairn. This, one may infer, was a shrine where prayer plumes and perhaps other offerings would have been placed, but the "home of the sun" in the pueblo, would have been at the big stone with the sun design where the rays of the rising sun first struck, if we may judge by the beliefs of today. The moon, important in dating and also because it supposedly exerts an influence upon sun, usually was represented by a crescent because time periods, each known as "the moon", began with the first appearance of the crescent moon, but half and full moons also were depicted. A crescent lying on its back was considered to be a bad omen because it was dry, but one with tips pointing downward or in vertical position was good because it was a wet moon. Parsons wondered if these interpretations could have been borrowed from white neighbors, but the concept extends from the eastern to the western pueblos and appears deeply ingrained. Sun and moon could be used together in dating and in foretelling the future by omens. Whether the 584-day Venus calendar which the Mixtec-Cholula people of Mexico thought important was of concern to the Pueblo priests we do not know, but Venus as the Morning Star certainly was very important. Although the most common Pueblo representation of a star was and still is an equal armed cross, the Great or Morning Star could be shown with five points or surrounded with spikes.

The hand design which accompanied the sun, moon and Great Star symbols in the Chaco group is rather frequently found in groups of pictographs over the southwest. This is a common Pueblo signature mark but with religious rather than secular connotations. Girls who have completed the replastering of a kiva finish by drawing cloud and lightning designs on a beam and add handprint, whether in Hopi or in the Rio Grande. Stephen (Parsons 1939, pp. 198, 202) was told that this symbolized the girls' prayer for rain. I have been told that at least in some pueblos the leader of a religious society which had completed a ceremony dipped his hand into wet clay and signed off by pressing it onto a wall to show that he had completed his duty. A footprint also could be used similarly by the Pueblos. Shrines in the open could be marked with a handprint so that the supernatural beings would know who was responsible for the prayers and offerings made there and hence could make accurate reciprocation (Ellis and Hammack 1968). Further, we have the intriguing note (Vaillant 1944, p. 197) mentioning the "imprinting of hand impressions" by captives who were to be sacrificed at the Aztec winter solstice ceremony (a southern use of the hand at this date possibly related to "the sacred hand" at Zuni) and the note that at a feast shortly preceding this and referred to as "the arrival of the Gods" (something of a prototype in concept to the Zuni Shalako known by the same term) the announcement that one or more supernaturals had arrived depended upon finding the print of a foot in a small cake of cornmeal set out for this purpose (Anderson and Dibble 1951, p. 21), again something of a signature mark.

We have spoken so far of crescent moons and solar horizon observations, but that which Cushing describes for opening spring agricultural work at Zuni was of a somewhat different sort. In the 1500's the Zuni ancestors were living in several pueblos in the Zuni Valley, all of which were abandoned in the Pueblo Revolt of 1680. Survivors constructed the Zuni of today across the river from the site of one of those old towns. Matsakya (Matsaki: Mat'saka) when first seen by the Spaniards in 1581 was a village of a hundred houses four and five stories high, arranged in compounds, but when Cushing was at Zuni in the 1880's Matsakya had been only a ruin on a hill at the northwestern base of To'wayal'lanne, Thunder Mountain, for two centuries. Yet it was not forgotten. When Cushing (1967) was living in Zuni in 1893, each morning at dawn in late February or early March he saw the Zuni Pekwin, their Sun Chief, walk three miles up the valley to Matsakya followed by the head Bow Priest, representative of the older War God (Morning Star). While the Bow Priest waited at a distance, the Pekwin entered what Cushing calls a small tower and sat down before a pillar marked with the symbols found in our Chaco complex and waited, praying and chanting, for the sunrise. His concern was not simply the point at which the sun rose over the horizon but on which morning "the shadows of the solar monolith, the monument of Thunder Mountain, and the pillar of the gardens of Zuni" lay in a single line. When this occurred, the Pekwin thanked the sun, the Bow Priest cut a notch "in his pine wood calendar", and both hurried back to announce that the time to begin farm work had arrived. To many this meant moving from the village out to field houses near the plots they would cultivate.

Cushing adds, "Nor may the Sun Priest err in his watch of Time's flight; for many are the houses in Zuni with scores on their walls or ancient plates imbedded therein, while opposite, a convenient window or small port-hole lets in the light of the rising sun, which shines but two mornings in the three hundred and sixty-five on the same place (See Fig. 2). Wonderfully reliable and ingenious are these rude systems of orientation, by which the religion, the labors, and even the pastimes of the Zunis are regulated." We may recall that on 7 March and six months later, but on no other days of this year, a beam of sunlight entered the dark inner room of the 15th century great house at Casa Grande in southern Arizona by passing through a narrow tubular hole in an inside and an outside wall of the building and probably a matching hole in the surrounding compound wall (Pinkley and Pinkley 1931).

For dating the summer solstice, which like the winter solstice is referred to as "the middle" (of time), the Zuni Pekwin, who is solely responsible for solstice dates and who may be impeached for what is believed to be an error in observation or a "bad heart" which will effect crops adversely, watches sunsets from the pillar at the sun shrine of Matsakya. The sun has been moving northward and passes the moon at a specific named spot. At about the end of May, when the sun descends behind a certain point on the mesa to the northwest, the Pekwin calls out for all Zunis to prepare prayer sticks for Sun, Moon, the dead, and the katcinas. General prayer stick planting is on June 22, after which the summer retreats for rain begin. The solstice is considered to fall on the last of the four days during which Sun sets at the same spot behind Great Mountain, northwest of Zuni. Winter observations are taken from a petrified stump (petrified wood which resembles bones being symbolic of the War Gods who slew the early monsters which preyed upon man) on the east side of the pueblo where sunrise points over Thunder Mountain are noted. When the sunrise reaches a certain point, he notifies the elder brother or head Bow Priest, his lieutenant, and the latter in turn notifies the first group of religious societies to hold a retreat. Two prayer sticks for Sun Father and two for Moon Mother, his sister, are placed in the sun shrine on Thunder Mountain and four for former Sun priests are deposited in a cultivated field. Four days later four more are placed in the sun shrine and after another four days four more are placed in the field. On the morning after the final planting the Pekwin announces that in ten days the sun will rise from "the Middle Place", his southern house, the solstice marker. During that period the Pekwin continues his sunrise visits to the stump for observations and prayers. In 1891 the actual ceremony began on 22 December, after general prayer stick planting on the 21st, and continued for eleven days (Stevenson 1904, pp. 108-141; Bunzel 1934, pp. 534-540).

Between the first and the fourth days, prayer stick offerings are made by members of the religious societies for the sun, the moon, the War Gods, the lightning makers, the ancestral dead (rain makers: katcinas) of the six cardinal points, and the Beast Gods who give aid to curing. We know that the Priestess of Fecundity (today no longer existent), their one female officer and representative of Mother Earth, a parallel to the Town Chief or cacique in the other pueblos, prepared "a number of cotton cord loops" which made up or represented a small sacred white blanket as a gift for the sun. All

2 *A stone plug for a small window of the type found in prehistoric Pueblo houses.*

prayer sticks except those made as personal offerings of the makers or their families are tied together, later to be deposited at the base of the knoll on which the sun shrine observatory stands in Matsakya. Most of the religious societies go into retreat on the second day after the Pekwin's announcement. For eight days there are ceremonies, including those intended to cure the sick, and visiting between the fraternities is prevalent. New War God images are carved from cottonwood, "all varieties of seeds" being placed in a hole representing the umbilicus in the body of the older god and covered by a serrated horizontal projection which is inserted and then decorated by attaching a miniature bow and arrow, shield, and war club, and plumes to symbolize clouds and lightning. An abalone shell pendant is hung around his neck, a belt of raw cotton is added at the waistline, and a netted shield is made for each image to stand upon. The figures are surrounded with prayer plumes, and gaming items for the older god are prepared. After the images have been honored by a kiva display and rituals they are placed in their respective shrines on Thunder Mountain. Among the many activities of this period is the plucking of one of the live captive eagles to obtain the needed soft eagle breast feathers, the selection of a man to make the new fire, the ceremonies held by members of every kiva, and further depositing of prayer plumes in other shrines and in holes dug in family fields. The Shalako, which celebrates the annual visit of the Council of the gods to Zuni and of the six great bird images who will carry the prayers of the people to the cardinal directions, is the most famous in the sequence of the many specific ceremonies (Stevenson 1904, pp. 108-141; Bunzel 1934, pp. 534-540).

Mrs. Stevenson (1904, fig. 3) has provided us with a photograph of the little "tower" at Matsakya sun shrine, which she describes as made up of a "stone wall, semi-circular in form, about 3 feet high, the inner space being 3 feet wide and opening to the east. A sandstone slab about 2 feet high and 14 inches wide, with a symbol of the sun 4 inches in diameter etched (pecked) upon it stands against the apex of the wall. A smooth-surfaced stone, on which are cut a number of lines, is inserted in each side of the wall about 8 inches above the base. Some of the priests declare that the lines on the south side of the wall indicate the number of years the previous Sun Priest held the office and the one on the north side the number of years ... (his successor) served. Nine concretions form a square on the ground before the etching of the sun, and there are three smaller ones in line in front of these. Concretion fetishes, valued as bringing fructification to the earth, are found in all the fields. A small flat stone rests on two of the larger concretions". That flat stone may have been the seat on which Cushing saw the Sun Priest place himself. The slab evidently was erected in front of the open side where observations were made. In the photograph one can make out the sun symbol on the face of the slab, four groups of projecting lines coming out from the circle to represent rays as in the Zia sun symbol copied by the state of New Mexico on its flag. Two eyes and a mouth provide features for the face. The other glyphs described by Cushing unfortunately are not visible in this dim photograph.

Very little has been written concerning calendric observations on the Rio Grande, but the word picture we have obtained from one Tanoan pueblo, which must remain nameless, adds appreciably to our understanding of the

Pueblo concept of Sun, coordinated celestial observations, and interpretation of shadow in relation to another "solar monolith". In the mythological system here, Sun's southern house is said to be Eagle Pueblo, the eagle being symbolic of Sun throughout the Southwest. There the Sun clan developed and there the arrangements for Sun's movements are made. In his manlike form, Sun is personified as Patiabu, Sun Man who was created in the underworld, the womb of Mother Earth, whence came mankind, plants, animals, and even the supernatural beings. Sun's body was covered with bright paint in all colors, and when he came up into our world his first gesture was to rub off some of this paint onto the plants so that man would have flowers. Flowers thus came to be symbols of Sun (probably a concept borrowed from Mexico). Before any man was permitted to emerge from the underworld, each was taken to an altar to be painted red, black, yellow, or some other flower color, like Sun Man. This was their initiation into the Sun cult and the hard and exacting life of ceremonial activities and precepts of old Pueblo pattern. Today young men may protest at the requirements imposed upon ceremonialists, but the elders believe that as long as one does not abandon the responsiblities imposed on his ancestors by the example of Sun Man and first accepted by Badger Man, he can be assured of good fortune through life.

Within a pueblo, the shrine or home of Sun Man is where the rays first strike in the morning. This also is the home of Moon, here a male spirit. The winter solstice ceremony is thought of as centering in offerings, which are spoken of as the "results of the ceremonies" such as prayer plumes and the new clothing for Sun, Moon, and Evening Star, all to be placed at the sun shrine in concluding the ceremony.

The top man in the little body of sun-watchers in this pueblo is the leader of the Koshare society. He, as Sun Chief, cares for the sun fetish, and wields more authority than any other man of the village. As representative of Sun, his presence in the local Catholic church is requisite to performance of the mass. It is he who gives orders for all the religious societies to perform the necessary ceremonies, including recognition of Sun, for the incoming secular officers just after their installation in January. The other sun-watchers are the Town chief or cacique, the War Priest (still an office of power here), the head War Captain who represents the older War God, and the leader of the important priestly group which represents Mother Earth and directs one of the two men's societies into which all youths are taken. It is said that at one time in the past this group of observers included a Star-watcher, presumably a representative of Big Star-Sky god as at Hopi.

When the time of the June solstice is known to be approaching, the sun-watcher scatters a corn meal offering and walks a short distance north of the pueblo to a stone slab on which he stands while noting the point of sunrise. Sun's "summer house" is reached when that body rises over a narrow peak known as "the heart", across a little arroyo from the watcher. When the cacique receives the water's report, he orders the sequence of religious society retreats due at this season. But there is yet another observation to be made. Late in the afternoon on 24 June, celebrated as San Juan's Day by all the Eastern Pueblos since Spanish times and close enough to 21 June to substitute for the Summer solstice in today's calendar, the head War Captain and the War Priest go to a solar monolith at the western edge of the pueblo. This is a white outcrop some two feet high and six inches thick, with an equal-armed cross representing a star cut into its east face. The shadows of the mountain range to the west are watched as they fall across this stone. If the shadow completely covers the stone, the omen is interpreted as foretelling a long summer; if the shadow is short the summer also will be short and early frosts will ruin some late crops. The watchers at once hurry to announce the result of their observation to the religious society which is in retreat on this day and its members rejoice or lament according to the shadow's length.

The watchers for the winter solstice are the War Priest and the head War Captain who, each evening, note the place of sunset and keep track of the appearance of the moon from the crest of the ridge on the east side of town. The moon is believed to travel between the north and the south just as the sun does but at opposite seasons so that their paths cross at one point. The matter of prime importance to these calendar priests, during the watch which precedes the date of the winter solstice, is to observe Sun and full Moon exactly when they most closely approach each other, a problem duplicating that of the Zuni Pekwin in trying to properly place the Shalako celebration of the same period. The meeting of Sun and Moon is said to be guarded by the Cacique's society. Should this conjunction not be seen, Moon, which competes with Sun, might steal a little of his time and the year thus be shortened by frosts before harvest. Another problem in some years is Sun's tendency to dally for three or four days in his "summer house", rising at the same point on the horizon repeatedly before starting southward. This also shortens the agricultural year and permits a freeze before Sun and Moon meet. The summer ceremonial cycle, which consists of two series of four day retreats for each religious society, now begins on the calendar date of 20 June and is intended to end about 15 August. If frost occurs the retreats are dropped, for their function is to provide constant aid for growth and maturation of crops during the agricultural season.

    The Winter Solstice ceremony begins about 12 December, the exact date being announced by the Town Chief after the observation of Sun and Moon's meeting. He opens the ritual with the request that the ceremonialists provide the Evening Star deity, the Morning Star deity, the Pleiades, the seven stars of Orion, the Sun and the Moon with clothing, food, and "medicine" so that they may have power to sustain themselves and bring about a long and fruitful year. All of the societies hold a four day retreat in their separate houses, which is followed by a night of dancing in each other's houses, much as in the Hopi Powamu and after the Zuni Shalako. One of the features of this celebration always is the making of new fire with a fire drill and a block of dry wood, another being held in reserve and a loop of cord at hand to make the twirling of the stick more effective in case a spark is not obtained at first. This new fire is guarded in the house of one of the women's societies for four days. At the end of the solstice ceremony, the deities are presented with miniature clothes. During the retreat, each society leader has set up a loom and begun weaving a small feather blanket (2 ft. x 15 in.), the feathers being laid across tightly twisted warp cords of home-grown cotton and held in place by a loosely twisted weft cord. The top strip, with eagle feathers, later will be clipped off to serve as a wrap for Sun. The second strip, with pale yellow feathers from the small "wild canary", will be clipped off for Moon. The third strip, with bluejay feathers, is explained as being for Fire and the stars. We already have mentioned the small white blanket prepared for Sun at the winter solstice in Zuni. Clothing also is offered to Sun in Acoma, Isleta, and probably through all the pueblos.

Summary

    In closing our hurried resume of what at present is known of the use of celestial observations in the Pueblo Southwest, we find the immediate concern to have been primarily the calendar and the foretelling of omens pertaining to the important economic matter of crops, both within the all-encompassing mantle of native religion. Basic concepts and even much of the detail originated in Mexican and Maya originals. The handling, like Pueblo culture itself, was simpler than that in Mexico, though it may have been appreciably more elaborate in late pre-Hispanic times than after its integrity and integration were appreciably broken by a long combination of Spanish and then of Anglo force and persuasive influence. Sun, the Morning Star, Moon, and a Sky God sometimes confused with Sun and sometimes with the Great or Morning Star were the chief recipients of attention in their own specific movements and in the inter-relationships of those movements and, understandably, in receiving compensatory rituals and offerings in return for their cooperation and services to man. The officers of the native religious system, who were responsible for keeping track of the movements of these heavenly bodies and in calculating the native calendar, were the local representatives of those bodies, especially in winter, and of Earth Mother, especially in summer.
    The year was cut into two equal divisions by the observed solstices, for which the associated ceremonial activities were intended to keep Sun functioning properly, for he and Mother Earth were responsible for fecundity.

That equal dual division was less used, however, than one into two unequal divisions comprising thirteen lunar periods which started with that of the winter solstice and was intended to separate the agricultural period from the non-agricultural period in which concentration was upon warfare, hunting, and curing. A more elaborate calendar system of the past is hinted in Stephen's early notes, but we have no evidence of interest in such matters as lunar and solar eclipses other than a brief note that such events were feared as possibly causing the death of children and even of unborn infants (Parsons 1939, pp. 86, 181 fn.).

Like the Aztecs, the Pueblos distinguished divisions of the night by the position of the moon and of Orion and the Pleiades, and to a lesser extent, of the Great Dipper. Their rising or their passage over the uncovered hatchway, which served as a ladder opening and smoke hole directly above the kiva fireplace, commonly was watched for timing rituals. The position of Morning Star, a patron of hunters and warriors, the recipient of prayers for propagation of domestic animals today, and most revered of all heavenly bodies other than Sun and Moon, also was considered in timing. Beyond this we really know little about man's relationship to the stars. Stephen (Parsons 1936, pp. 857-863) reports that Aldebaran, the Broad Star, received prayers in a Hopi curing ritual. The Galaxy or Milky Way is revered by all the Pueblos. On Zuni and Acoma masks and altars it is represented by a white band or a band of black and white squares or a ladder, for like the rainbow it was believed to provide a bridge from earth into the heavens, but the occasional addition of eyes and mouth (Parsons 1939, pp. 182, 340 fn.) indicates that it was personified like the stars as a whole which are spoken of as Night People, "our fathers and mothers", and "little priests". There are statements about stars of cardinal directions, but Parsons found no recognition of the North Star, perhaps because it was too dim to have served as a northern marker in the time of their prehistoric ancestors. The orientation of prehistoric pueblos and of kivas is a subject which may provide some insight into Pueblo beliefs. If we examine the orientation of existent Pueblo towns (Stubbs 1950) we find that all but Sandia, Zia, Zuni, and all the Hopi villages other than Mishongnovi are oriented with

marked accuracy to the cardinal points. This undoubtedly
has to do with intent of the leaders who originally laid
out the sites and properly buried a shrine offering in
the plaza and established a directional shrine on each
periphery. Rio Grande kivas characteristically were
oriented to the east in historic and prehistoric sites
but in the Mesa Verde, Chaco, and Kayenta areas, the
preferred orientation was to or toward the south. Why
the difference?

   Could the much discussed "towers" of the prehistoric
Southwest have been observatories from which designated
calendar priests kept track of the movements of heavenly
bodies? This was my proposal at the meeting of the
Society for American Archaeology in Mexico City in 1970,
and I still believe that their structure, which raised
them somewhat, if not above the roofs and trees within
a village area, together with their common association
with kivas in the Mesa Verde and Chaco areas, makes this
a very possible explanation, though they probably also
were used for smoke and fire signals. The fact that
structures dedicated to Quetzalcoatl especially in his
guise as Ehecatl, the Wind God, in Mexico were circular
and that this deity eventually arose to become the
Morning Star would fit into such thinking, and the
little "tower" at Matsakya still used for at least solar
observations by Zuni suggests a residue from something
more important in the past. This certainly does not,
however, proffer a hypothesis that all observation of
heavenly bodies formerly was done from towers, the distribution of which was relatively limited, and which may
have been used primarily for watching stars even within
the limits of that geographic distribution. The presence
of towers certainly points to a greater sophistication
of culture in some prehistoric Southwestern districts
than in others, but with our present knowledge we hardly
can say more. We reiterate our statement that the group
of petroglyphs near Penasco Blanco in the Chaco quite
certainly was a sun-watcher's station, pre-dating that
at Matsakya by 300 years. About 15 miles northeast of
Zuni is the site known as the Village of the Great Kivas,
contemporary with and very closely related to the big
Chaco pueblos, and on the cliff face just above the talus
which rises behind that site are petroglyphs. One pair
consists of an equal armed cross or star and a horizontal

crescent moon with tips turned down. Near it is the
simplified drawing of an owl and a long zigzag line
which Roberts (1932, p. 151, plate 62b) was told by Zuni
workmen referred to a tale told to children by the War
Chief, the owl coming at night to lead the chief to the
camp of their enemy, the Navajo. Roberts carefully
explains that he cannot say whether the explanations he
received were or were not those of the prehistoric people
who made the designs. Nor can we, but we are dubious of
the moon and star symbols having pertained to the story
given. That association of crescent moon and star, we
would say, is more likely to have marked a sun and moon-
watching station. We have another example in the Chama
drainage of the Upper Rio Grande where, in the center of
an open cave, the hand, the sun symbol of concentric
circles, a Great Star and a lesser star (Morning and
Evening Stars?) have been painted onto the natural rock
wall. In 1955 Miller suggested that two northern Arizona
pictographs showing a vertical crescent moon and a circle on a rock surface, one at White Mesa and the other
in a tributary of Navajo Canyon, depicted some special
celestial event, possibly the supernova of A.D. 1054.
I would be inclined to interpret these as probably markers for sun-watcher's stations, the sun in this case
being a circular plaque. These complexes lack the hand
and the star, and the Chama drainage example lacks the
crescent moon, but from our knowledge of the white stone
outcrop merely marked with the cross-star symbol but used
as one of the sites for calendric divination and dating
by members of a Rio Grande pueblo today, we can be sure
that the complex of four features need not be found together on a station. Miller has illustrated the possibility of his hypothesis by records of the crescent moon
having been in the sky on 4 and 5 July of that year, but
we would quote the Pueblo people again in saying that
although the supernova may have been seen and discussed,
there is no reason to think it would have been recorded.
The long and very serious concern of Pueblo people with
the solstices and the lunar periods, however, is amply
documented.

5

Early Navajo Astronomical Pictographs in Canyon de Chelly, Northeastern Arizona, U. S. A.

Claude Britt, Jr.

Chinle, Arizona

General Description of Canyon de Chelly

Canyon de Chelly is a spectacular sheer-walled, red-rock geologic phenomenon located in the center of the Navajo Indian reservation in northeastern Arizona. The mouth of the canyon is at an elevation of 5500 feet, while the canyon rim at the deepest point is approximately 7000 feet in elevation. Canyon de Chelly National Monument comprises a network of tributary and side canyons. Canyon del Muerto, the major tributary canyon of de Chelly, was so-named after the discovery in 1882 of natural mummies in dry caves. Wild Cherry Canyon, Spring Canyon, and Monument Canyon are other well-known tributary canyons of the labyrinth of the Canyon de Chelly network. In depth, the canyon ranges from approximately 30 feet near park headquarters to 1020 feet at its deepest point. In size, the national monument encompasses approximately 84,000 acres.

The canyon system is noted for its numerous pre-Navajo cliff dwellings. It is estimated that there are more than 1000 archaeological sites within Canyon de Chelly National Monument. When the first Navajo people came to this area, they saw numerous ruins all around them. Thus, they were aware that people had lived there before their arrival. They did not know who constructed these ruins, so they simply referred to them as the "Anasazi", meaning "the ancient ones" in the Navajo language. Today, archaeologists use the term Anasazi to refer to the prehistoric culture who inhabited this portion of the southwestern United States from approximately A.D. 1 to circa A.D. 1300. For reasons not fully understood, the Anasazi abandoned the canyons by the year A.D. 1300 (Britt 1971).

There are tremendous difficulties that must be overcome in doing field work in the canyons, especially recently. At the time the abstract of this paper was submitted in November, 1972, it was not possible to enter the canyons even by four-wheel drive vehicles. The greatest problem last fall was deep, dry, loose sand in the bottom of the canyon. It was planned to do library research during the winter months when the canyon was closed due to snow and all the foot trails were icy and slick. Field work was to continue in the spring of 1973. However, again vehicles were unable to enter the canyons, this time due to high water and resultant quicksand which came from melting snow. Finally, the writer resorted to descending steep, precipitous, primitive foot trails into the canyon to study some of the more important sites. Unfortunately, the only time a traditional Navajo will talk about these planetaria or many religious matters is during the winter between the first frost in fall and the first thunder of spring. This makes it almost physically impossible to obtain on-site interpretations.

Significance of the Canyon to the Navajo

Archaeologists are not in agreement as to when the Navajo people first arrived at Canyon de Chelly. The earliest documentary evidence comes from Cuervo Y. Valdes when he records the Navajo inhabiting Canyon de Chelly in 1706 (data in the National Park Service files, Chinle, Arizona). Because the Navajo were already established in 1706, it is quite likely that they first entered this area by 1700, or perhaps before the end of the Dinetah phase in New Mexico.

In the Canyon de Chelly area, Navajo chronology can be divided into two periods: (1) The Navajo Period and (2) The Historic Period. No absolute dates can be set for the earliest Navajo occupation of the Canyon de

Chelly area. It is definite that by 1706 the Navajo had been occupying this area for some time. Therefore, it is likely that the earliest phase would correspond to the Dinetah phase in New Mexico. The Dinetah phase ended at A.D. 1700. The Spider Rock phase in Canyon de Chelly started at A.D. 1700 and ended at A.D. 1864. It should be noted that the Spider Rock phase overlaps and corresponds to the Gobernador phase in New Mexico. It was during the Gobernador and Spider Rock phases that Navajo rock art developed. The Navajo culture was virtually unchanged by white contact prior to the "The Long Walk" which started in 1864. As evidence of this, the Navajo people, themselves, consider their modern Navajo nation as beginning in 1868 when they returned from "The Long Walk" (Link 1968).

Canyon de Chelly has, since 1700, been the stronghold of the Navajo people, as attested by Navajo fortresses in the canyon and by historical accounts. There are some manmade Navajo fortresses in the canyon as well as numerous natural fortresses. In the 1700's and early 1800's the Navajo, themselves, left records of important battles with other Indian tribes in the form of pictographs on the canyon walls. In Canyon del Muerto, Massacre Cave is named for a massacre of Navajo people which occurred in 1805 under the command of Lt. Antonio de Narbona (Document No. 1792 in the Spanish Archives of the New Mexico State Records Center and Archives in Santa Fe, New Mexico). In 1864, the Navajo people again used the canyons as a stronghold when Kit Carson was ordered to force the Navajo to surrender and sign a peace treaty (Frink 1968). Because of the numerous natural fortresses in these canyons, Kit Carson's campaign would have been unsuccessful had he not destroyed the Navajo's crops thus forcing them to surrender or starve to death.

The contemporary Navajo people make the canyons their summer home. Many families move into the canyon in the springtime to herd sheep and to plant crops. Their basic crops are corn, beans, squash, and melons. In addition, they have peach orchards. Much of this food is dried for use in the winter. Because of the general lack of firewood, the difficulty of getting supplies into the canyons, the necessity to enroll their children in school, and the fact that the canyons are bitterly cold in the winter, the canyons are practically abandoned from about 1 November to about 1 April.

Any place where people live, one finds their places of worship. Accordingly, Canyon de Chelly has many such sacred places. It is said by local Navajo people that there are nearly 100 sacred places within Canyon de Chelly. All Navajo medicine men must visit each of these places at least once during their lifetime. Watson (1964, p. 22) discusses four types of Navajo sacred places: (1) a location mentioned in a legend, (2) a place where something supernatural has happened, (3) a site from which plants, herbs, minerals, and waters possessing healing powers may be taken, and (4) where man communicates with the supernatural world by means of prayers and offerings.

Many types of areas in Canyon de Chelly are sacred to the Navajo. Some are merely unusual rock formations; some are rocks or other objects that have been struck by lightning; some are areas in which meteorites struck or other unusual phenomena occurred; some are Anasazi ruins; some are natural caves; some are stone cists; and some are springs (Museum Notes 1938). From what all 20 Navajo informants told the writer, there can be little doubt that the planetarium sites or star ceilings in Canyon de Chelly are truly sacred sites, the significance known only to Navajo medicine men.

Description of the Planetarium Sites

DeHarport (1951) reports ten star ceiling sites in Canyon de Chelly. This writer now knows of fourteen such sites. Navajo informant C states that he has discovered 32 planetarium sites within Canyon de Chelly National Monument, with relatively few in Canyon del Muerto. Informant C also states that he is still discovering more planetarium sites as he explores remote areas of the canyons.

These star paintings are precisely painted on the roofs of high overhangs or on the roofs of cave-like rock shelters. Site CWC 3 is characteristic of these cave-like rock shelters. Site CWC 3 is located in Cottonwood Canyon, a small tributary of de Chelly. The ceiling of this cave is about 12 feet above the present floor of the shelter. Site CC 95 or M 27 is another well-known cave site with a planetarium panel. It is located opposite the mouth of Spring Canyon. Mindeleff (1897, pp. 97-98) describes CC 95 or M 27 as follows:

"...at this point a large rock stands out from the cliff and in it there is a cavity shaped almost like a quarter sphere. Its greatest diameter is 45 feet and its height about 20 feet. The bottom land here is 10 or 20 feet above the stream bed and slopes up gradually toward the cliff, forming the bottom of the cave, which is perhaps 18 or 20 feet above the stream and some distance from it...."

Many of the star paintings occur on high overhangs, some of which are as much as 150 feet above the canyon floor. Most of these sites are now inaccessible. They may have been reached by hand and toe holds, although it is possible that some of them were reached by use of ropes.

Navajo informant H states, "because four and seven are the sacred numbers of the Navajo, I would expect planetarium panels to be grouped in series of fours and sevens." Although DeHarport (1959) reports only one planetarium panel in Cottonwood Canyon, this writer now knows of six in that canyon. When telling Navajo informant H this, he said, "there must be one more in Cottonwood Canyon you haven't discovered yet." At the junction of Canyon de Chelly and Canyon del Muerto, there occurs a series of small deep caves with openings to the west. There are three planetarium panels at these sites (Sites CC 70, CC 71, and CC 72), and there may have at one time been an additional panel in this area which has been destroyed by exfoliation. It is interesting to note that the next closest site, Site CC 73, contains excellent examples of other pictographs of Navajo origin: an antelope, a row of European horsemen, as well colorful yeis (Navajo representations of supernatural beings).

Nine types of design elements which most commonly represent the stars are illustrated in Fig. 1. Other design elements are rarely seen in Canyon de Chelly. These crosses, representing stars, vary in size from about one

1 *Nine of the most common types of design elements currently recognized in the planetarium sites of Canyon de Chelly.*

inch to approximately six inches, with the average painting being approximately three inches in length. Size variations are often present within a single panel. Many times it is evident that these size variations correspond with the brightness of the actual stars being depicted (Navajo informants A, C, H, I, K, and M). Each design element is basically a cross with variations. At present the significance of these variations is unknown by any of the Navajo informants. Thus, it is not possible to interpret the design elements at the present time. Informant D thought it was possible that the different design elements might have been executed by different artists. The sites contain varying numbers of crosses.

There are only ten crosses at Site CC 70, while the writer counted 73 star paintings at Site CWC 3.

These star depictions are executed in black pigment, thickly applied, the edges of which are remarkably straight and well-defined. This black pigment has not been analyzed, but it is quite likely charcoal in an oil base because this type of paint is still used by the modern Navajo. It has been reported by a park ranger that a few crosses in Canyon de Chelly were executed in white pigment. The star paintings in the Navajo Reservoir District were executed with red and orange pigment (Schaafsma 1962, pp. 209-210). Some are also black, white, and gray at the Cuba site in New Mexico (Schaafsma 1972, p. 48).

Canyon de Chelly apparently has the largest concentration of Navajo star paintings in the southwestern United States. Other areas where planetarium sites occur are in the Navajo Reservoir District, Site IA 4398, on the San Juan River (Schaafsma 1963). In addition, star paintings occur at Dulce and in Delgadito Canyon, as well as at a site near Cuba, New Mexico (Schaafsma, written communication, 1973).

Age and Cultural Affiliation of Planetarium Sites

Although in the past some park service personnel have attributed these star paintings to the Anasazi culture of Canyon de Chelly, there is ample evidence that they are unequivocally early Navajo as opposed to Anasazi. The writer believes the star paintings in Canyon de Chelly are early Navajo and most likely date between circa A.D. 1700 and approximately A.D. 1864, during the Spider Rock phase. During that time, Navajo rock paintings were a fine art as evidenced by polychrome yeis, cattle, Spanish horsemen, and antelope executed in sharp detail. After the Navajo people returned from "The Long Walk" in 1868, their rock art, for the most part, underwent rapid deterioration. The fact that the star paintings were so precisely executed is evidence that they were painted between approximately A.D. 1700 and A.D. 1864. More recent Navajo rock art is of an entirely different style, resembling modern graffiti.

The fact that 17 Navajo informants have either stated or implied that the planetarium sites are of Navajo origin is in itself good evidence that the star paintings cannot be of Anasazi or Hopi origin. Also, Navajo informants A, C, E, F, G, I, J, L, M, and N have even named certain constellations in some of the sites by use of photographs.

Schaafsma (1966, p. 12) states that "small crosses or star depictions precisely painted on the roofs of rock shelters are as characteristic as the yei of early Navajo rock painting". This writer has found no star paintings in Canyon de Chelly which were in direct association with yeis, horsemen, or other pictographs which would prove the star paintings to be of unquestionably Navajo origin. Consequently, he wrote Polly Schaafsma. In a letter from Schaafsma, dated February 25, 1973, she states, "direct association (of stars) with the yei figures of Todosio Canyon was quite evident....Crosses of a similar type in stenciled form were found surrounding a horned mask in Delgadito Canyon in the Largo drainage, the mask and stars being a unified design....Finally, there is a planetarium near Cuba, New Mexico, with not only the usual stars or crosses making up the ceiling decoration, but with a variety of other figures: dragonflies, birds in flight (cross-like), thunderbird type birds, circles, an occasional animal, and also tiny horsemen carrying lances!" Therefore, it should be apparent that these planetarium sites cannot be of pre-Navajo origin due to their direct association with Navajo yeis and horsemen.

Star lore is prevalent in Navajo mythology and crosses depicting stars occur on some ceremonial equipment. The Pleiades star group occurs on some Navajo rattles used in various curing-of-the-ill ceremonies. A kehtahn is a hollow reed about the size of a cigarette which is filled with tobacco, bits of shell, and other materials which are precious to the Navajo. The writer has in his possession a kehtahn decorated with four crosses (Fig. 2). According to McAllester (1956), such items are used in the Great Star Evil Chasing Chant.

An interesting example of the use of star paintings as a charm against evil is found on a 19th century Navajo war shield which is on display in the Canyon de Chelly Visitor Center Museum. A Navajo singer, upon looking at the war shield (Informant J), stated that the stars were "good medicine", especially if the warriors were engaged in battle at night. This shield depicts a crescent moon, the North Star, Pleiades, Orion, and Hyades among many other stars.

Crosses representing stars are common on Navajo rugs. The writer was curious as to whether these crosses woven into Navajo rugs were merely some type of design, or whether they actually represent stars. Navajo informants B, C, G, and K were consulted and they said that in some rugs they do represent stars, while in others they are only designs. According to these informants, if a rug is picturing a sandpainting, the sky, or a Navajo story, the crosses would represent stars. In addition, crosses representing stars are used in a number of sandpaintings of the Big Star Chant (McAllester 1956, Plates I, II, VI, VII, IX, and X). The sandpainting of Father Sky (Fig. 3) depicts stars in the form of crosses. The Father Sky sandpainting includes several constellations which are currently recognized by astronomers.

Still more evidence that these planetarium sites are of Navajo origin comes from a statement by Navajo informant O. When questioned as to what constellations the Navajo recognize, this informant said, "some represent horses and rabbits, but most people have forgotten most of the constellations." The fact that some of these star paintings depict horses is in itself evidence that these sites cannot be of pre-Navajo origin.

The fact that planetaria follow that of early Navajo occupations and crosscut a number of Anasazi sub-regions or districts with differing rock art styles in the Plateau Anasazi tradition is more evidence that the star paintings cannot be pre-Navajo (Schaafsma, written communication, 1973).

It should be noted that the Hopi, who occupied the canyons intermittently between A.D. 1300 and A.D. 1700 did not use caves as shrines. There is ample evidence that the Navajo did use caves as sacred places. One of the best known Navajo shrines in Canyon de Chelly is "Ceremonial Cave" located near Spider Rock (Museum notes 1938). Although this sacred cave is closed to the public and non-Navajo people are not permitted to visit it, Navajo informant K states that there are star paintings on the roof of Ceremonial Cave. He also states that

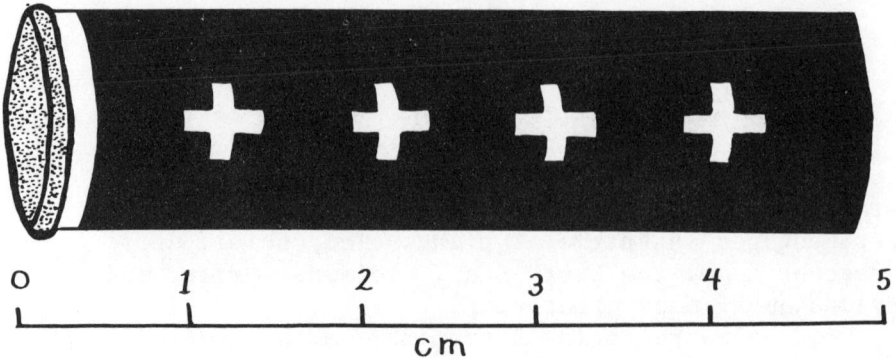

2 *Navajo kehtahn illustrating the use of crosses to represent stars. Drawn from a specimen in the author's collection.*

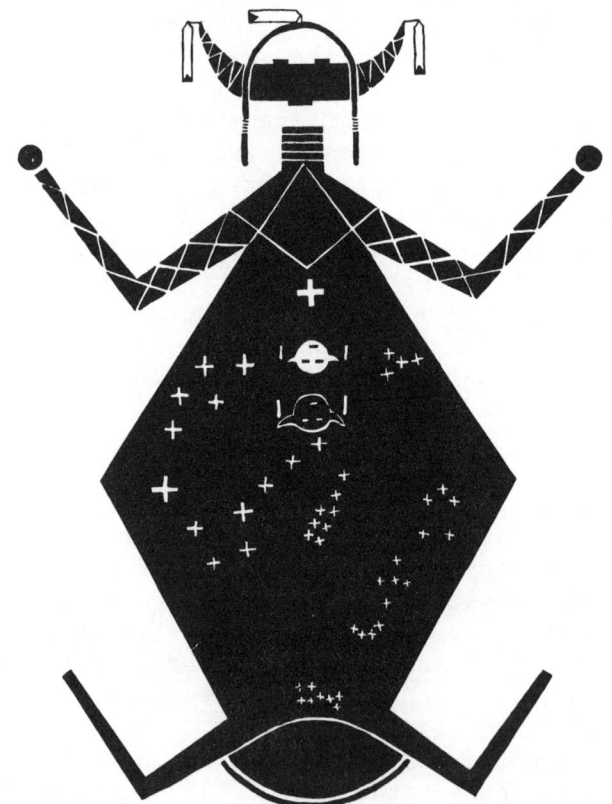

3 *Navajo depiction of Father Sky derived from a sandpainting. It depicts the sun, the moon, the Milky Way, and the more important constellations.*

Navajo medicine men visit the star ceilings in Canyon de Chelly to leave offerings in the form of kehtahns.

From the foregoing discussion, it should be evident that the planetarium sites in Canyon de Chelly are of unquestionable Navajo origin. Also, it is not meant to imply that all the star paintings date prior to A.D. 1864. On some panels many stars are faded while others are sharp and clear. Some crosses appear to have been painted over the top of previously-existing ones. Some of these crosses could be fairly recent, although there appear to have been no new additions since the park service established Canyon de Chelly as a National Monument in 1931.

Interpretation

The problem of getting on-site Navajo interpretations in the winter has already been noted. In addition, this writer encountered several other problems in obtaining interpretations of the planetarium sites by the Navajo. Few Navajo know the significance of the sites. One young Navajo, informant N, said, "these places (planetarium sites) are so highly secret that the old people will not even tell me anything about them. The older generation has stopped telling the young Navajo people their legends." One elderly traditional Navajo, informant B, said that the significance of the star paintings is so secret that he couldn't tell anyone. As further evidence of the highly sacred nature of these star paintings, this writer gave a young Navajo man a drawing of Site CC 95 to take home to his elderly father for an interpretation. Upon questioning the young man several days later, he said, "when I showed the drawing to my father he burned it. My father said that place is so highly sacred that it would be disrespectful to the yei for us to have a copy in the house. I am sorry I don't have your drawing to give back, but perhaps you can draw another one." In Navajo religion, many things are not supposed to be looked upon without the proper medicine. Evidently the elderly Navajo who burned the drawing of Site CC 95 must have felt that he did not have the proper medicine to look upon it.

Many of the planetarium sites undoubtedly depict constellations. This presents a problem in on-site interpretation also, because nowadays, for the most part, only Navajo medicine men know the constellations and the mythological laws they represent. Five Navajo informants state that an interpretation of the planetaria in Canyon de Chelly could only be accurately given by a Navajo medicine man. These same informants also state that it is very unlikely that a non-Navajo could get such information because there are so many things in Navajo religion which are so highly secret.

Another factor which tends to increase the secretivness of Navajo star stories is financial consideration. A medicine man may be paid from $100.00 to $1000.00 in cash, livestock, or merchandise for a single ceremony. Each medicine man is known by the ceremonies which he is able to perform well. Thus, his knowledge is jealously guarded and considered to be his own personal property. For this reason, it is necessary to keep his knowledge highly secret, otherwise he would have too much competition. Even if a medicine man does teach a Navajo apprentice, he will never cover the subject completely until he himself is near death. One hundred years ago medicine men who could perform the Blessingway ceremony were highly respected and wealthy. Since the recording of the Blessingway in several publications (Wyman 1970), the stories and chants have become almost commonly known throughout the Navajo reservation and beyond. Almost any medicine man can now perform the Blessingway ceremony. One of the causative factors in the secrecy of the planetarium sites is the fact that Navajo medicine men are afraid the data will be published, as was the Blessingway.

The heart of all Navajo star lore is in the creation story. According to the legend, after the Navajo people emerged from underground onto the surface of the present world, certain beings began putting stars into the sky. Some references refer to these beings as First Man and First Woman (Yazzie 1971, p. 21), or Fire Man and First Woman (Newcomb 1967, p. 88), or Black God (Haile 1947, pp. 23-28). These beings planned to arrange all stars in constellations representing every animal on earth. Before they were finished, Coyote came along. Seeing how slowly the work was progressing, Coyote picked up the blanket on which the remaining stars were lying and flipped them into the sky. He thus created the Milky

Way and myriads of stars that are scattered among the constellations. Coyote's own star is one of these and most likely corresponds to what we know as Canopus.

According to Newcomb (1967, p. 88), First Woman, after all stars were placed in the sky, said, "now all the laws our people will need are printed in the sky where everyone can see them....The commands written in the stars must be obeyed forever." According to Navajo informant C, each constellation represents a law which all the Navajo people must obey. He further states that when the Navajo stop obeying the laws which are written in the sky, the Navajo tribe will come to an end. In 1946, few medicine men knew the constellations, as noted by Haile (1947, p. 6) when he states, "...and while there are not many constellations, and some appear in the winter, others in the summer, some early in the night, others late towards dawn, few Navajo know anything about it at all, because it is too difficult to learn."

The Navajo recognize approximately 37 constellations (Haile, 1947). The three most important and central ones are atse'ets'oozi (Slim First One) which corresponds to Orion, so'ahots'i'i (Pinching Stars) which corresponds the Hyades, and Dilyeehi (no English equivalent) which corresponds to the Pleiades. Other constellations of the Navajo which correspond roughly to those recognized by North American modern astronomers are: (1) "Man With Feet Ajar" - Corvus, (2) "The Big First One" - Scorpio, (3) "Revolving Male" - Ursa Major, and (4) "Revolving Female" - either Ursa Minor or Cassiopeia. The other constellations represent animals, insects, or supernatural beings.

The sandpainting of Father Sky (Fig. 3) illustrates the use of constellations in Navajo ceremonies. The sun and moon are illustrated with the Navajo sun and moon face figures. These are facing a large star which is close to the Milky Way, represented by white intersecting lines across the arms and chest of Father Sky. This star may be the Coyote star (Canopus). The large star on the left side of the figure represents the North Star (Polaris). Above it is a group of five stars representing Revolving Female (Cassiopeia). Below it is a group of seven stars representing Revolving Male (Ursa Major). To the right of the sun and moon is a group of four small stars probably representing Hyades. Directly below the sun and moon is a group known as "Rabbit Tracks" (English equivalent unknown at this time). There are three remaining constellations, two of which are unknown. The center one is "The Big First One" (Scorpio). The remaining two could represent Orion and the Pleiades. It is obvious that these constellations are not executed in correct sky positions, nor are the sizes of the stars related directly to the brightness of the stars. The reasons for these apparent discrepancies are not known, but some suggested explanations are: (1) Sandpaintings, unlike star ceilings, are not permanent works and therefore are subject to modern changes and individual artistic license; (2) The constellations in the sandpainting are likely grouped or arranged in accordance with their relationships to one another in Navajo legends or their importance in these legends. This grouping does not seem to appear in star ceilings; (3) A Navajo medicine man must make four deliberate errors in the execution of any sandpainting that is not part of an actual ceremony.

As noted by Schaafsma (1965, pp. 7-12), pictographs were painted for a variety of reasons. Grant (1967, pp. 28-39) lists five purposes of pictographs: (1) ceremonial, (2) mnemonic, (3) records of important events, (4) clan symbols, and (5) doodling and copying ancient designs. It is obvious that the planetarium panels in Canyon de Chelly are primarily ceremonial and secondarily mnemonic to the Navajo.

It is beyond the scope of this paper to discuss all the known planetarium sites in Canyon de Chelly.. For this reason, only Sites CWC 3 and CC 95 will be discussed, as these are two of the better known ones. There appear to be three separate panels at Site CWC 3. Portions of two are shown in Fig. 4 and 5. Of the 73 figures depicted at Site CWC 3, 30 star paintings and one man-like figure are painted on one panel of this site (Fig. 4). The central group of stars in Fig. 4 probably represents "The Big First One". This constellation corresponds roughly to the one known to astronomers as Scorpio. The individual represented is the author of witchcraft according to Navajo legends (Haile 1947, p. 7). The human form of this constellation is definitely indicated by the small anthropomorphic form in direct association with it. This could indicate that this site was used by

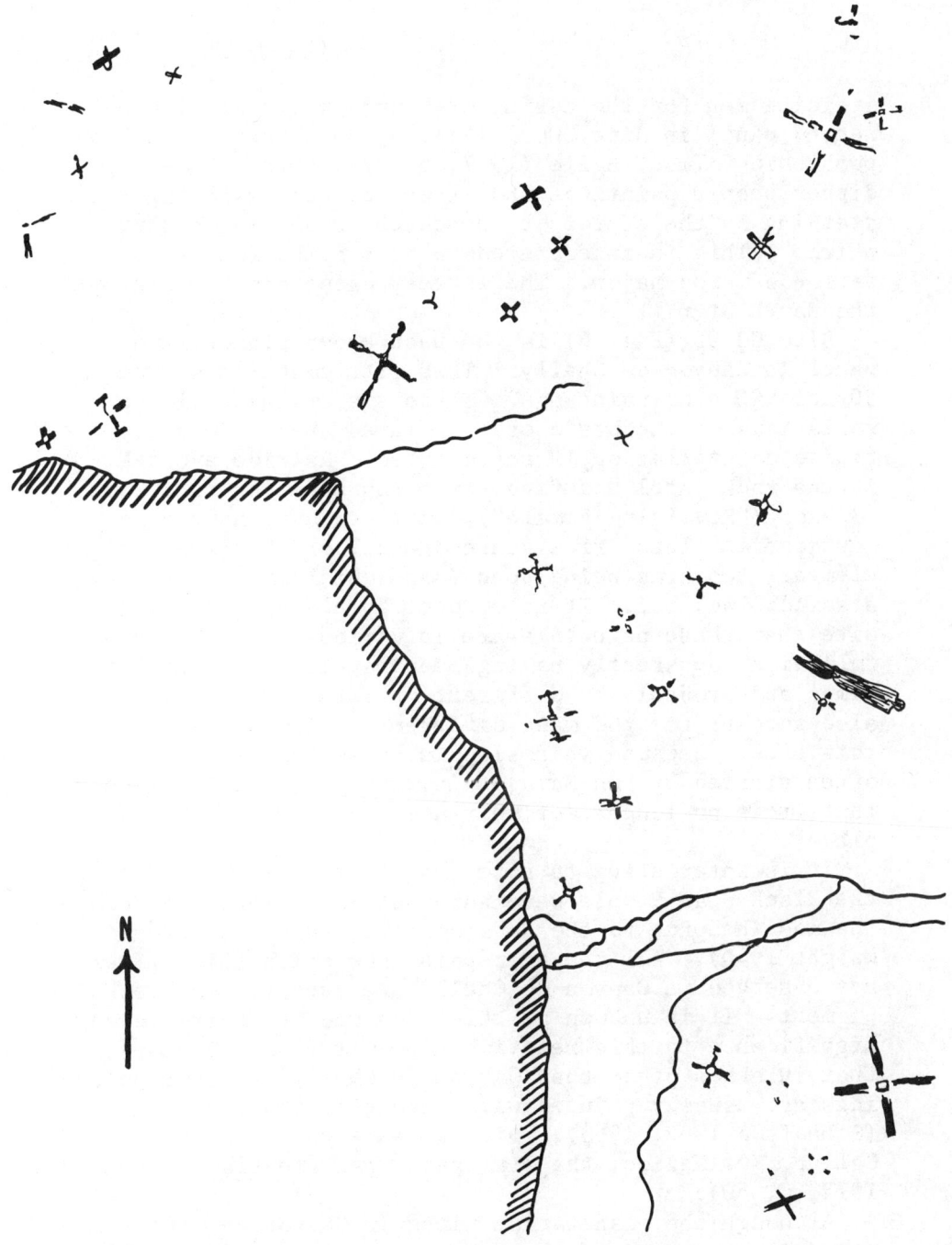

4 *Portion of Site CWC 3, redrawn from a photograph. Largest stars are approximately 6 inches long. Directional arrow is approximate.*

medicine men for the curing of Navajo witchcraft. The second panel in Site CWC 3 (Fig. 5) depicts four anthropomorphic forms. Haile (1947, p. 37) states, "the big dipper people paintings show four (or two) male figures standing on the center of the earth around which they whirl." This is interpreted to be a symbolic representation of Ursa Major. The largest cross might represent the North Star.

Site CC 95 (Fig. 6) is the best known planetarium panel in Canyon de Chelly. This site contains between 90 and 100 star paintings. There are at least eleven variations of the basic cross painted there. One definite constellation is represented. Astride a crack in the rock panel are five stars representing nahookoos bi'aadi ("Revolving Female"), known to astronomers as Cassiopeia. These five stars are all of the same design element, two arms being open loops which are crossed by a solid black bar. It is evident by close study of the site that these paintings are in various stages of exfoliation, apparently having been painted at different times and probably by different artists. This would also account for the many different design elements at this site. Because this site is so well-known and is often visited by non-Navajo visitors, it is quite likely that it is no longer actively used as a Navajo sacred place.

It is interesting to note that in the Big Star Chant the Black Star People were supernatural beings who were the most helpful to the Navajo people on earth (Wheelwright 1940). All the star paintings which this writer has observed in Canyon de Chelly are executed in black pigment. It is unkown at this time whether there is any significance to this relationship. It should be noted that in areas other than Canyon de Chelly the star paintings are usually painted with orange or red pigment (Schaafsma 1962, 1963), while at a nearby site near Gallup, New Mexico, the star paintings are black (Schaafsma 1972, p. 50).

Although the planetarium sites in Canyon de Chelly are primarily ceremonial, they also, to a lesser degree, serve mnemonic functions. The changing of the seasons in the past were noted by positions of the constellations in the sky. Some of these sites may have helped remind people of this use of the constellations. Each

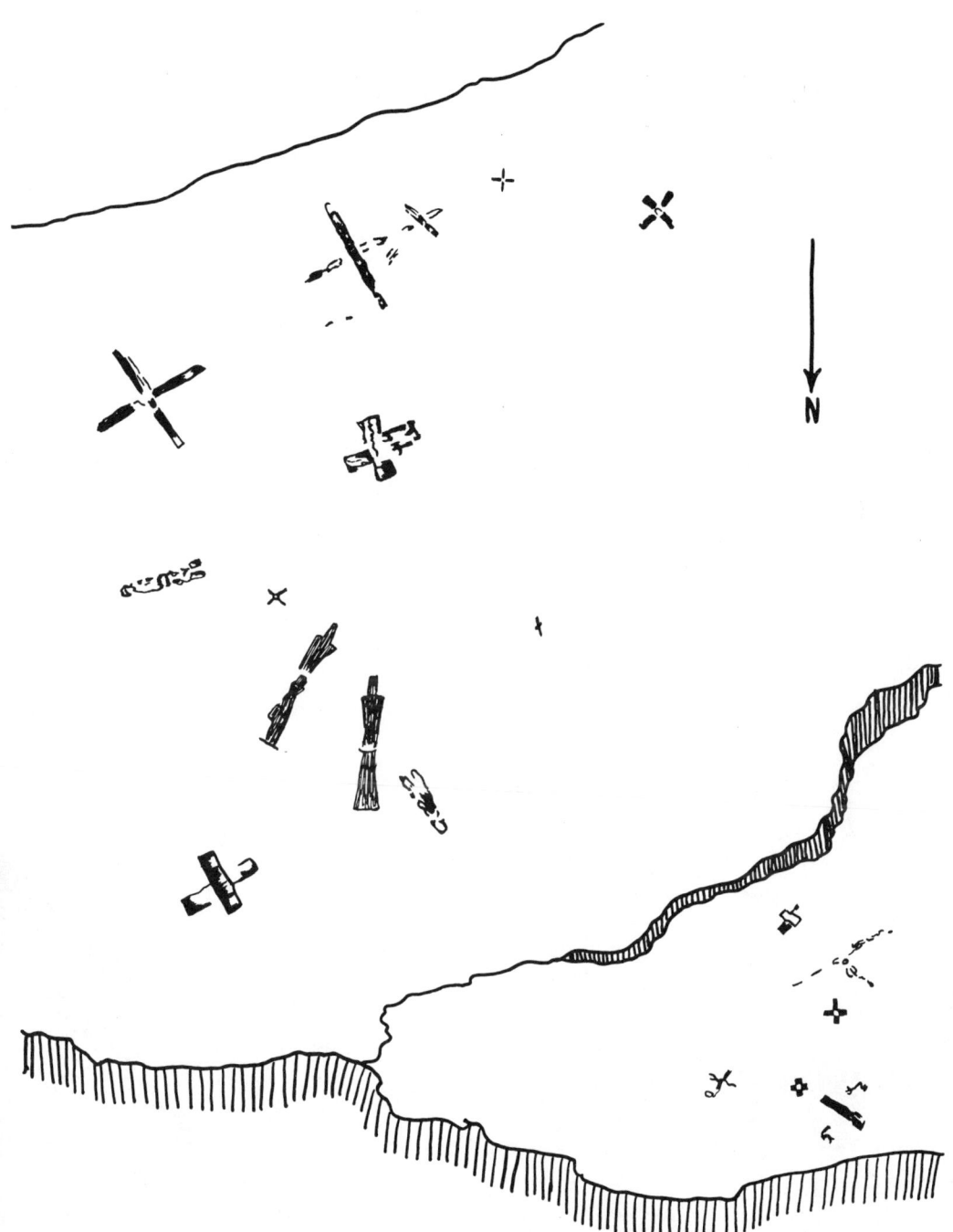

5 *Portion of Site CWC 3, redrawn from a photograph. Largest stars are approximately 6 inches long. Directional arrow is approximate.*

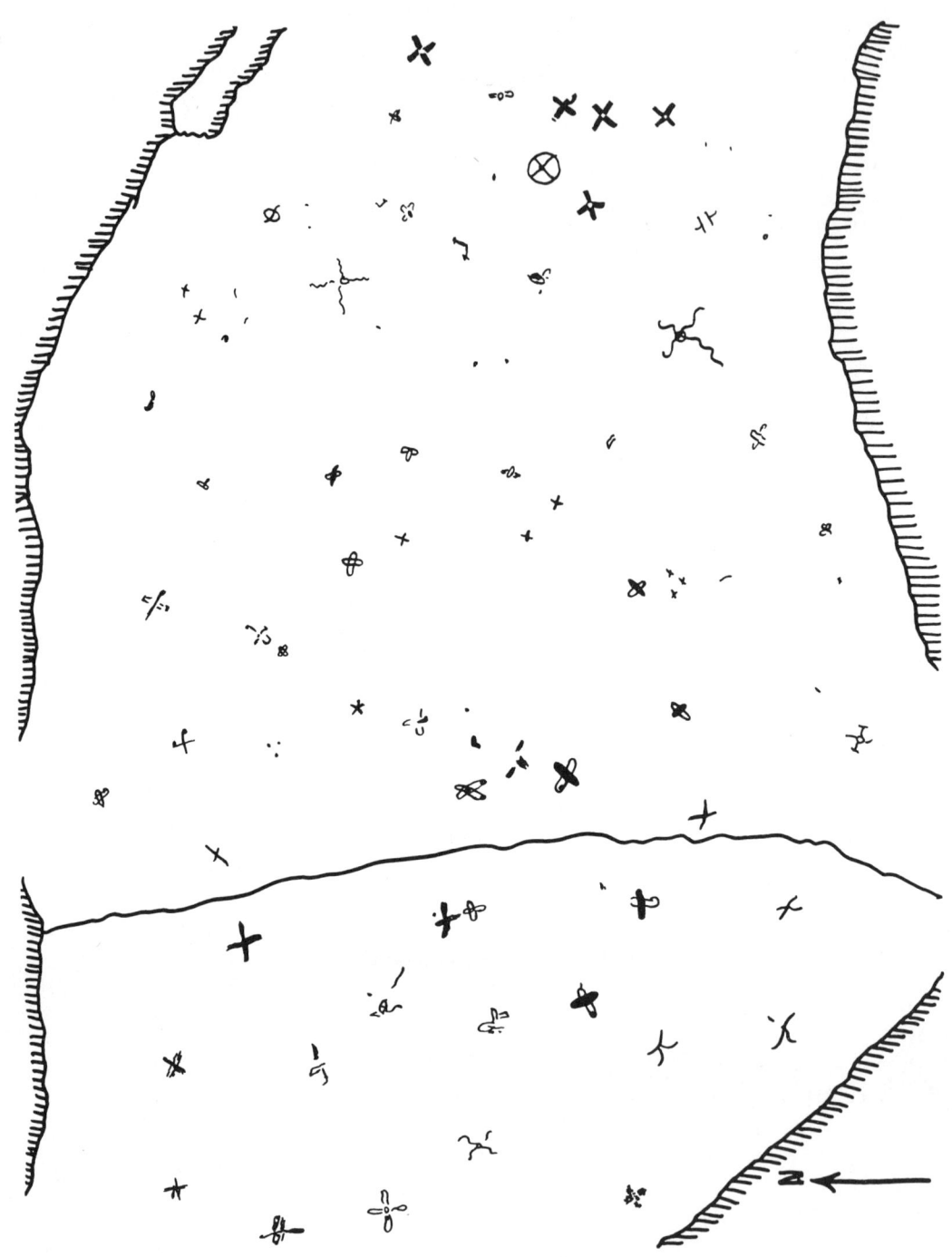

6 *Portion of Site CC 95, redrawn from a photograph. Stars average four inches in length. Directional arrow is approximate.*

constellation has a story or legend connected with it. These star paintings could remind medicine men of the constellations and the story they depict. In this capacity, they can also be used to train apprentice medicine men.

Conclusions

The planetarium sites in Canyon de Chelly are highly sacred places to the Navajo people and there can be little doubt that these star paintings date from approximately A.D. 1700 to A.D. 1864.

This is the first study of this nature in Canyon de Chelly. From this brief survey, it is apparent that much more work needs to be done. This writer plans to survey, photograph, and record all known planetarium panels in Canyon de Chelly National Monument. He hopes to publish this information as a monograph.

From the information obtained in this brief study, it is quite apparent that the stars were and still are very important to the Navajo people.

Acknowledgments

First, I would like to thank my Navajo informants for their help. Their names are: Cozy Brown, Kellywood Harvey, John C. Harvey, Agnes Harvey, Wilson Davis, Jerry Begay, Dee Dee Bia, Kee Begay, Billy Joe, James Gorman, Ben Miller, Stella Goldtooth, Johnson Hunter, Janeway Wheeler, Rose Begay, Johnson John, Thomas Chee, Alvin Draper, Larry Yazzie, and Ben Henry. In addition, appreciation is extended to the Navajo medicine man who sketched the drawing of Father Sky for me. He told me I must never publish his name. Finally, appreciation is also expressed to my wife, Shirley, for correcting my dangling modifiers and for drawing the illustrations.

6

Star-Patterns in Great Basin Petroglyphs

Dorothy Mayer

Berkeley, California

In this paper I present the hypothesis that some of the Indian petroglyphs of the Great Basin region are representations of stellar patterns. Fifty-three petroglyphs have been correlated with these patterns (Table 1), and examples of the correlations are given in Figs. 1 - 11; further examples are available in Xerox form, since space prohibits publishing them all. All the petroglyph numbers refer to pages in Heizer and Baumhoff's Prehistoric Rock Art (1962), herein abbreviated as PRA.

I have not attempted to find calendric displays. Also, I have not looked for designations of planetary and lunar periods.

My procedure was as follows: I initially formed the hypothesis by noticing petroglyph figures which seemed to be closely correlated with the visual distribution of the fixed stars. Of course, lines drawn at random among the stars may produce almost any pattern; so it is important to begin the procedure with asterisms which have a very definite visual pattern, such as the semi-circle of Corona Borealis, the "V" of the Hyades in Taurus, the curve of the "tail" of Scorpio. If such a figure appeared on a glyph, I checked the surrounding design for figures corresponding in form and position with other definite stellar patterns. If this succeeded, I tried to correlate other parts of the design with less well-determined structure.

This procedure is necessarily relational. At this point in the investigation a simple figure cannot be used, since whatever resemblance it has to stellar patterns may be by chance. There is a probability that the glyph is stellar only if a quite complex single-figure glyph, or if the several figures of a multiple-figured glyph, show a resemblance to stellar patterns and have elements in the same spatial position relative to one another as the asterisms.

There are a number of instances in which individual elements on a stellar glyph are precise mirror-imaged representations of asterisms (No.'s 99g, 105h, 107f, 114e, 340d). Also, occasionally some elements are rotated within their position (No.'s 105h, 303m, 340d). I also have found that faint asterisms in significant positions such as along the ecliptic or near the poles, are sometimes given as much prominence as more conspicuous patterns. There are also some indications that the petroglyph-markers may have constructed the celestial equator (No.'s 114h, 118a, 183g, 372).

A rather remarkable result of the investigation is the high proportion of petroglyphs oriented southward. Twenty-nine of the fifty-three analyzed glyphs are oriented directly to the South, (Table 2). **Here** "orientation" means roughly "the direction the observer faced to look at the sky". A Xeroxed appendix giving details is available from the author.

Some of these southward oriented petroglyphs seem to depict stellar patterns which are too southerly in declination to be in view at the present day from their latitude (Table 3). It is an interesting possibility that they may be the work of individuals who have travelled from the South: they may be the record of a journey ("I have travelled to the place where the sky looks like this."), or perhaps they were a teaching device or had a ritualistic use. But others of these glyphs may reflect a change in the sky itself; for due to precession the declination of the fixed stars changes by a noticeable amount in several centuries, a situation which brings portions of the present-day "southern" sky into prominence in northern latitudes. If precessional changes can be distinguished from journeys, the precessional change would provide a rough chronology for the site.

Another way of assigning an approximate chronology to the site would be to notice the positions of colures depicted in the glyphs, and any marker denoting the intended solstice or equinox. Such an abstract marker

Star-Patterns in Great Basin Petroglyphs

Table 1

Ch-3	Cl-4	Inv 208
(Grimes site)	(Cane Springs)	343 c
96 b	124 c	344
98 c		348
98 e	Do-22	
99 g	(Genoa)	Inv 210
100 a	141 c	350 b
102 d		350 c
103 b	Lv-1	
103 k	(East Walker River)	Inv 269
	147 k	352 m
Ch-20	155 k	
(Fish Cave)		Inv 272
105 h	Nv-2	358 e
	169 f	Plate 19 b
Ch-71		
107 f	Wa-5	Las 32
317 e	(Spanish Springs)	371 a
319 f	Plate 14 a	371 b
323 a		371 g
323 e	Wa-69	
328 d, 329 b,	(Massacre Lake)	Las 39
Plate 23 f	183 g	372 b
340 d	183 m	(section)
		373 e
Cl-1	St-1	
(Valley of Fire)	(Lagomarsino)	Las 76
114 e	302 i	376 a
114 h	302 s	(section)
118 a	302 v	
118 k	303 c	Mi 4
121 c	303 d	Plate 11 a
122 h	303 i	
	303 m	

may have been used to indicate the galactic equator (No.'s 114 e, 96 b), so its presence is not out of the question. This idea should be used conservatively, however, since it adds another degree of freedom to the hypothesis.

Table 2

Orientational Distribution of 53 Stellar Petroglyphs

Orientation	% of Petroglyphs
North	5.7%
North-East	0
East	17
South-East	5.7
South	54.7
South-West	5.7
West	7.5
North-West	3.7
	100.0%

Table 3

Petroglyphs Depicting Possible Stellar Patterns Too Southerly in Declination to be Visible at Present from the Latitude of Their Site

# of glyph	Site	Approximate latitude of site	Present most northerly latitude of stellar patterns (approx.)
141 c	Do-22	39° N	20° N
183 m	Wa-69	41.7° N	Equator
302 s	St-1	39.3° N	20° N
303 c	St-1	39.3° N	Equator
303 i	St-1	39.3° N	20° N
303 m	St-1	39.3° N	20° N

The depiction on some glyphs of asterisms around the positions of the present equinoxes and solstices is probably fortuituous, (Table 4). These positions would have significance as equinoctial and solstitial points only: (1) recently, with the last 2,000 years; (2) very long ago, the total precessional cycle of 25,800 years b.p. (before present), when the present position of the colures would have obtained; (3) at the quarters of the cycle; at 6,450 b.p. the equinoctal colure would have been in the position of the present solstitial colure, at 12,900 b.p. the vernal and autumnal equinoxes would have exchanged positions, etc.

In PRA, Heizer and Baumhoff present evidence that most of the petroglyph sites in the Great Basin are situated along game trails, and suggest that they are connected with hunting magic. This is consonant with the present hypothesis, for both the flow of game and the aspect of the night-sky may have been observed in a divinatory or calendric sense to predict the flow of game. Or perhaps at crucial moments of a person's life, such as at initiation or death, both phenomena were regarded as highly significant omens. Again, game-flow could have been used by shamans to predict the appearance of the sky.

One of the most striking aspects of most of the stellar petroglyphs is their initially cryptic quality, combined with a few well-defined clues which can lead to elucidation. Although as abstract visual patterns they are often precise, pleasing, or amusing, they do not seem to be repeated in the sense that our "constellation figures" are repeated, as permanent culturally-bound patterns. Rather, each stellar glyph seems to be the product of a unique and ingenious "vision of the universe". Whoever the petroglyph makers were, they seem to have placed a high premium on two values which appear incompatible to us: precision in recording objective phenomena, and individuality of expression.

I wish to thank the Sky Publishing Corporation for permission to use hand-copies of star-charts from Sky and Telescope magazine and Norton's, and the U.C. Press for permission to reprint the line-drawings from PRA.

Fig. 1. The Petroglyph figures of 103 k (site Ch-3) correspond to configurations of stars in the constellations of Hercules and Lyra. β Lyrae is a binary variable (magnitude 3.4 - 4.1, period 12.91 days) shown here almost as prominently as Vega (α Lyrae, magnitude 1); perhaps its prominence refers to an event, such as sudden nova brightening. It is interesting that the Ring Nebula, M 57, is quite near to β Lyrae.

Table 4

Petroglyphs whose Design Correlates with Stellar Patterns around the Present Solstices and Equinoxes

# of glyph	Site	Present solstice or equinox
96 b	Ch-3	Summer solstice
98 e	Ch-3	Summer solstice
100 a	Ch-3	Autumnal equinox
105 h	Ch-20	Summer solstice
114 h	Cl-1	Autumnal equinox and Summer solstice
124 c	Cl-4	Autumnal equinox
155 k	Ly-1	Summer solstice
183 g	Wa-69	Vernal equinox
302 i	St-1	Autumnal equinox
303 m	St-1	Summer solstice
348	Iny-208	Autumnal equinox

Two sky maps of the Hercules region are shown; the more detailed map (Fig. 1 C) shows a closer correspondence to the glyph. Lyra lies outside Norton's map No. 11, but its position is shown on the Sky and Telescope map (Fig. 1 B). Corona Borealis does not appear on this glyph, but its position is shown on the charts for purposes of reference.

The hand-copies of star-charts in these illustrations should be used only in conjunction with a good star-atlas.

Fig. 2. Petroglyph 99 g (site Ch-3) has a good correspondence to the asterisms Hydra, Corvus, Hercules, and Corona Borealis. The two objects above Hercules on the glyph did not seem to correspond well to anything on the S & T star map. However, reference to a more detailed atlas (Norton's) showed that the lower of these two objects has a good correspondence in shape to Ursa Minor, mirror-imaged, (Fig. 2 C). Its position on the glyph in relation to Hercules and Hydra seems definitely to be displaced, though account must be taken of the curvature of the rock surface, and this could not be

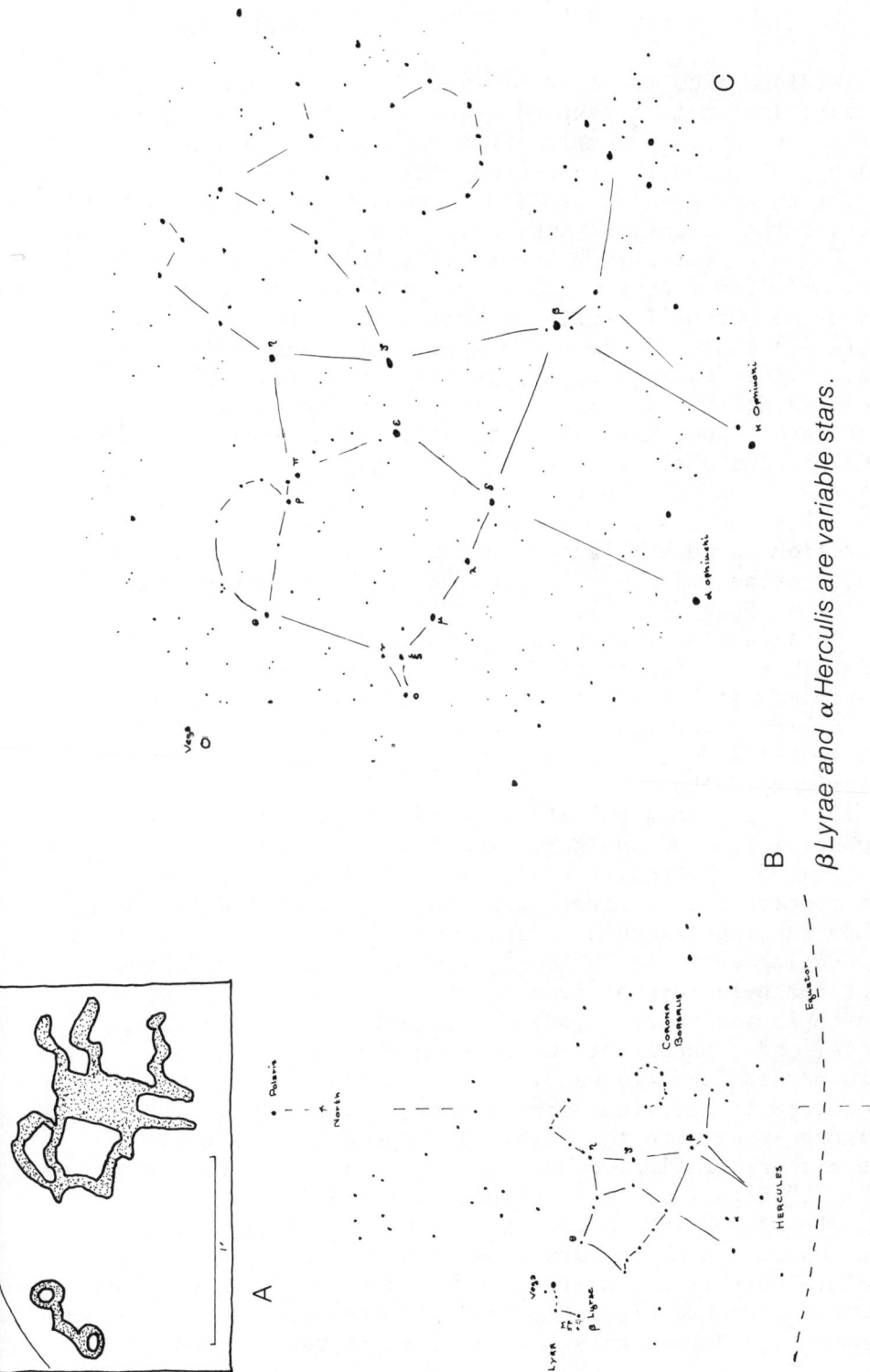

1 A *PRA 103 k* (site Ch-3).
  B *Sky & Telescope* (hereinafter abbreviated "S & T") July, Hercules region.
  C *Norton's Star Atlas* (hereinafter abbreviated "Norton's") map 11, Hercules region.

βLyrae and αHerculis are variable stars.

determined from the line-drawing. The upper "double-ax" figure does not correspond well to any asterism, though it may refer to a doubled Ursa Major. It may also refer to a past colure. The two objects above "Hydra" may refer to the head of Leo (left), and to Cancer reflected around the ecliptic (right).

Fig. 3. Petroglyph 107 f (site Ch-71) is a curious seven-limbed figure which corresponds to the contellations of Corona Borealis, Serpens Caput, Hercules, and Lyra. A photo of the glyph appears in PRA, plate 24 c. Three line-drawings appear in PRA: pp. 107 f, 326 g, and 327 d; the line drawing on p. 107 f seems to be most accurate. Two charts have been given, the more detailed (C) showing a closer fit with the glyph.

The "head" of the figure seems to depict both Corona Borealis and its mirror-image, perhaps because the composition is more balanced this way. One other case in which an asterism may be doubled is the possible Ursa Major of 99 g (Fig. 2).

This stellar configuration is presently on the meridian in early August at 9 p.m., late December at 12 ., early May at 3 a.m., etc.; and the current solstitial colure runs through it. The equinoctial colure bisected it around 4500 B.C., and it was then on the meridian at midnight of the Spring equinox.

Fig. 4. The glyph 317 e (Ch-71) corresponds to the constellation of Ophiuchus and its environment.

Fig. 5. Petroglyph 105 h (Ch-20) depicts the sky as it appears facing first in a roughly easterly direction, then turning around and facing west.

Facing east and following the galactic equator from the north-eastern horizon to the meridian (Fig. 5 B), there is good correspondence to Auriga, Perseus, Cassiopeia, and Cepheus. At Cepheus on the meridian, turn around and face the west. There is now a good correspondence to Hercules, Corona Borealis, Ursa Major. A figure appears to the right of "Cepheus" which could be either a reflected Ursa Minor, or a reflected Draco. The "Y" figure may be a highly stylized Draco.

The star-chart is for late October, 8 p.m. The same sky would appear in late June, 4 a.m.; that is, just before sunrise around the present summer solstice. Therefore the spoked circle to the left may refer to the rising sun at the summer solstice; also it is well-placed in

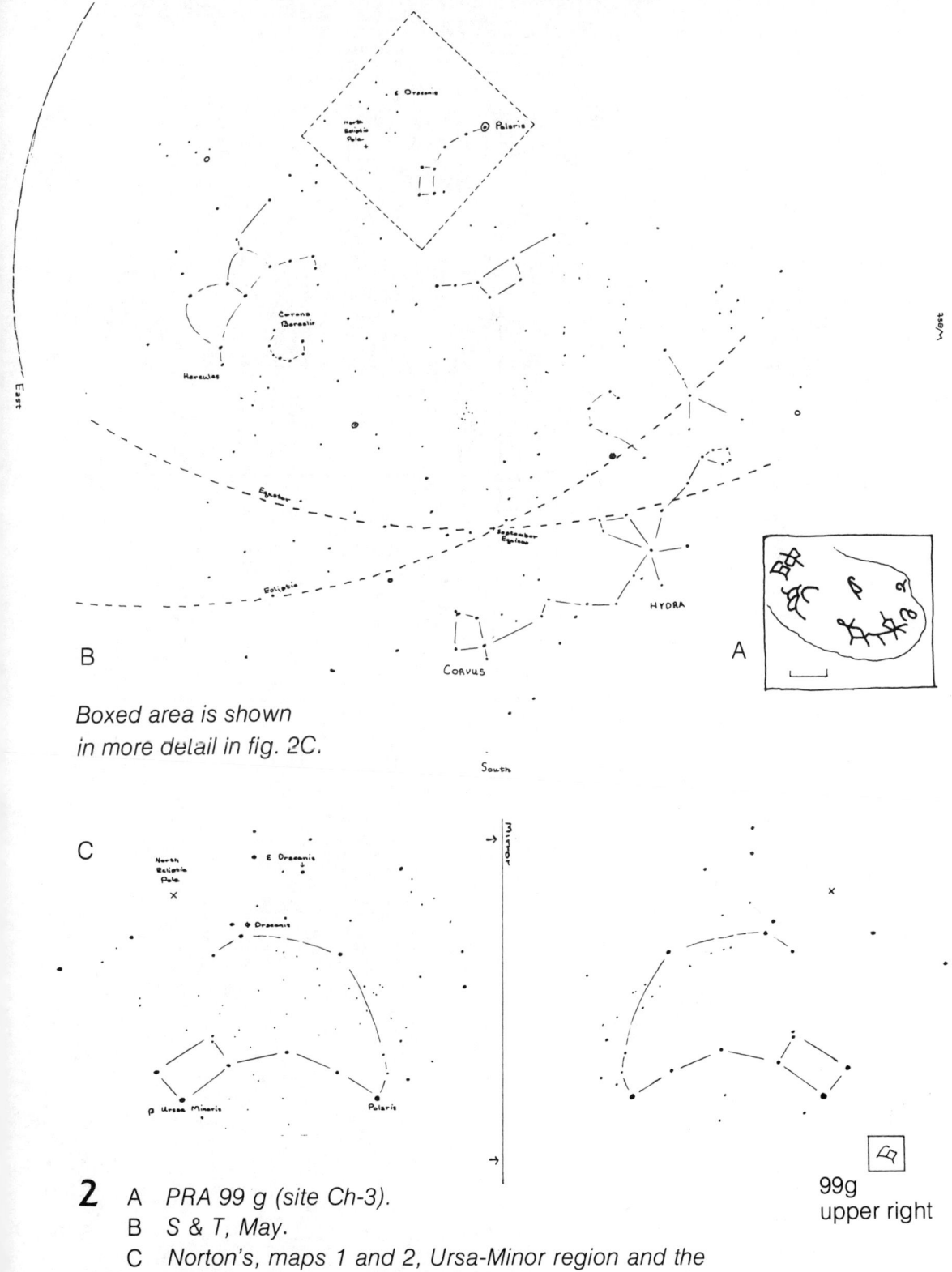

2 A PRA 99 g (site Ch-3).
B S & T, May.
C Norton's, maps 1 and 2, Ursa-Minor region and the mirror-image of this region, compared with a section of 99 g.

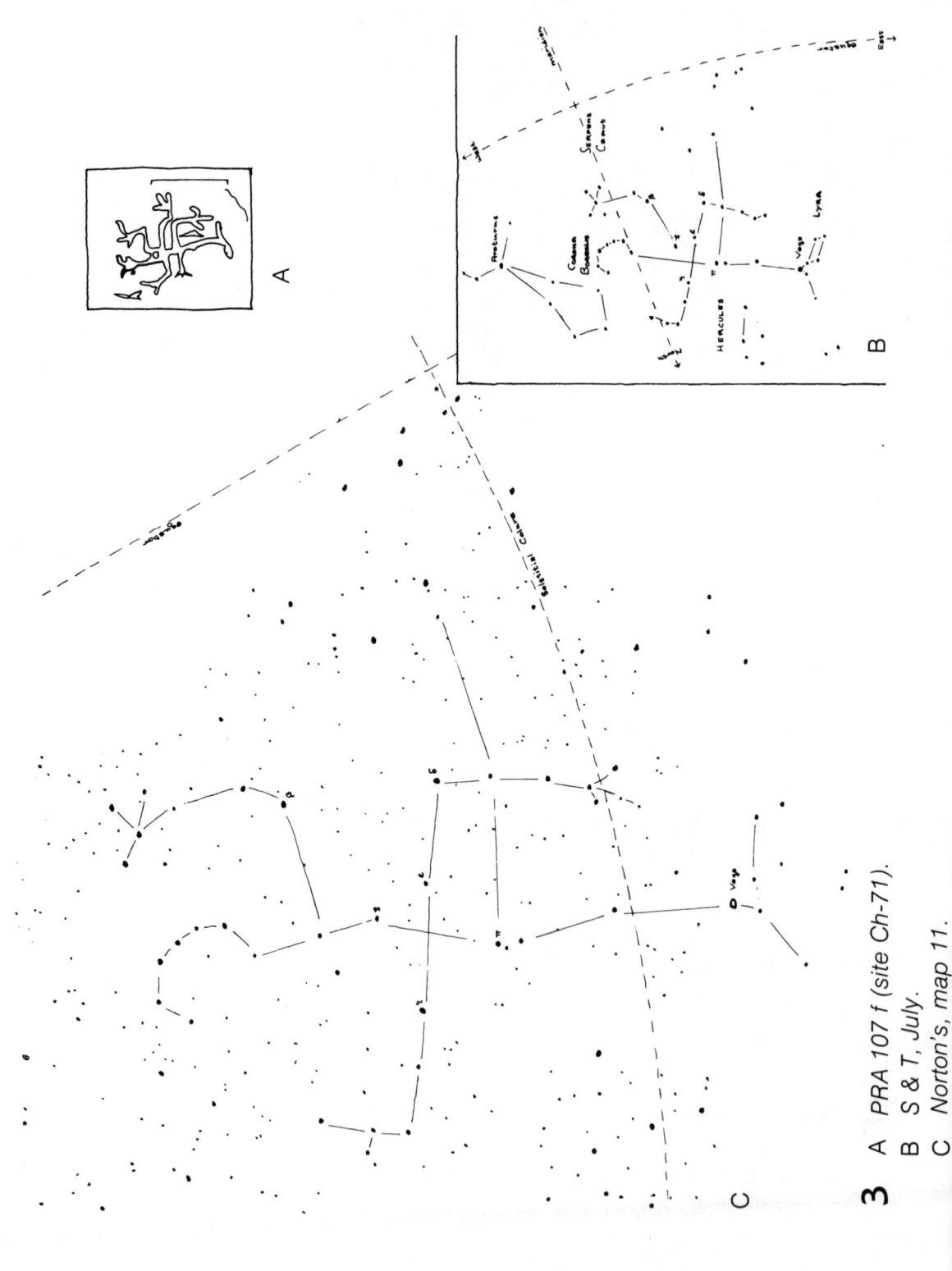

3 A *PRA 107 f* (site Ch-71).
B *S & T*, July.
C *Norton's*, map 11.

Star-Patterns in Great Basin Petroglyphs 119

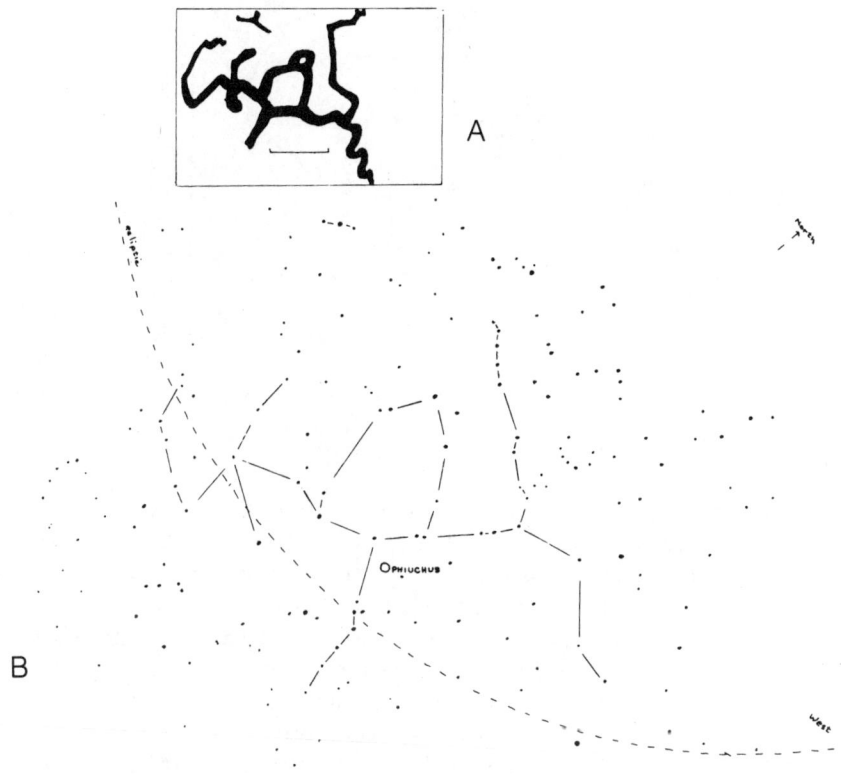

4 A *PRA 317 e (site Ch-71).*
   B *S & T, July, Ophiuchus region.*

relation to the other figures to be on the ecliptic at the position of the present summer solstice.

Fig. 6. Petroglyph 114 e (site Cl-1) appears to depict a bolide or comet with a train of about 50°, falling from Northeast to Southwest between Ophiuchus and Capricorn-Aquarius-circlet of Pisces (R.A. 20-21 hrs.).

Ophiuchus is depicted clearly, but the eastern arm is exaggerated (Fig. 6 B). Cygnus seems to be dovetailed with Draco to the west. Draco is repeated again at the upper right in conjunction with part of Ursa Major (Fig. 6 F). Camelopardalis appears rather clearly, carrying Auriga, though probably only α, β, and δ Auriga would have been visible. At the right of the glyph, Ursa Major is depicted three times, once mirror-imaged; this

5   A   *PRA 105 h (site Ch-20).*

section is analyzed in detail in Fig. 6 C-G. The "raised-arm" figure below them correlates with α, ν, η, δ, and ρ Bootes, mirror-imaged around the axis shown in Fig. 6 G. Corona Borealis is also mirror-imaged.

The cervids do not correlate with anything in the sky. Their appearance may refer to a seasonal migration. The two figures "φ" line up, but do not point north to form a colure; rather they seem to express quite exactly the Galactic Equator. The comb "⋈" is on the ecliptic. The "bear paws" correlate in position but not in shape with prominent patterns: Perseus, Aquila, Sagittarius, tail of Scorpio, and probably Gemini (counterclockwise from upper left). (Fig. 6 C) shows the same configuration of stars in Ursa Major used three different ways in three contiguous figures (Figs. 6 D, E, F). Bootes may also be shown mirror-imaged (Fig. 6 G), represented by the "raised-arm" figure on the glyph.

Fig. 7. The pattern of this glyph corresponds to the distribution of the fixed stars along the celestial equator from about R.A. 8.5 to 15 hrs., approximately half the hemisphere. This region holds the constellations Leo, the head of Hydra, Crater, stars of Virgo, α Coma Berenices, and ν and δ Bootes. It is centered on the present autumnal equinox, though this is probably fortuitous. The arrow points to φ Leonis, a 4th magnitude star appearing in Norton's, but not in the Sky and Telescope star map; the region is of course frequented by planets.

5 B *S & T, October.*

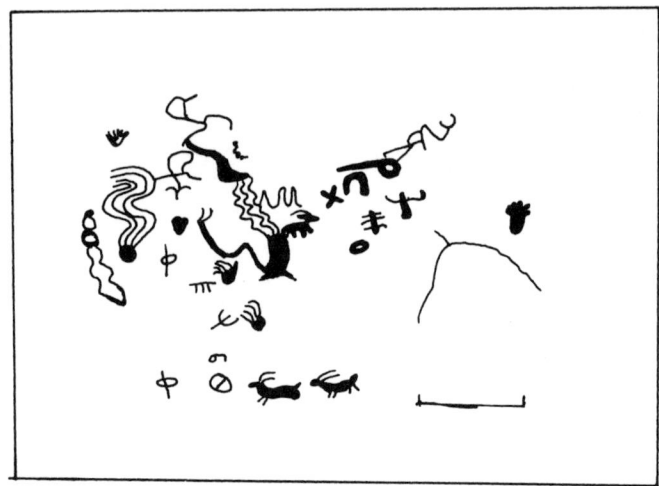

6   A   *PRA 114 e (site Cl-1).*

Fig. 8. Petroglyph 141 c (Genoa site, Do-22) corresponds to the galactic equator from Scorpius to Orion. The head is on or near the ecliptic, apparently about $20°$ west of the current winter solstice. From $20°$ north latitude this can be seen at present along the extreme southern horizon from east to west.

Fig. 9. This glyph (183 g, Wa-69) corresponds to star-patterns around the Great Square of Pegasus, Pisces, Cetus, and Aquarius; the correspondence is closer on the more detailed map (Fig. 9 C). It is interesting that the "ruled" line on the glyph is quite precisely correlated with the equator for about $10°$ west of the present spring equinox, and with the ecliptic for about $10°$ east of the equinox (At the equinoxes the path of the sun is of course very close to the equator.).

Fig. 10. The peculiar glyph 302 i (St-1) of a two-headed "serpent" attached to a cup corresponds rather well with the constellations of Hydra, Crater, and Corvus. It is on the verge of being too "simple" to be included in the data, and was finally included because of the improbability of the design.

Fig. 11. Petroglyph 303 d (St-1) appears to depict a glancing occultation of Antares (α Scorpio), by the crescent moon. Since Scorpio is "head downward", it is probably being shown near the western horizon.

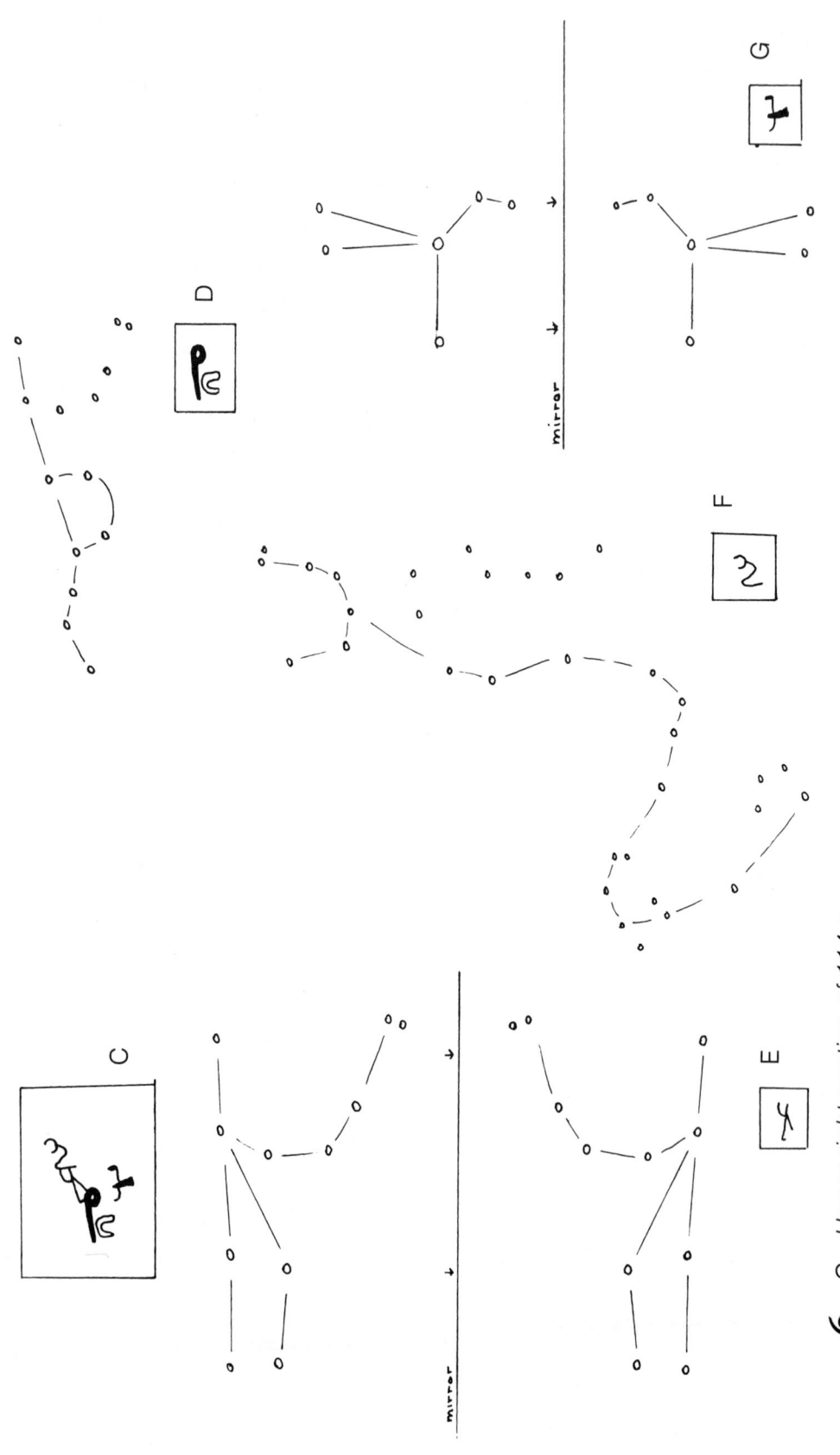

6 C Upper-right section of 114 e.
   D S & T, September, Ursa Major region.
   E S & T, September, Ursa Major region and its mirror-image.
   F S & T, September, Draco and Ursa Major region.
   G S & T, September, Bootes region and its mirror-image.

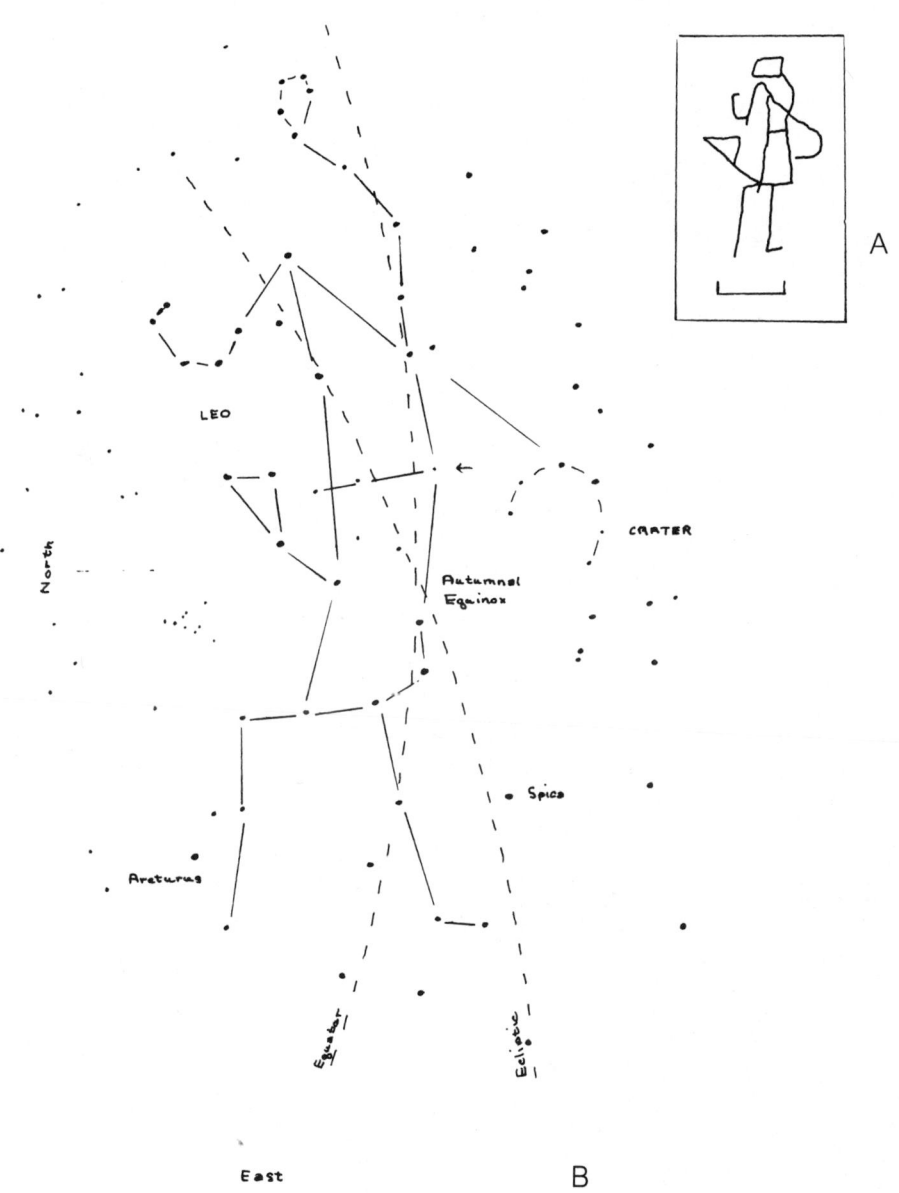

7 A *PRA 124 c (site CI-4).*
B *S & T, May.*

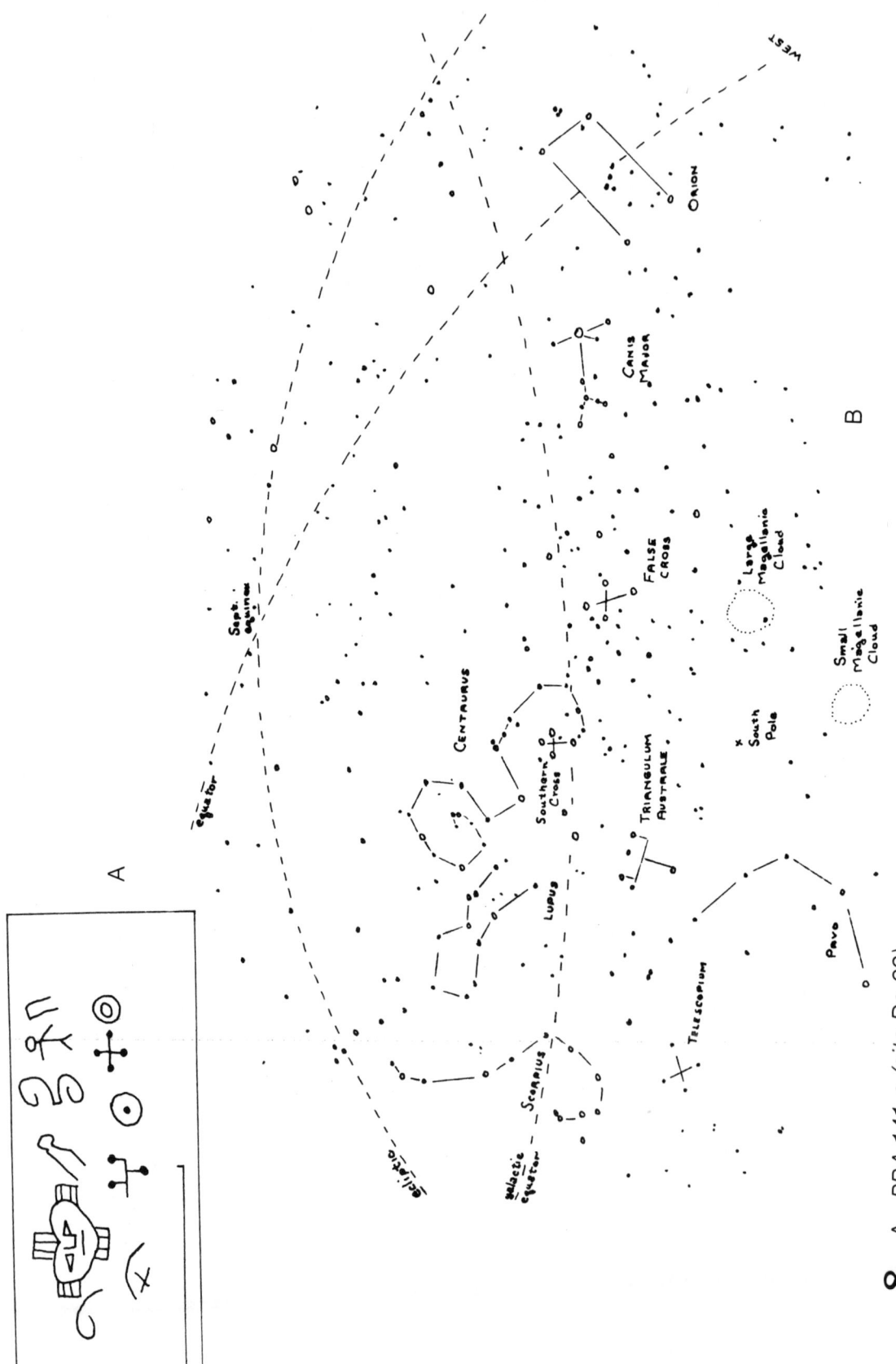

8 A *PRA 141 c (site Do-22).*
  B *S & T (southern hemisphere chart), March.*

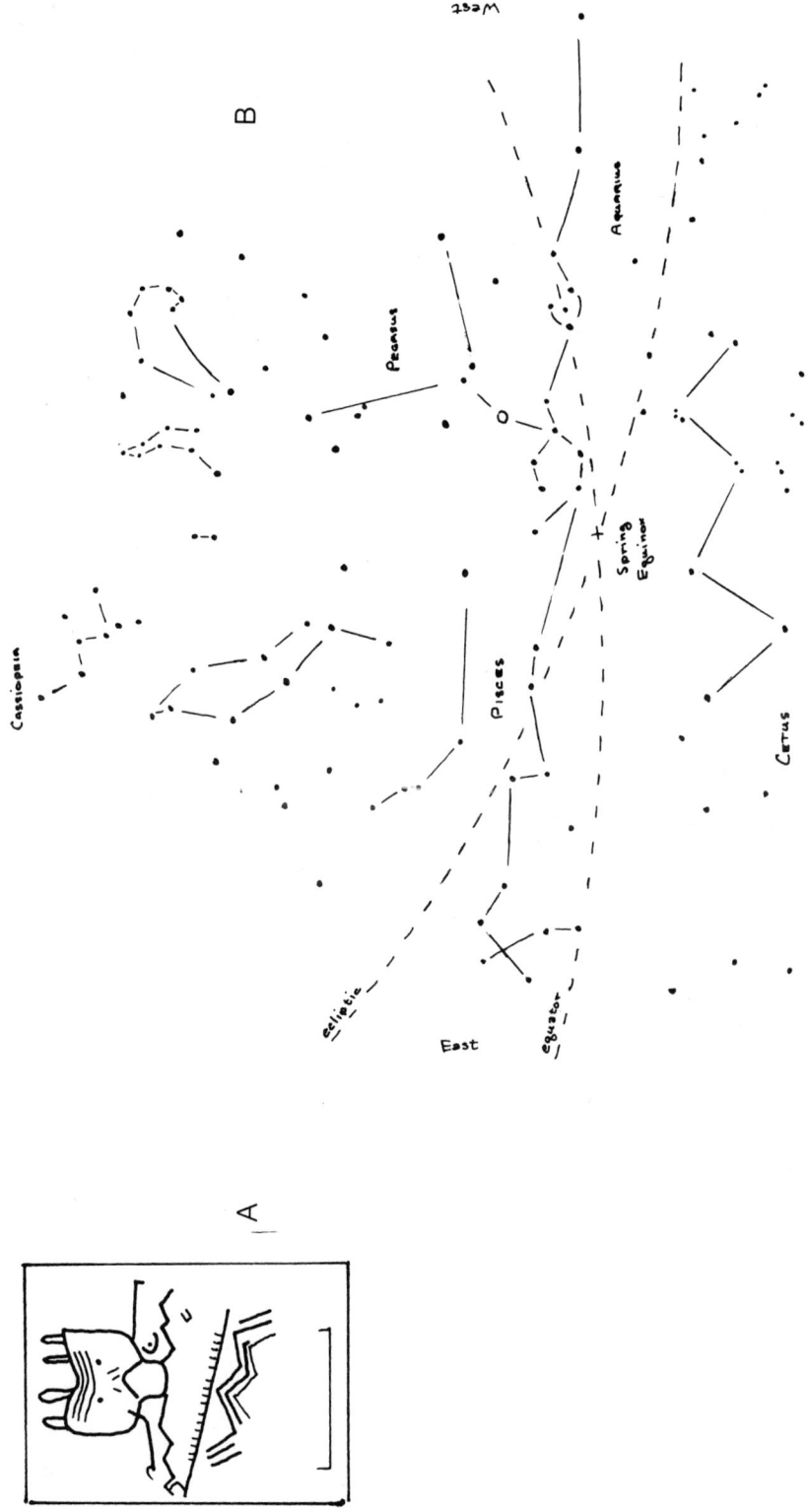

9 A *PRA 183 g (site Wa-69).*
   B *S & T, November, Pegasus region.*

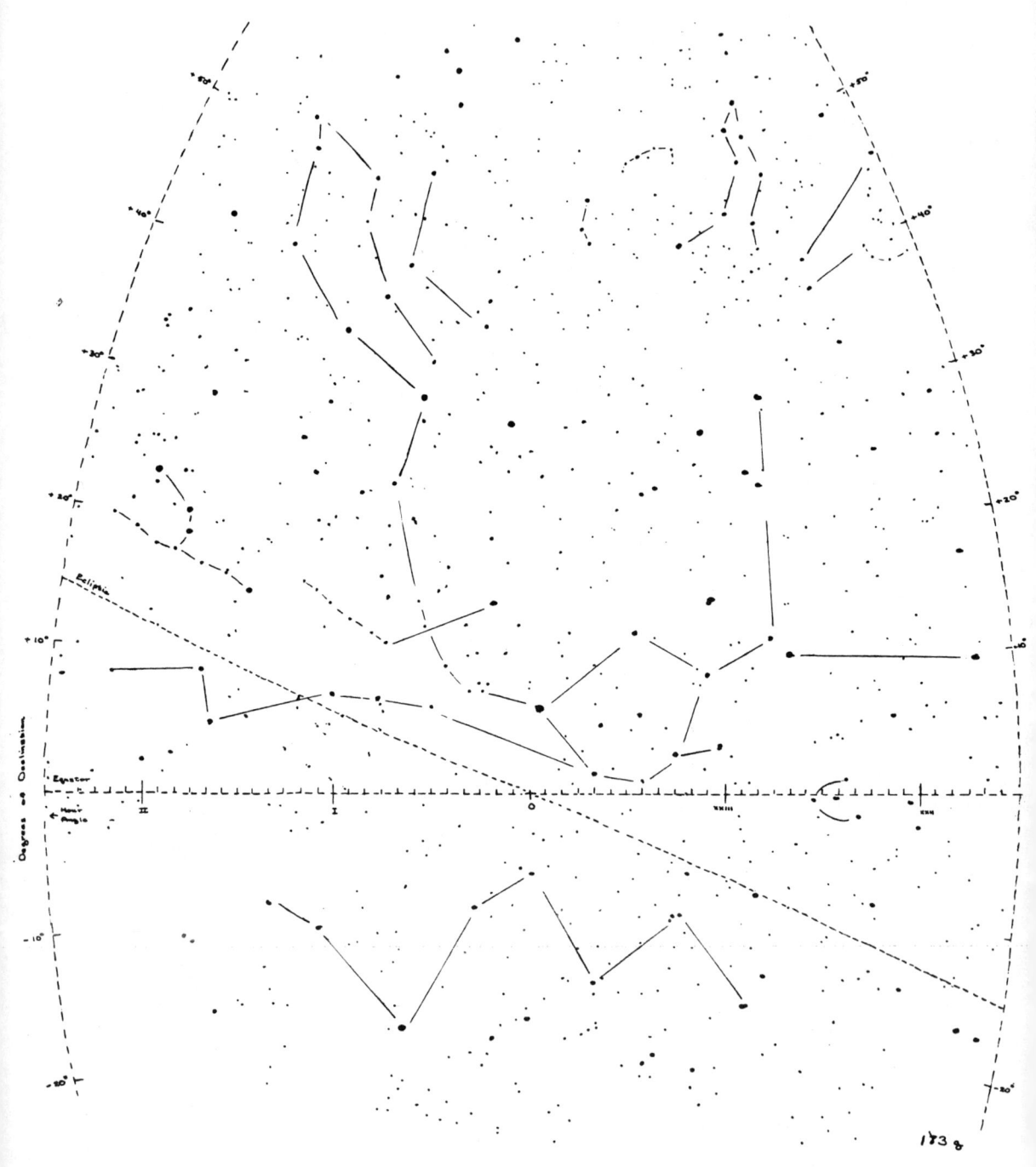

9 C *Norton's, maps 3 and 4.*

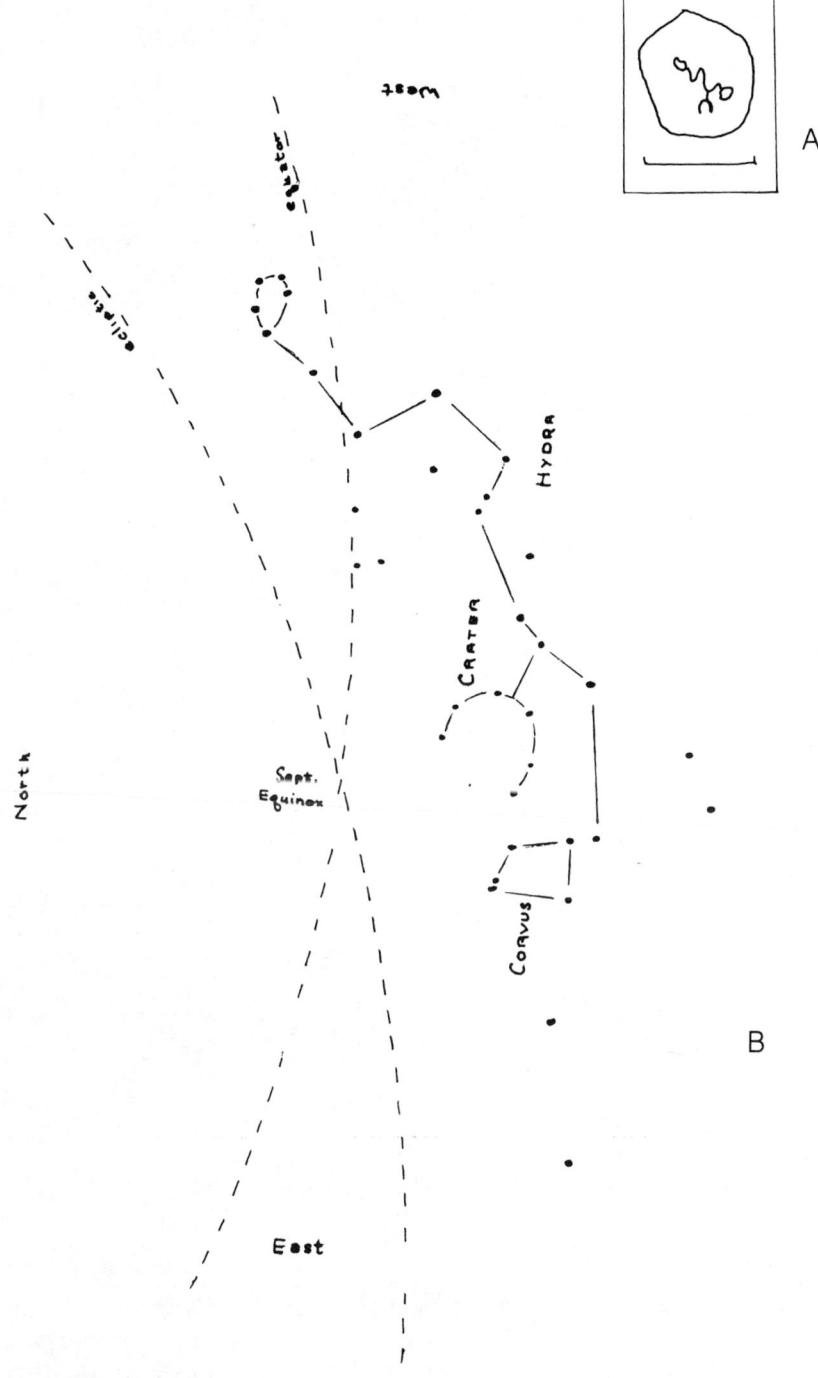

10 A *PRA 302 i (site St-1).*
B *S & T, May, Hydra region.*

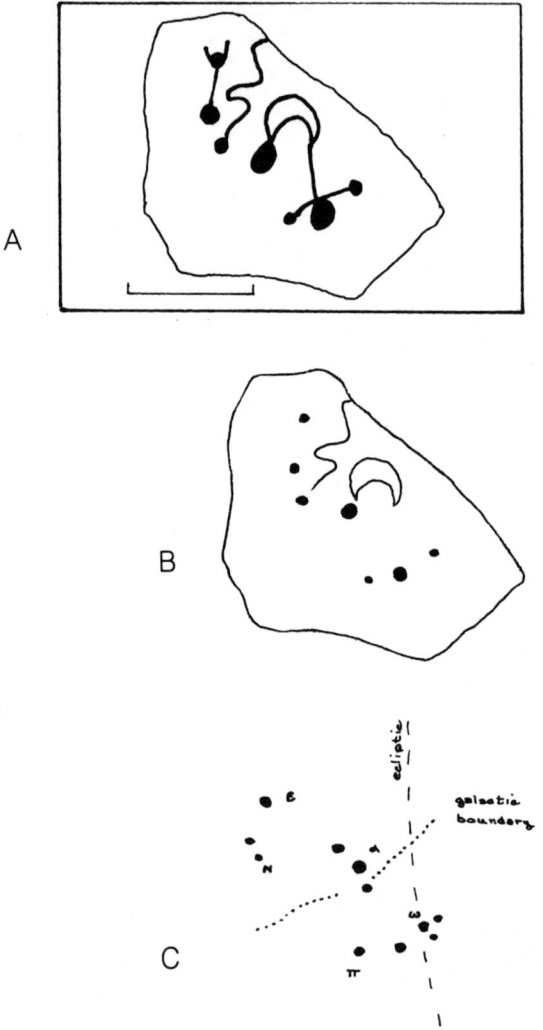

11 A *PRA 303 d (site St-1).*
  B *303 d without some of its connecting lines.*
  C *S & T, July, Scorpio region.*

Therefore the crescent moon would be waxing and about 3 or 4 days old. The cusps are pointing in the wrong direction, a very common error. The wavy line attached to N Scorpius could depict the galactic boundary, which is of course quite indistinct.

7

Astroarchaeology: The Unwritten Evidence

Gerald S. Hawkins

Smithsonian Astrophysical Observatory

Cambridge, Massachusetts 02138

Introduction

At a discussion meeting of the Royal Society and British Academy in London, in December 1972, the process of extracting information on prehistoric man's knowledge of astronomy and mathematics was recognized as a new field of research, with a subtitle of "The Unwritten Evidence." Astronomical information has been derived from the siting and alignment of buildings and crude megalithic structures with celestial objects on the skyline or in the zenith. Numerical information relating to astronomical periodicities can be found in the disposition of stones and holes in circles, in otherwise ornamental carvings, and in bone markings. Nor is this process limited to a study of nonliterate cultures. Within the ancient literate cultures, there was a tendency for astronomical knowledge to be regarded as secret, or at least not to be written down in any explicit form. Also, some early writings not yet fully understood may be clarified by the quantitative results revealed by this unwritten evidence.

This type of study therefore has some aspect of detective work and unavoidably involves a degree of speculation. R. J. C. Atkinson regarded the discovery of sun and moon alignments at Stonehenge as mere speculation and reviewed the work as "Moonshine". On the other hand, Fred Hoyle (1972) said:

It is not a speculation to assert that we ourselves could use Stonehenge to make eclipse predictions. We could certainly do so without making any substantive changes in the layout. While this does not prove that Stone Age man did in fact use Stonehenge for making eclipse predictions, the measure of coincidence otherwise implied would be quite fantastic. How does one prove any incident belonging to the past? Historians argue by documentary evidence. But how if their documents are false? A plethora of documents belonging to the present day are false, many of them made so deliberately. It is not possible to argue that Stonehenge was falsified deliberately, to maintain a facade of astronomical subtlety by a people ignorant of astronomy. It will probably be hard for the historian to accept the idea of geometrical arrangement of stones and holes providing evidence much stronger than a document, but I believe this to be true.

Between the viewpoints of Atkinson and Hoyle, I think it is agreed that speculation is a necessary prerequisite to any advance in research and that as the body of evidence grows, the speculation becomes a working hypothesis and finally is established as an acceptable explanation of the behavior pattern and level of knowledge of an ancient culture in terms of our modern concepts and vocabulary. In this paper, I will summarize the methods of astroarchaeology and illustrate the type of result that may be obtained, with examples from the Old and the New Worlds.

Method

The analysis is more secure if one works with an archaeological site, the construction date of which has been determined by other methods. Otherwise, one is postulating an alignment and then deducing a date that will make the alignment work. This approach tends to beg the question. Also, the analysis should be comprehensive, with all similar markers and all related positions for the sun and moon being examined; otherwise, a selectional bias may occur and other alignments be overlooked.

If λ is the geographic latitude of a site and A the azimuth East of North of a line in the structure, then that line points to declination δ, given by the equation

$$\sin \delta = \sin \lambda \sin h + \cos \lambda \cos h \cos A , \qquad (1)$$

where h is the geocentric altitude of the astronomical object. Now, h is not the observed altitude at the site; rather, it is the angle as viewed from the center of the earth above a plane drawn parallel to the horizontal plane at the site, with atmospheric refraction not taken into account. Let us suppose that an object (sun, moon, or star) is exactly on the horizon, that is, at altitude h_s. For an extended object such as the sun, of semidiameter q, three conditions must be considered: the tangent on the skyline, the disk bisected, and the tangent with the whole disk below. The geometrical altitude of the center of the object is related to the skyline altitude by the relation

$$h = h_s - r + p \pm q , \qquad (2)$$

where the + sign is taken for the condition of the tangent with the disk standing on the horizon, p is the parallax, and r is the total atmospheric refraction, tabulated as a function of h_s in Table 1. Thus, knowing the skyline altitude h_s, one can substitute for h in equation (1) and calculate δ directly. Values for q and p are given in Table 2.

From the date of construction, it can be ascertained, by reference to Table 3, whether or not the chosen line was directed toward the extreme declination of the sun or moon. For a test of possible stellar alignments, reference should be made to the 5000-year star catalog (Hawkins and Rosenthal, 1967), and for planetary extrema, to the tables of Tuckerman (1962). Generally, there will be a difference $\Delta\delta$ between the tabulated declination and the value computed from the data at the site. This can be converted to a horizontal "error" of pointing of

$$\Delta A = \Delta\delta \, \text{cosec} \, B \qquad (3)$$

and a vertical "error" of

$$\Delta h = \Delta \delta \sec B \, , \qquad (4)$$

where

$$\sin B = \frac{\cos \lambda \sin A}{\cos \delta} \, . \qquad (5)$$

Further details and refinements are given by Hawkins (1968).

Table 1

Refraction as a Function of Observed Altitude

Observed Altitude h_s	Refraction r
0.00	0.58
0.10	0.56
0.20	0.54
0.30	0.53
0.40	0.51
0.50	0.49
0.60	0.47
0.70	0.45
0.80	0.44
0.90	0.43
1.00	0.41
1.50	0.35
2.00	0.30
2.50	0.26
3.00	0.24

Examples

Stonehenge and Megalithic Britain

Stonehenge, a complex structure in southern England, was built and rebuilt through the end of the Secondary

Table 2

Semidiameter and Parallax of the Sun and Moon

Object and Condition	q	P
Sun, tangent	0.267	0.002
Moon, tangent	0.259	0.951
Star	0	0
Sun or moon bisected	0	---

Table 3

Extreme Declination of the Sun and Moon

Year B.C.	Sun	Moon
4000	24.11	29.26 and 18.96
3500	24.07	29.22 and 18.92
3000	24.03	29.18 and 18.89
2500	23.98	29.13 and 18.83
2000	23.93	29.08 and 18.78
1500	23.87	29.02 and 18.72
1000	23.82	28.97 and 18.67
500	23.76	28.91 and 18.61

Neolithic period and into the Bronze Age. It was the first structure to be comprehensively studied by this method. A pattern of alignments emerged in which lines between stones and archways pointed to the turning extrema of the sun and moon on the horizon. This pattern, shown in Fig. 1, is not exhibited in an obvious manner with, say, a central post and surrounding stones with a north-south axis, and this is perhaps why the underlying astronomy was hitherto unnoticed (Hawkins 1963). Nevertheless, as Hoyle (1966) pointed out, the main alignments are related to the geometrical regularities and the arrangement is not random. Subsequently, alignments were found between the outer stones of the earliest structure, I, and the trilithons of III (Hawkins 1973). The number of archways in the sarcen circle, 30, was suggested to be a first approximation to the lunation (29.53 days), and the Y and Z holes, 29 and 30, a second approximation (Hawkins 1965). The number of Aubrey holes, 56, was correlated with a cycle containing two 19-year periods and one 18-year period, which fitted the azimuthal swing of the moon as shown in Fig. 1, and possibly with the occurrence of "danger" years for the eclipse of the sun or moon at midsummer and midwinter (Hawkins 1964).

This was unwritten evidence for astronomical knowledge of high order in a prehistoric culture. At first, it was contested and debated, but now it is generally accepted as having some validity (Hawkins 1973). It led to the discovery of lunar lines in more than 50 other sites in Britain (Thom 1967, 1971) and to the realization that this was a significant component in the indigenous Neolithic culture. The latter works showed, in general, a single line at each site in the form of a row of stones or an outlier to a circle, whereas Stonehenge contained the comprehensive pattern. Also, the single-alignment sites show evidence of a greater pointing accuracy. Notwithstanding the difficulty of surveying between irregular stones, the misalignments, or "errors", at Stonehenge are considerable, amounting to more than a lunar diameter. During discussions, it has been suggested that Stonehenge might be a repository, a central filing system, in which all the components of the pattern of extrema were brought together and incorporated in a symmetrical, and pleasing, architectural structure (C. Renfrew, private communication). In forcing the asymmetries of the sky directions to a symmetrical architecture, some misalignments would necessarily result.

"Sky gods" were important in ancient Egypt, and it is of interest to examine the temples for possible alignments. Some egyptologists are of the opinion that the ancient Egyptians were not orientation-conscious with regard to the sky, the temples being built to face the banks of the Nile (private communication, anonymous

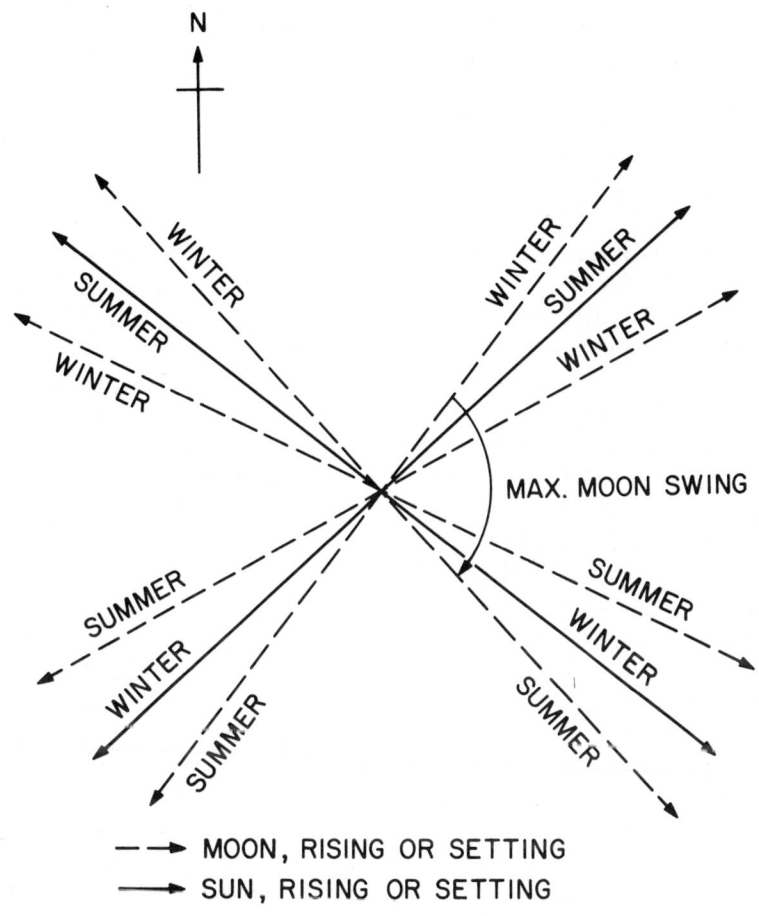

1 *The pattern of solar and lunar extremes in azimuth.*

referee of paper submitted to <u>Science</u>). One writer, commenting in the <u>Encylopaedia Brittanica</u> (1960 edition), said:

The worship of the physical sun (or perhaps of a god behind the sun) as the giver of life was an eminently simple and rational one but far too much so for the Egyptians, who, like other people, preferred irrational mysteries to such rational simplicity....The further we go back the more complicated, the more self-contradictory, the more ritualistic and the more barbarous it is, as is the case among all nations.

Various Temples at Thebes

The greatest existing temple complex of ancient Egypt is at Karnak, on the east bank of the Nile within the boundaries of Thebes, now the city of Luxor (latitude 25.73 N). It comprises some 15 major temples and chapels in an area of 3 sq. km. Some of these temples interlock as a group, parallel or perpendicular to a chosen axis.

The main group was dedicated jointly to Amon, the "unseen god" of Thebes, and the sun god Ra. The main temple structure runs a distance of about 530 m from the quay at the old course of the Nile to the eastern gateway. It was extensively rebuilt with the addition of the Hall of Festivals by Thutmosis III circa 1480 B.C. The temples of Seti II and Rameses III within the forecourt are perpendicular to the axis, and the High Room of the Sun and the edifice of Taharqa are parallel. The old sanctuaries of Ipet Sout, the temple of "Amon who hears the prayers" and the temple built by Rameses II to Ra-Horakhety, are collinear with the axis. The group thus conforms to a rectangular grid, and the orientation of each temple is defined by the main axis. The azimuth of the axis was measured by ground survey (Richards 1932), and horizon altitudes were measured by theodolite under the direction of the author. A photogrammetric survey was made of Karnak by the Franco-Egyptian Research Center, and azimuths of other temples were measured from this plan by using the Amon-Ra axis as a reference.

The temple of the moon god Khonsu was probably built circa 1160 B.C. The representation of Khonsu, Ibis-headed, is the horizontal crescent moon with the full disk between the cusps. Next to the temple of Khonsu, with its axis perpendicular to it, is the temple of Opet, the two conforming to a rectangular grid rotated by $88°.1$ to Amon-Ra.

The southern group at Karnak contains the temple of Mut surrounded by a crescent-shaped lake. Temples in this group are built to a rectangular grid rotated $82°.0$

Astroarchaeology: The Unwritten Evidence

to the Amon-Ra axis. The temple of Rameses III was built parallel to Mut, and the temple in the northeast precinct was built with its axis perpendicular. All temples here are in a ruined condition. Figure 2 shows the axis of the northeast temple and the distant skyline. The god of this temple, Khonspekherod, is translated as Khonsu-the-child by Barguet (1962).

Azimuths of the Khonsu and Mut grids were read from the photogrammetric charts, and the skyline altitude for the Amon-Ra and Khonsu lines was measured by theodolite. The skyline for the transverse axis of Mut was scaled from the photograph of Fig. 2, allowing for the $0°.5$ parallax in the hill as seen from Mut. Computed declinations are given in Table 4. The uncertainty in the Amon-Ra declination is $±0°.05$, and for Khonsu, $±0°.2$. At the temple of Mut, the uncertainty is larger, $±0°.4$, because of the difficulty of defining the axis of the ruined temple and because of the steep slope in the distant hills. A theodolite measurement at Mut would reduce this uncertainty somewhat, but unfortunately a skyline traverse has been made only at the Amon-Ra site.

It can be seen that the temples of the gods of the Theban triad pointed at the time of construction to sun and moon extremes. Amon-Ra and Khonsu were aligned with their celestial namesakes, and Mut, the consort and/or mother of Ra, was aligned with the moon. The accuracy of the Amon-Ra axis, pointing to the sun tangent on the horizon on the shortest day of the year, was equal to or better than $0°.05$ in azimuth. This surprising degree of precision would have persisted as an observational phenomenon from about 2000 to 1000 B.C.

Examined in more detail, the Khonsu and Mut lines pointed to the first crescent moon following the day of the summer solstice. The moon extrema moved from the transverse axis of the main temple of Khonsu to the Mut line and back again, swinging like a pendulum along the hills of Thebes with a period of 18.61 years. Observation of the first crescent was a stochastic process - if it was too close to the sun for visibility, then the next chance for observation was 24 hours later. Let x be the fraction of a day that must elapse before it is possible to see the crescent. The average delay for visibility over several periods of observation is then $(x + 1 + x)/2 = x + 0.5$. This is also the most probable

2 *Axis of ruined temple in Mut complex, showing cliff at the end of Theban Hills.*

Astroarchaeology: The Unwritten Evidence

value for a single observation. Persistent observing, of course, could reduce the minimum lag to x. The declinations at Khonsu and Mut are systematically lower than the extrema of Table 3. In Table 4, these values are converted into the time taken for the moon to move from the extreme to the aligment declination, which is effectively the "age" of the new moon when so aligned. The crescent "new" moon was important in fixing the beginning of the month, and the first crescent seen after the summer solstice was used in regulating the calendar of the seasonal year. The Khonsu and Mut temples were placed broadside to the extreme positions of this crescent.

Are these alignments accidental, or were they known to the builders? By dismissing them as accidental, one returns to the viewpoint quoted earlier in this paper, which accepts ancient Egyptian thought as an irrational mystery. But the alignments are precise and argue for acceptance as new data to be correlated with our previous knowledge of this extinct cultural system.

In a definitive work, Barguet (1962) found astronomical connotations throughout the Amon-Ra complex, so much so that he put forward the working hypothesis that the structure had some singular cosmic significance. He felt that the temple worked backward, pointing away from the Nile. There were inscriptions about the sunrise and references to the horizon, and he remarked on the orientation of the subsidiary temples of Ra-Horakhety, "Amon who hears the prayers" and the High Room of the Sun on the roof of the Hall of Festivals toward the rising of the sun. Table 4 shows that the alignment was toward a specific sunrise, the extreme azimuth of midwinter.

The extensive rebuilding by Thutmosis III apparently adhered to the construction axis of earlier sanctuaries. The northeast wall of the Hall of Festivals was decorated with murals showing the pharaoh attending an Amon-Ra, placing foundation stakes with the goddess Sechat, and making and laying foundation stones. This wall could be the original survey line, with observations of the midwinter sunrise made before the foundation ceremony. The High Room of the Sun is on top of this wall at the eastern end. It was originally built by Thutmosis III and dedicated to Ra-Horakhety, the sun god of the "two

Table 4

Alignment of Ancient Egyptian Temples

Temple	Line	Azimuth	Skyline
Amon-Ra	Main axis	$116°.9$	$0°.58$
Khonsu	Transverse axis	298.8	2.6
Mut	Transverse axis	288.9	2.0
Colossi Memnon	Main axis	117.3	0.5
Abu Simbel	Main axis	100.55	0.5
Abu Simbel side chapel	Pylons	116.	0.8

Declination	Object	Age of crescent
$-23°.87$	Midwinter sunrise	
$+27.4$	Midsummer moonset	1.56 days
$+18.2$	Midsummer moonset	1.1 days
-24.3	Midwinter sunrise	
-9.6	Sunrise October 18	
$-24.$	Midwinter sunrise	

horizons." The room was approached by two stairways, one of which remains (Fig. 3), and Barguet concluded that it was open to the sky with a single window opening to the southeast (Fig. 4). Murals show a pharaoh, gods, and the rekhyet (common folk) facing toward the window and worshiping. The inscription runs: "Make acclamation to your beautiful face, master of the gods, O Amon-Ra, primordial god of the two lands...."

It would seem that this roof temple was, in modern parlance, an observatory. The position of the sunrise disk was observed as it stood on the horizon at the time of the winter solstice. A "High Room" was needed at that epoch because the enormous axis of the temple complex, though still in alignment with the solstice sun, was blocked by the cross wall of the Hall of Festivals.

Astroarchaeology: The Unwritten Evidence

Engraved over a window in the Hall was a poetic and somwhat enigmatic statement from a priest, Hor, in the reign of Takelot II: "One goes towards the Hall, horizon of the sky, one climbs to the <u>aha</u>, lonesome place of the majestic soul, high room of the ram who sails across the sky; one there opens the doors of the horizon of the primordial god of the two countries to see the mystery of Horus shining."

Barguet favors a literal translation of the hieroglyph <u>aha</u>, rendering it as "place of combat." The astronomy supports this interpretation - the sun battles with the powers of the winter solstice at the lowest point on the ecliptic in its annual journey. The "Ra" and horizon connotations found ubiquitously throughout the Amon-Ra complex seem to carry more meaning than a purely religious one. What, for example, are the "two horizons" of the sun? In my opinion, this does not refer to rising and setting, east and west. It indicates the two eastern extremes of midwinter and midsummer, the solar lines of Fig. 1. However, no alignments have been found at Karnak or elsewhere in Egypt to the extreme northerly direction.

It is interesting to see whether the astronomical results aid similarly in the interpretation of the Khonsu temple. There is a plethora of murals and inscriptions that can be studied at the site, although no copy nor translation has yet been published. There is a roof temple placed over the southeast wall with a ruined window pointing outward (Fig. 5). Another window points inward across the roof of the structure to the hills of Thebes (Fig. 6). Both windows have inscriptions. The inscription to the right of the east-pointing window was translated by Dr. L. Habachi:

Utterance! O Amon-Ra-Horakhety-Atum-Horus, who crosses the sky, the great falcon who causes the body to be happy as a man at a festival, beautiful of face with the two great feathers, you rise beautifully in the dawn while the gods and all the people say to you 'Jubilation!' In the night the stars adore you while you sleep in the pregnant (mother). You rise at your birth and join your mother everyday. You rise and your enemies die. You are stable; rebels fall, O Amon-Ra....

3 *High Room of the Sun, Temple of Amon-Ra.*

4 *The remains of a window in the eastern wall, High Room of the Sun.*

The main axis is too near the north-south direction to align with the sun, moon, or planets, nor did it align with any bright star. The transverse axis through the outer window pointed 1°9 to the south of midwinter sunrise, and it cannot have been built as a repeat of the Amon-Ra alignment. (Incidentally, this deviation, which is almost imperceptible as one looks from the Khonsu to the Amon-Ra temple, serves to illustrate the degree of precision of the astronomical alignments that are under discussion.) In keeping with the inscription, general observations of sunrise each dawn were made through this window.

The west window pointed in a non-functional way across the expanse of the roof, and it was this abnormality, and the inscription, that prompted me to compute for the transverse alignment. The inscription to the left reads as follows: "Live the good god beloved of Amon! (...text damaged...) who illuminates the benbeni house like the horizon house of the sky of (Ra?, text damaged)."

Short and cryptic, the sentence does not make complete sense, although there seems to be an astronomical connotation. The hieroglyph benben causes difficulty. It is usually translated as "the pyramidion on the top of an obelisk" or "the tall stone on which the falcon Horus alighted," yet the word recurs throughout the inscriptions of Khonsu so that, as one egyptologist remarked, "the whole temple is a benben!" Over in the Amon-Ra temple, Barguet (1962, p. 203) found evidence to suggest that the High Room of the Sun was called a benben; he speculated whether or not at one time it had a pyramidal roof.

The astronomy suggests a more scientific meaning for benben. It might refer to the highest and lowest declinations of the sun and moon. The temple of Khonsu was coupled to the turning of the moon on the horizon by its astroarchaeological alignment. The inscription, which clearly refers to a god other than Amon, would then carry a functional meaning. The moon god illuminates its extreme turning point on the horizon, as does the sun. Similarly, the Great Temple of Amon-Ra might be regarded as the benben of the winter sun.

5 *Southeast wall of Khonsu, with person standing in roof temple.*

6 *Window pointing across roof to hills of Thebes.*

Astroarchaeology: The Unwritten Evidence

Colossi of Memnon

Across the Nile from Karnak, the temple (now gone) flanked by the two Colossi of Memnon pointed at the time of construction to the midwinter sunrise to within $\pm 0°.5$ in azimuth. The uncertainty is mainly in the conversion of the magnetic azimuth, $119°.12$, to true. The bearing was obtained in 1934 (G. Haeney, private communication), and interpolation of magnetic data between 1909 and 1965 gives a variation of $1°.79$ W in 1934. In folklore, the Colossi have always been associated with the sunrise as a general phenomenon. This result shows it to be a specific sunrise.

Abu Simbel

This temple (Fig. 7), with its facade of gargantuan effigies, is considered by some to be the finest monolithic sculpture in the world. Statuettes of the crowned hawk line the terrace, Ra-Horakhety again stands over the door (Fig. 8), and a frieze of baboons surmounts the structure to "greet the dawn". The brilliant Nubian sunrise appears across the Nile, and the light glides down the rose-colored massif, penetrating the 60 m length of the temple to the inner sanctuary, where it illuminates a statue of Rameses II with Amon and Ptah on his right and Ra-Horakhety on his left. There seems to be significance in this play of light, which takes place on only two days in the year – in early spring and late fall – but no rationale is given in the inscriptions. With commendable foresight, UNESCO engineers preserved the alignment when the temple was cut into blocks and raised above the waters of Lake Nasser.

As determined by the Center of Documentation at the Department of Antiquities, Cario, the azimuth is $100°.55$; the latitude, $22°.1$ N; the elevation, 123 m; and the skyline altitude, 0.5. At the original position of the temple, on the day when the sun's azimuth was $100°.55$, the sun shone through a notch in the cliffs of the east bank of the Nile. This notch was approximately $0°.3$ deep.

7 The rock-hewn temple of Abu Simbel with side chapel at the lower right.

8 *"Ra-Horakhety stands over the door."*

It has closed up at the new site from the change of perspective. The declination for the sun's disk tangent on the horizon was $-9°.6$. This is not part of the extrema pattern (Fig. 1); it falls into the category of a singular astronomical alignment (Hawkins 1973). The first example of this was the Plataforma Adosada, Teotihuacan (Marquina 1951), where the alignment yielded that day when the sun was in the zenith. But the situation at Abu Simbel is somewhat different.

The sun was at declination $-9°.6$ on 18 Oct. and 22 Feb. by the 1970 Gregorian calendar, and the dates would be the same to within a day at the epoch of Rameses II. (This calculation is confirmed by photographs available at the Center of Documentation. Christophe (1965), however, reports by visual observation a date of 20 Oct., which may correspond to illumination by the "first flash.") The calculations confirm the suggestion of Christophe that it is the Oct. date that is operative. It coincided with the beginning of the Egyptian civil year, 1 peret I, at the time of Rameses' first Jubilee. The civil year, of course, slipped by one day every four years against the true, tropical year, and 1 peret I fell on 20 Oct. (Gregorian) in 1256 B.C. and 18 Oct. in 1248 B.C. Rameses' 30th-year Jubilee fell at the beginning of this span, circa 1260 B.C.

This simple result radically alters the interpretation of the temple: it was carefully site-selected and astronomically planned ahead of time so that the flash of sunlight would alight on the pharaoh-god effigies, bringing life and rebirth to Rameses and starting a process of deification. The few seconds of this celestial ceremony would be reenacted each year, although the date would thereafter fall progressively later in the civil calendar. These are almost inescapable conclusions, yet there is no explicit mention in the hieroglyphics. Egyptologists have suggested, however, that the evidence points to Rameses II becoming equal to the gods during his lifetime, with the change beginning at his 30th year.

To confirm further that the ancient Egyptians had an exact awareness of the azimuthal swing of the sun, it is interesting to note the exterior side chapel, Figs. 7 and 9. Dedicated again to the sunrise god, Ra-Horahkety, this chapel is fronted by two dummy pylons and originally

Astroarchaeology: The Unwritten Evidence

contained four baboon statues, obelisks, and scarab – all symbols of the sun god. There are the remains of a raised altar and ascending stairs. This chapel does not integrate aesthetically with the architecture of the facade. Its function is to pick up the turning of the sun on 23 Dec., some 64 days after the critical day for the main temple in Oct.

Desert Markings, Nasca, Peru

The above select examples show how positive astronomical results have contributed to the understanding of ancient cultures. I would like to conclude with an example from South America, an investigation that produced a negative result with regard to astronomical alignments.

On the desert at the foothills of the Andes, Kosok (1949) and Reiche (1949, 1968) reported lines and geometrical figures drawn on a vast scale, revealed by scraping away the black pebbles to expose the light-colored subsoil. Both workers suggested that the lines pointed to the rising or setting sun or stars and called the network an "astronomical calendar". I selected a site near Nasca (longitude 75°08' W, latitude 14°42' S) for study by the astroarchaeological method. A portion of this site is shown in Fig. 10. The large rectangle is approximately 840 m long; the black line is the Pan-American Highway. A photogrammetric survey was made, a portion of which is shown in Fig. 11. Azimuths were determined from the plan, and horizon altitudes were obtained by panoramic photography. A total of 186 azimuths were obtained (93 linear features) for testing for alignment with astronomical bodies.

An attempt was made to determine an archaeological date by searching for artifacts. The area is barren and dry, and objects dropped or placed on the surface stay in situ for millenia. We noted many pottery shards scattered over the entire area, and in some groupings, the shards represented a complete vessel. These were photographed and kindly, though tentatively, identified by J. H. Rowe and G. R. Willey. The results of the identification are shown in the histogram, Fig. 12,

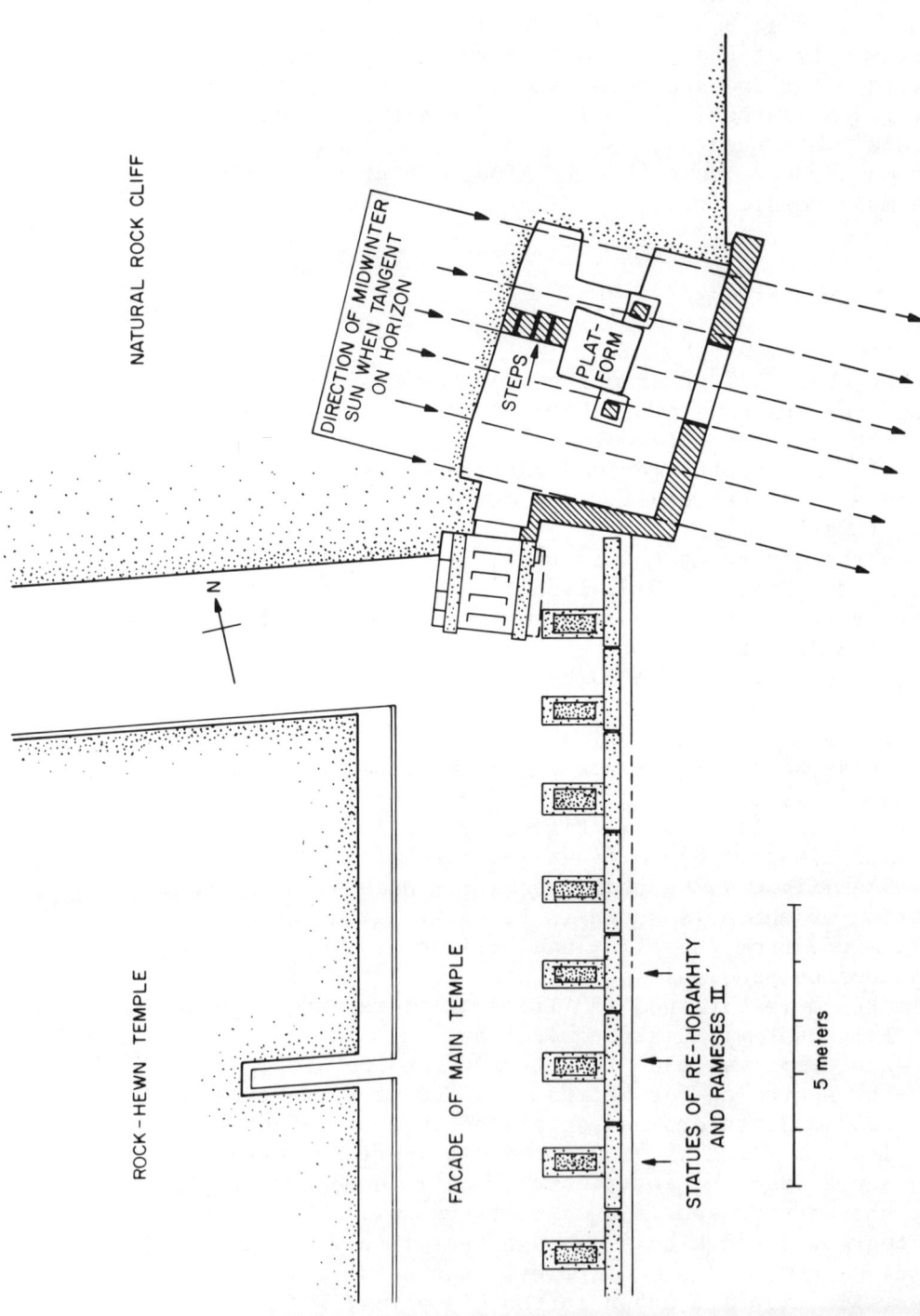

9 Side chapel of Ra-Horakhety, showing midwinter solar alignment.

10 Desert markings near Nasca, Peru.

where it is seen that the peak occurs for pottery of types 3 and 4 of the Nascan culture. Rowe and Menzel (1967) place the date of this pottery between 100 B.C. and A.D. 100. However, this is subject to some uncertainty because of internal inconsistencies in the radiocarbon data. Nor can we be sure that the pottery is contemporaneous, although it is scattered over the lines and placed on mounds nearby the drawings.

Computations were made for the archaeological date. Since only a few seconds of additional computer time were required, the calculations were extended to cover the period from 5000 B.C. to the present.

Allowing an error of up to $1°$ in declination, we would expect about one direction in 10 to correspond, purely by chance, with one of the 18 sun-moon extrema or equinox positions as shown in Fig. 1. Actually, 39 alignments matched the sun or moon positions, although only a few of these were with significantly differentiated lines. Thus, a sun-moon hypothesis failed to account for the majority of the lines, even if we agree to reserve judgment on the few that appear to fit. Interestingly enough, the large rectangle pointed to the extremum of the winter moon at declination $18°.5$, but there are no matching rectangles for any of the seven other extrema taken up by the moon in its 18.6-year cycle.

Alignments were computed for the 45 stars brighter than magnitude +2 and for the Pleiades cluster, an asterism of Pre-Columbian interest. The number of alignments found for the 93 linear features per century is plotted in Fig. 13. Allowing for precession (Hawkins 1973), we would expect approximately 20 "alignments" for the group of features per century. This is confirmed by the mean value actually found (17.3) over the 7 millenia. The number for the two centuries of archaeological interest is actually less than expected, although the significance is too small to postulate that the Nascans were pointing lines so as to avoid the stars! In no period of one or two centuries did the majority of the lines fit the stars, and thus the stellar hypothesis is unsupported. Similarly, an analysis for planetary extremes would also fail.

We did confirm the calculation of Reiche that Eta Tauri, the brightest star in the Pleiades, rose along the

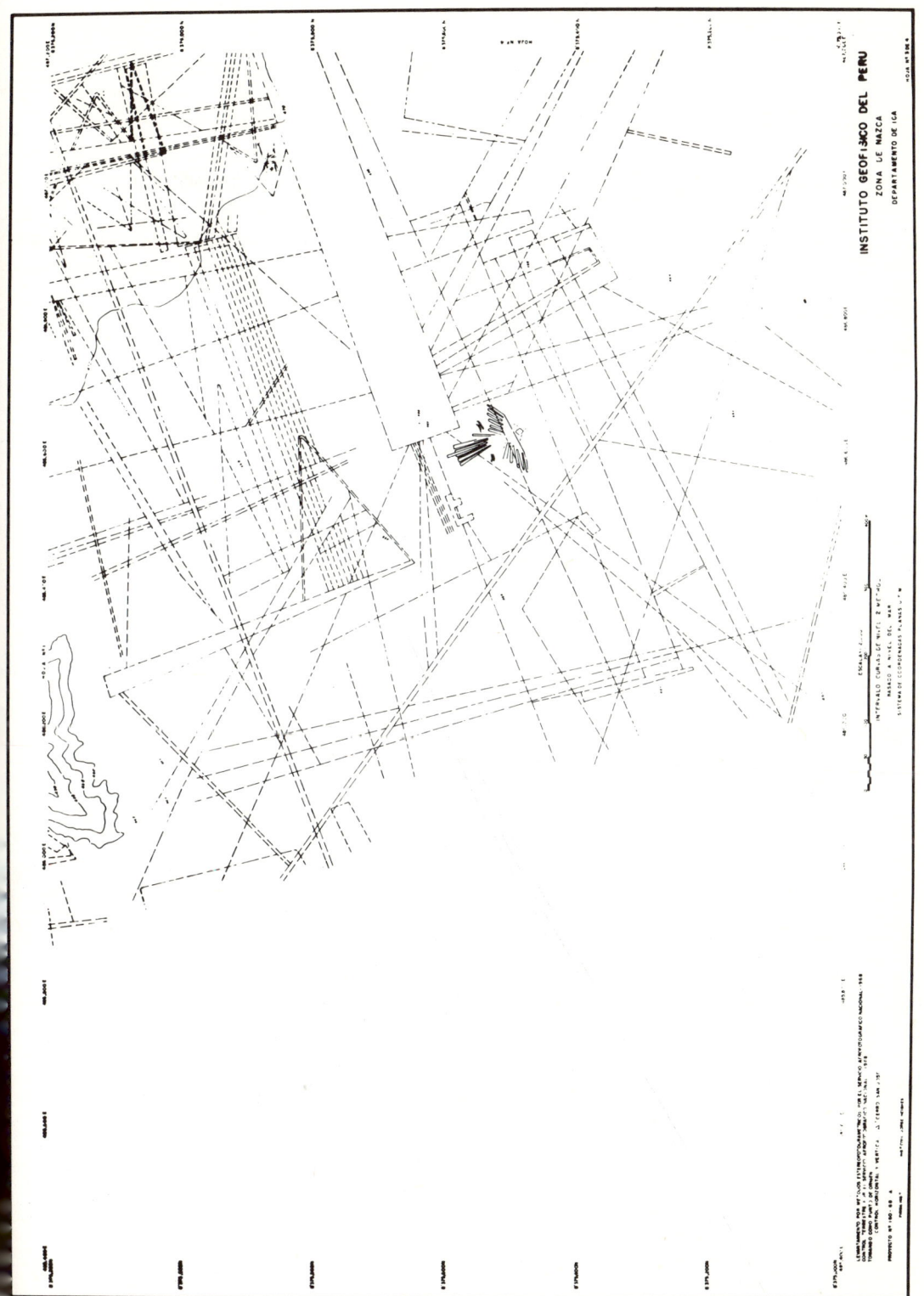

11 *Portion of photogrammetric survey.*

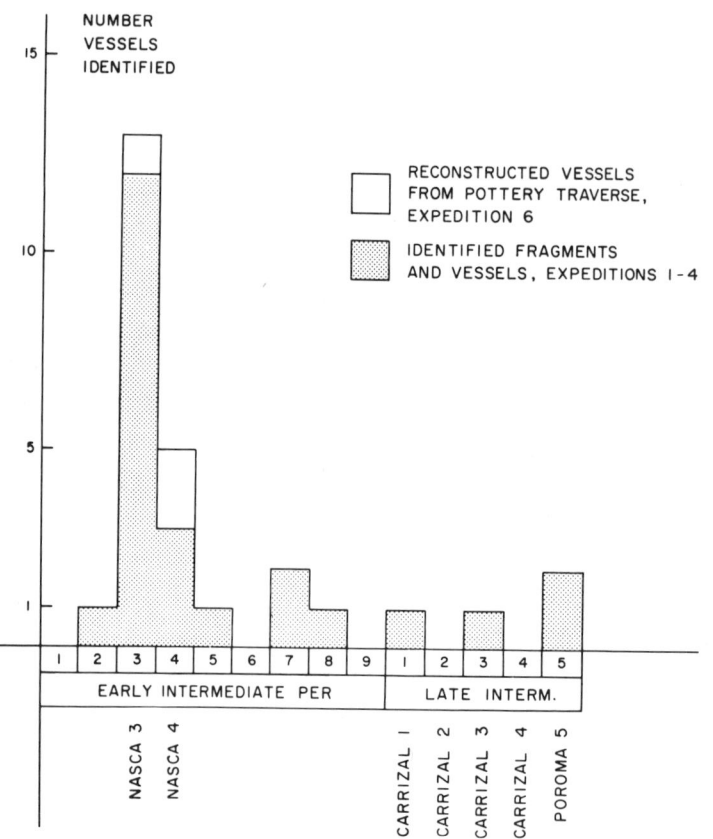

12 *Pottery shard analysis.*

central line of the main rectangle in A.D. 610. But so did Regulus in A.D. 410, and there were 16 other alignments over the complete period of study. Reiche's statement must now be discounted in light of the redetermination of the radiocarbon dates and the fairly specific identification now made by means of pottery shards with Nasca 3 and 4. The Pleiades seem to be no more than a part of a series of chance alignments that must be expected with any linear feature.

Thus, the line complex was not built to point to the sun, moon, stars, or planets. Astronomically speaking, the system is random. Perhaps a clue to its existence lies in the figurative drawings also found at the site, a few of which are shown in Fig. 14, 15, and 16. These

Astroarchaeology: The Unwritten Evidence

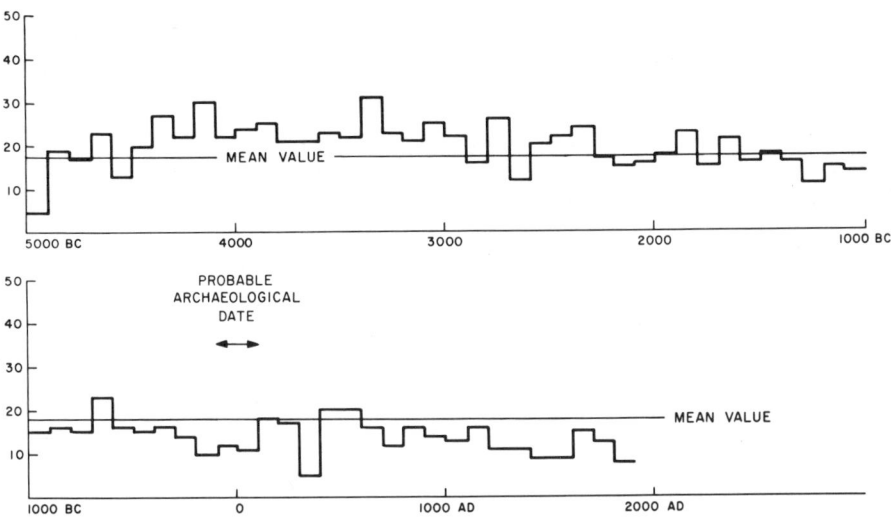

13 *Number of alignments with stars per century.*

drawings, some 200 m in length – so large as to be unrecognizable on the ground – were made with a continuous, reentrant line. Speculations have been given elsewhere (Reiche 1968; Hawkins 1973). This somewhat unfruitful exercise in astroarchaeology tends to negate the lurking feeing that alignments can be found with any old structure, whether intended by the builders or not!

Discussion

At the Royal Society and British Academy discussion meeting, a questioner from the floor humorously asked if astroarchaeology was the digging for artifacts on the surface of the stars. It was pointed out that "astro-" is an adjectival prefix and the work is a subfield within the discipline of archaeology. It is an application of the principles of the oldest of the exact sciences to the study of the past. An alignment or a number in a circle or on a bone is a hard fact, as real as any artifact. The new result must be considered in the light of what was previously known about the culture and the information assessed and evaluated. If the weight of the evidence argues for acceptance, then the new information becomes part of the broad field of archaeoastronomy and is of importance in our understanding of the nonliterate culture, or helps clarify the nonexplicit phrases of a literate culture. These and other similar fields of study contribute to what might be called the prehistory of science.

Because of the great difficulty of reaching firm conclusions, akin to detective work without interview or document, more than normal caution must be exercised. In the investigation of an alignment, basic criteria have been recommended (Hawkins 1968). The work should not be attempted if an archaeological date is unavailable; tumuli should not be lined up with megaliths or other inhomogeneous markers; and an isolated notch or horizon feature should not be postulated unless there is a clear and accurate man-made marker as well. By ignoring these and other simple criteria, the evidence becomes unnecessarily compromised. J. Norman Lockyer (1884, 1906) published in this field and caused regrettable controversy - so much so that astronomical alignments were totally disregarded. But, as a recent biography notes (Meadows 1972), Lockyer was impetuous and pugnacious and thrived on controversy, even though he "in his scientific controversies at least, generally came out the loser." Lockyer, and to some extent Thom (1967) after him, computed star alignments with an assumed construction date. Because of the effects of precession, various bright stars are brought into line through the passage of the centuries, and in any one century, the probability of a random "hit" with one of the 45 brightest stars is 0.11 (Hawkins 1973). Since Lockyer was apparently willing to assign a construction date anywhere within a thousand-year time slot, alignments were inevitable. Subsequent revisions and refinements of archaeological dates have tended to put Lockyer's timing "out of joint" and invalidated the theoretical superstructure pertaining to the pantheon of Egyptian gods, the regulation of the ancient British and Egyptian calendars, and other matters. However, his work was totally rejected soon after his death because of its lack of compatibility with other archaeological and egyptological evidence. In the case of solar alignments, the work of Lockyer is more likely to find confirmation,

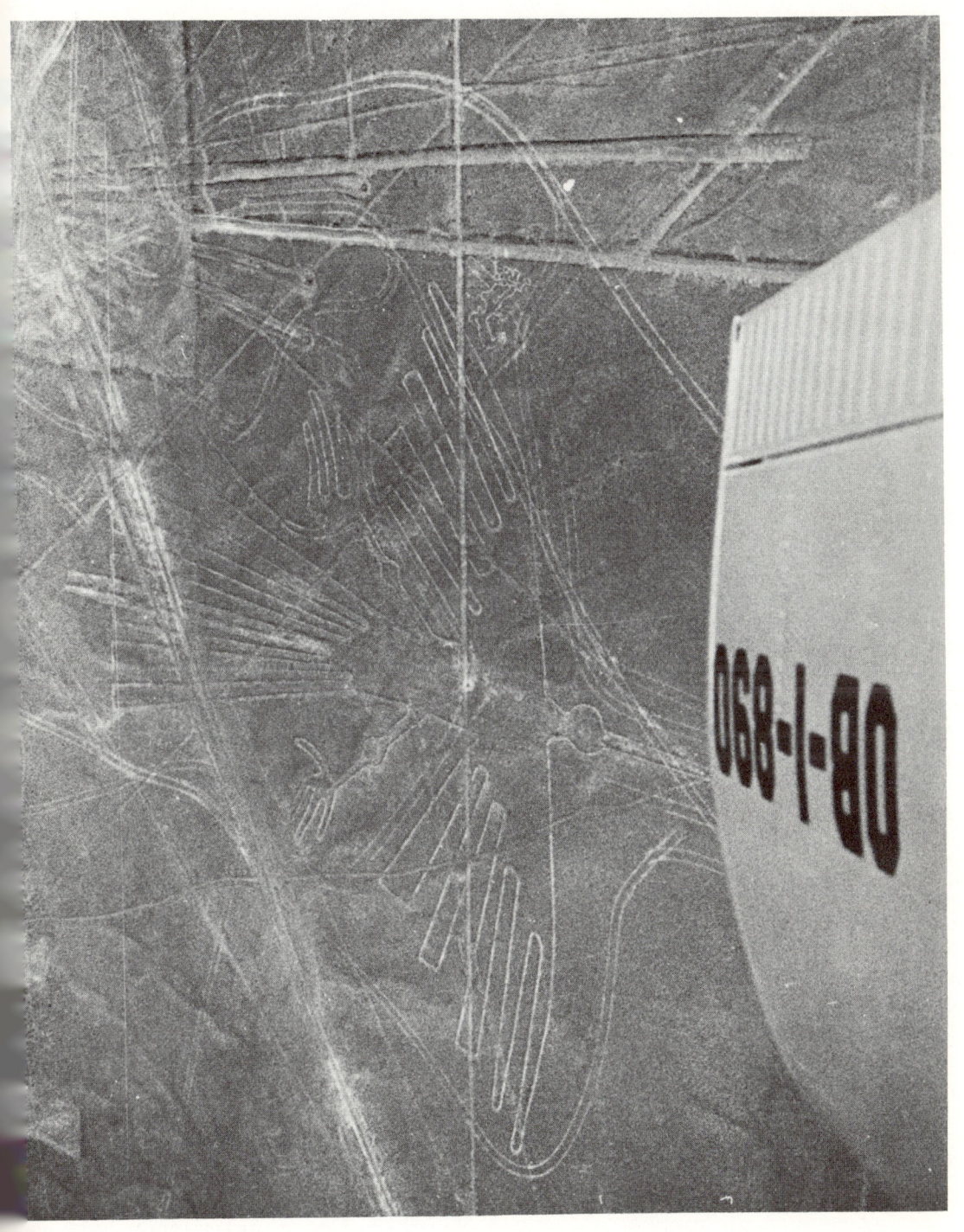

14 Ancient drawings in the Nasca desert. Condor bird.

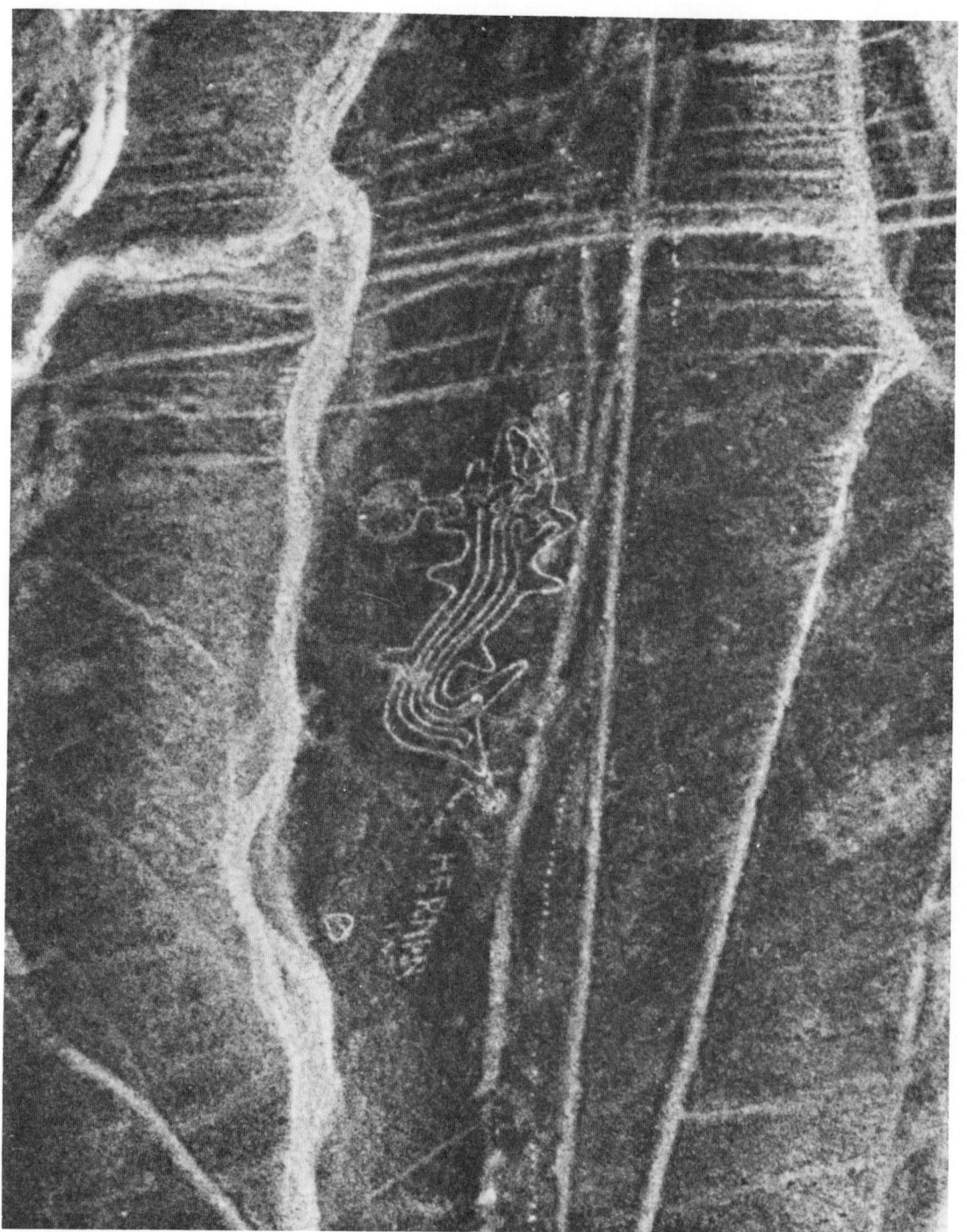

15 Ancient drawings in the Nasca desert. Fish.

16 Ancient drawings in the Nasca desert. Beaked creature.

because these alignments are not so dependent on a knowledge of construction date. However, Lockyer generally tended to work from Ordnance Survey maps and inadequate site surveys. A subsequent accurate survey at Karnak, for example, showed that his putative sunset alignment was blocked from view by the hills of Thebes. It is hoped that present and future astroarchaeological work will avoid these pitfalls.

The research was supported in part by the National Geographic Society and the Smithsonian Foreign Currency Program. Photogrammetric plans and data were kindly made available by the Center of Documentation at the Department of Antiquities, Cairo, the Franco-Egyptian Research Center, Luxor, and Servicio Aerofotografico Nacional of Peru. Assistance was provided by the American Research Center in Cairo. Dr. Labib Habachi has offered helpful suggestions with regard to the egyptology.

8

Possible Astronomical Orientations in Ancient Mesoamerica

Anthony F. Aveni

Department of Physics & Astronomy

Colgate University

Hamilton, New York   13346

Introduction

The planned nature of most Mesoamerican cities and ceremonial centers has been recognized for a long time, and a few investigators have considered the possibility that astronomical considerations may have played a role in the planning, e.g. Ricketson (1928), Marquina and Ruiz (1934), Dow (1967), Fuson (1968), Hartung (1971) and Aveni and Linsley (1972).

In view of the great degree to which the ancient people of this region seem to have practiced astronomy, it is somewhat surprising that at this late date no organized study of the possible extent of astronomical orientations throughout ancient Mesoamerica had been undertaken until the author initiated such a study in 1971 with the support of the National Science Foundation. The investigation involves direct measurement with a transit instrument of particular alignments at the archaeological sites and their subsequent matching with local astronomical rise-set phenomena utilizing a set of computerized tables (Aveni 1972). In all cases the guidelines set up by Hawkins (1966) and Reyman (1973) have been followed and sun and star shots have been employed in determining alignment azimuths. Particular attention has been paid to alignments which might have served a functional purpose, e.g. the azimuths of sunrise and moonrise at their extreme northerly and southerly declination limits and the rise-set positions of bright objects for which heliacal rising and setting could have served to mark important dates in the calendar. The investigation has been undertaken, not in the hope of supporting the notion of a pre-existing cosmic grand design for Mesoamerican architecture, but rather to serve as a basis for formulating hypotheses about the influence astronomy might have had upon native American architecture.

Though only six one-month seasons of field investigation have been undertaken to date, it is already clear that orientations gleaned from site maps available in the literature ought to be given low weight since most maps are too inaccurate. In particular, the direction of true north on many maps often is laid out erroneously; frequently, it is confused with the direction to which the needle of the magnetic compass points, and in many cases it is not clear whether a magnetic declination correction has even been applied. When an adjustment to true north has been made through the use of magnetic declination tables, one must still question the accuracy of the result, since rapid time variations in the direction of the magnetic compass pointer are well documented (Fuson 1969). The contents of Table 1 illustrate the necessity for each investigator to make direct measurements at the sites under study: the table displays the large differences between azimuths measured with the transit and those read from published site maps for a random sample of sites (in cases where magnetic north is the only north indicated on a site map, a $5\ 1/2°$ correction has been applied). In a few instances, e.g. Yagul and Ikil, site map readability is perhaps accurate enough to warrant an attempt to seek a preliminary astonomical match-up for a given alignment.

In this paper, the author presents results of possible astronomical orientations at sites in Central Mexico, Oaxaca and the Maya region of Northern Yucatan. In the next section, we examine the directions of principal axes of Mesoamerican cities and ceremonial centers and in the section following, buildings are considered which have a peculiar shape and/or orientation relative to other buildings at a given site.

Table 1

Azimuths of Alignments Read from Site Maps and Measured with a Transit Instrument

Site	Alignment	Site Map Azimuth	Transit Azimuth	Map Source
Labna	Str. 14, axis	5 1/2°	12° 02'	Marquina (1964) pg. 750
Acanceh	Stucco Palace, axis	5 1/2°	13° 17'	ibid. pg. 803
Izamal	Kinich Kakma, axis, upper portion	216 1/2°	197° 11'	ibid. pg. 807
Yagul	Axis of ballcourt	102°	101° 48'	ibid. pg. 996
Monte Alban	Axis of main ballcourt	0°	6° 42'	ibid. pg. 313
Chichen Itza	Castillo W. steps	17 1/2°	23° 22'	ibid. pg. 895
Uxmal	Governor's Palace, to front face	119°	118° 05'	Hartung (1971)
Ikil	Axis	22 1/2°	23° 08'	Andrews IV and Stuart (1968)

The Axial Distribution of Mesoamerican Cities and Ceremonial Centers

Macgowan (1945), in a pithy note to American Antiquity, seems to have been the first person to suggest that the plans of a large number of Mesoamerican cities exhibited an east of north axiality. Among those sites which evidenced some orderly arrangement, he observed that the orientations fell into three groups: true north, about 7° east of north, and about 17° east of north; he noted that few sites were oriented west of north. In the 17° group were Teotihuacan, Cholula, Tenayuca, Mexican period buildings at Chichen Itza, Tula, and the pyramid adjacent to the Zocalo in Mexico City. A number of sites in the Peten District seemed to belong to the 7° group. Macgowan suggested that a historical pattern might emerge in the sense that early structures such as Cuicuilco possessed a nearly true north axiality while the 17° east of north orientation showed up in later buildings.

Though there is much fanciful speculation in Macgowan's note, some of his ideas can be tested objectively. Fig. 1 clearly establishes his suggestion that an asymmetry about the true north direction exists and that, for whatever reason, people throughout Mesoamerica did, indeed, tend to lay out many of their cities along axes oriented slightly east of true north. The figure is a histogram showing the distribution about astronomical north of the primary axes of all sites the author has measured to date with the transit. In each case several walls of primary buildings at the sites were measured and the results averaged. Fifty of the fifty-six sites examined align east of north.

The histogram appears to support the existence of a 17° "family" of orientations though no axial trend through time has yet been discerned. The 17° orientations seem to be prevalent in the valley of Mexico, but the extent to which the idea was employed in this region will not be known until data at several more sites have been collected.

Dow's (1967) analysis of the orientations at Teotihuacan, carried out as part of the Teotihuacan Mapping

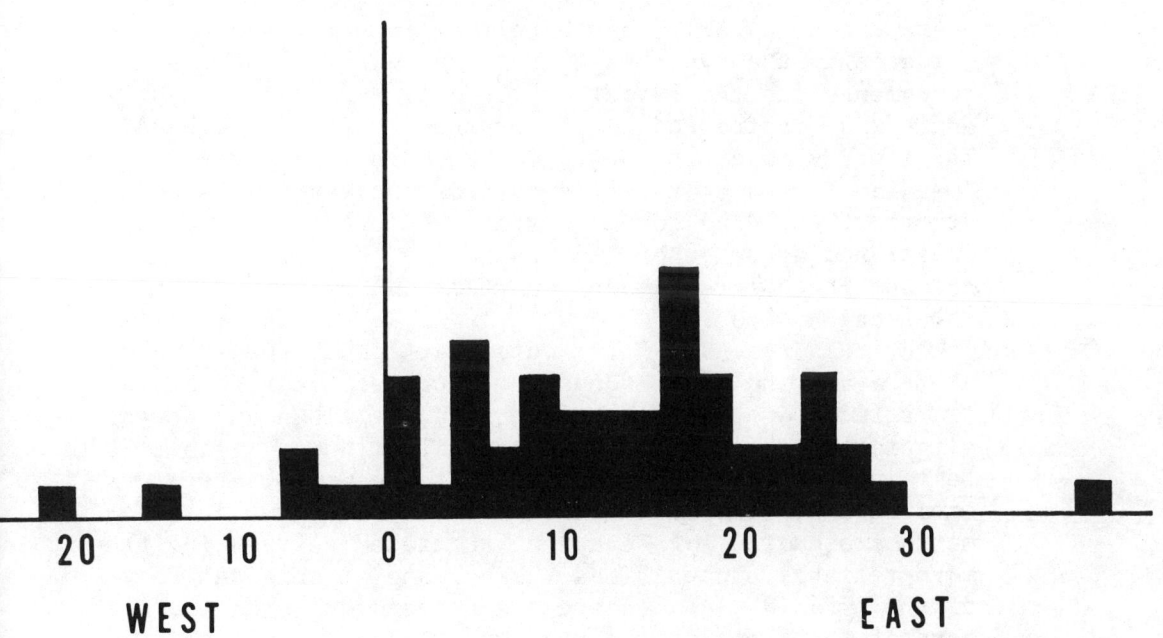

1 *Axial Distribution*

Project, reveals two sets of alignments there: an east-west orientation of 16°30' south of true east and a north-south orientation of 15°25' east of true north. He also notes a 16°55' south of east orientation at the Ciudadela. Dow mentions the existence of a long baseline running between a marker located about 100 meters north of the Viking Group adjacent to the Pyramid of the Sun and a similar marker on the southeast slope of Cerro Colorado, 3 km to the west of the Street of the Dead. Each marker consists of two concentric circles surrounding a pecked cross (See Hartung's paper this volume, Fig. 1, for a sketch). In January 1974 we measured the azimuth of this baseline with a transit and found it to be 285°21' or 15°21' N of W, about 4 minutes of arc different from a perpendicular to the direction of the Street of the Dead. As Dow has already suggested, the near perpendicularity of these two lines lends strong evidence to the hypothesis that the E-W baseline was fundamental in the layout of the city and that a right angle was constructed to orient the Street of the Dead. The clear view of the Cerro Colorado and the horizon immediately above it as viewed from the marker at the Street of the Dead suggests that the latter may have functioned as an observing post while the former marked the setting point of some important celestial body on the local horizon.

Our measurements of the outer portion of the structure at Tenayuca and at the House of Tepozteco near Tepoztlan, two structures built much later than Teotihuacan, reveal alignments close to those found at Teotihuacan (Table 2). Buildings at Tula, the ancient Toltec capital, also exhibit a similar orientation. In Table 2, two measurements are quoted for Tenayuca, the later entry being the more reliable since it was taken along an original face of the pyramid. At Tepozteco a perpendicular to the base of the steps of the structure, also an original baseline, is quoted, while at Tula the average azimuth of the original baselines of Temple B and the Ballcourt behind it are given.

Marquina and Ruiz (1934) allude to the setting position of the Pleiades as a possible stellar orientation point in setting up the axis of the Tenayuca pyramid, while Dow also mentions the Pleiades as a strong possibility at Teotihuacan. At the time the Street of the

Dead was constructed, the Pleiades would have touched the horizon point above the Cerro Colorado marker at azimuth 284°40', or 14°40' N of W, but the situation is not so simple. It is difficult to determine precisely where the Pleiades would have "set" since they may have vanished from view before touching the western horizon; though they appear as a prominent compact group of stars even to the untrained eye, their brightest member is a third magnitude star. Modern observations (Thom 1967) reveal that under the best observing conditions for viewing objects along the horizon, the extinction angle (the altitude above the western horizon at which an object would vanish from view) in degrees is approximately equivalent to the magnitude of the stellar object. Thus, the Pleiades should be invisible well before they reached the horizon of Teotihuacan, e.g., if the group could no longer be discerned at a 2 1/2° elevation, then the "setting point" is shifted westward by 20 minutes of arc. On the other hand the precessional motion of the Pleiades is so rapid that a mistake of 100 years in the dating of the baseline is equivalent to a shift of 1/2° in the azimuth of the setting point.

The Pleiades are worth considering as a possible astronomical motive behind the 17° plan, since they are mentioned repeatedly in myth and legend throughout all of tropical America and are strongly linked with the seasons. Levi-Strauss (1969) devotes almost an entire chapter to Pleiades myth in native America. Even among modern tribes the group receives prominent mention, e.g. he tells us that the Sherente:

count the months by lunar periods and their year begins in June with the appearance of the Pleiades, when the sun is leaving the constellation of Taurus....When these stars appear in the morning, it is believed to be a sign of wind....Their heliacal rising (before the sun) and also their cosmic rising (with the sun) is observed. Between two such risings the Sherente count 13 moons and this forms a year. They divide the year into two parts: (1) four moons of dry weather, more or less, from June to September (2) nine moons of rain from September to May. In the first two dry months the large trees of a piece of forest land are felled, to free it for cultivation. In the following two months the ground is cleared by burning the scrub, and then the seeds are sowed to profit by the rains at the end of September and October.

Furthermore, the Pleiades could have served the function of "announcing" the first annual passage of the sun through the zenith of Teotihuacan, since their heliacal rising and the passage of the sun through the zenith occurred approximately on the same day (58 days after the vernal equinox). While these two events did not coincide at the other sites, as Table 2 illustrates, one must not overlook the fact that these structures, nevertheless, have nearly the same orientation, though they are quite removed from one another. Can the similarity of these orientations be a mere coincidence or is there an underlying relationship? The magnitude of the shift in the setting position of the Pleiades due to precession of the equinoxes is far too large to explain the observed directional difference between the layout of Teotihuacan and that of the other sites. Perhaps the orientation of the sacred city of Teotihuacan was marked by other celestial reference points and transferred to Tenayuca, Tula, and Tepozteco at the time the latter were erected.

Drucker (unpublished) has suggested an alternate hypothesis for the orientation of Teotihuacan utilizing, instead of the Pleiades, solar positions with respect to local topographic features; however, it is not known whether his system is adaptable to the other sites since elevations and depressions along the local horizons have not yet been measured.

It is tempting to conclude that the layouts of all sites in the $17°$ group are simply non-functional imitations of the great city of Teotihuacan. Still it is not known how the orientation of these sites was actually accomplished and what techniques might have been used in laying out the plans of the later buildings to correspond so closely to the axiality of Teotihuacan.

Individual Structures with Peculiar Shape and/or Orientation

While most Mesoamerican cities exhibit a planned appearance, frequently one or more buildings at a given site seem

Table 2

Orientation of Principal Axes at Four Sites in Central Mexico

	Teotihuacan	Tenayuca (Last Epoch)	Tepozteco	Tula
Adopted Construction Date	A.D. 150	A.D. 1500	A.D. 1500	A.D. 1000
Site Latitude (North)	19°41'	19°32'	19°11'	20°05'
Horizon Elevation	1 1/2°	2°	3°	1 1/2°
Azimuth of Axis	286°30' 285°25' 286°55'	287°42'±30' 286°27'±10'	288°40'±30'	287°10'±20'
Azimuth Pleiades Set	284°40'	293°16'	292°48'	291°28'
Heliacal Rise Pleiades	T+58^d	T+76^d	T+76^d	T+76^d
First Solar Zenith Passage	T+58^d	T+58^d	T+58^d	T+58^d

out of line relative to neighboring structures. Whenever an ancient building having peculiar shape or orientation is encountered, the question of what motivated its designers becomes especially prominent. One possibility is that astronomical events occurring on or near the horizon could have determined the way a building would face. Maudslay (1912), quoting Motolinia, suggests that great care was paid to the correct astronomical orientation of the Temple of Huitzilopochtli at Tenochtitlan: "The festival called Tlacaxipeualiztli took place when the sun stood in the middle of Huicholobos, which was at the equinox, and because it was a little out of the straight, Montezuma wished to pull it down and set it right." He suggests that the priests faced east to watch the sun rise precisely between the oratories at the top of the building at the equinox.

Even without historical evidence, the possibility that peculiar structures might have served an astronomical-ritualistic function is so straightforward that such buildings are often called "observatories" by the excavators and site mappers. In this section we report the results of astronomical tests made on three unusual structures: Building O at Caballito Blanco in Oaxaca, the Caracol at Chichen Itza, and the Palace of the Governor at Uxmal.

Building O, Caballito Blanco and Its Resemblance to Building J, Monte Alban

One of the most peculiar buildings in Mesoamerica is Building J at Monte Alban. With its arrow-shaped ground plan set about an axis deviating by 40° from the predominantly north-south rectangular grid structure of the site, it stands out noticeably even to the casual visitor (Fig. 2). The presence of a tunnel in the rear of the building suggested to Caso (1938), who excavated the site, that astronomical observations might have been made there. Aveni and Linsley (1972) have tested the structure with the transit for possible astronomical orientation. We found two horizon events which conceivably could have been manifested in the architectural plan:

1) In the period of construction (Monte Alban II, about 275 B.C.) 5 of the brightest stars in the sky would have set at a position close to the axis of the arrow extended outward to meet the horizon.

2) At the same time, a perpendicular to the steps at the front of the building would have pointed close to the rising position of the bright star, Capella. This object is interesting since at this particular location and time it would have made its first annual appearance in the pre-dawn sky on about the same date as the first passage of the sun through the zenith of Monte Alban. The possibility that the observation of Capella could have been used to "announce" the zenith passage is further enhanced if we note that an extended perpencidular to the doorway of J passes over a narrow vertical tube built into the steps of Structure R (also called Structure P) on the east side of the plaza. I am indebted to H. Hartung for singling out the doorway-sight tube from the stairway-Capella relation; both alignments are illustrated on our site map (Fig. 2). Could the tube have been employed to spot the sun on its course through the zenith? The hole is wide enough to admit the solar image even a few days before and after the zenith passage event and we know that the dates of solar zenith passage were of considerable importance in the calendars of native American people (Dobrizhoffer 1822, Nuttall 1901, Marquina 1964) and were frequently occasions for celebration and religious festival.

The peculiar shape of Building J is duplicated approximately in Building O at Caballito Blanco, a site contemporary with and about 50 km east of Monte Alban. Paddock (1966) has suggested that while the two buildings have different absolute orientations, their arrangement relative to the other buildings at their respective sites is similar. Having mapped Monte Alban and determined the orientation of Building J, we decided it worthwhile to perform the same type of investigation at Caballito Blanco. Fig. 3 is a plan of the site. It was drawn by R. M. Linsley from measurements obtained with plane table and alidade, true north having been determined from corrected sightings of Polaris made with a transit. The similarity in form of Buildings J and O is immediately obvious upon inspection of the maps. A closer comparison

174   Anthony F. Aveni

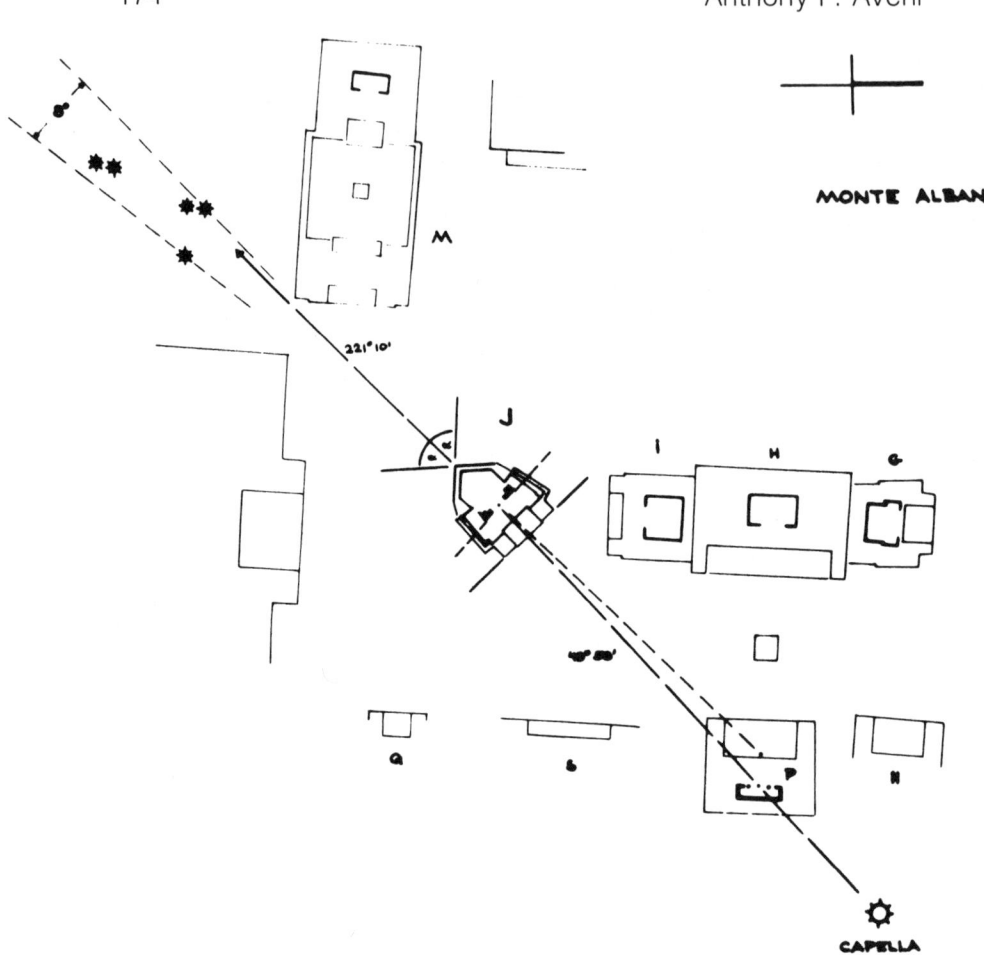

2 *Southern portion of Monte Alban showing principal alignments associated with Building J.*

of the two reveals that O is considerably smaller and has more pronounced extensions on the north-north-west and south-south-east sides. Also the "arrow-shape" is more evident in O, the outside angle at the west-south-west corner of the building being 82° as opposed to 96° for J. The similarity of the orientation of J and O relative to principal buildings at their respective sites turns out to be quite evident as Paddock suggested. At Monte Alban the angle formed by a perpendicular to the front of J and the axis of Buildings G, H, and I is found to be 44°33'. For Caballito Blanco the corresponding measurements yield

Possible Astronomical Orientations

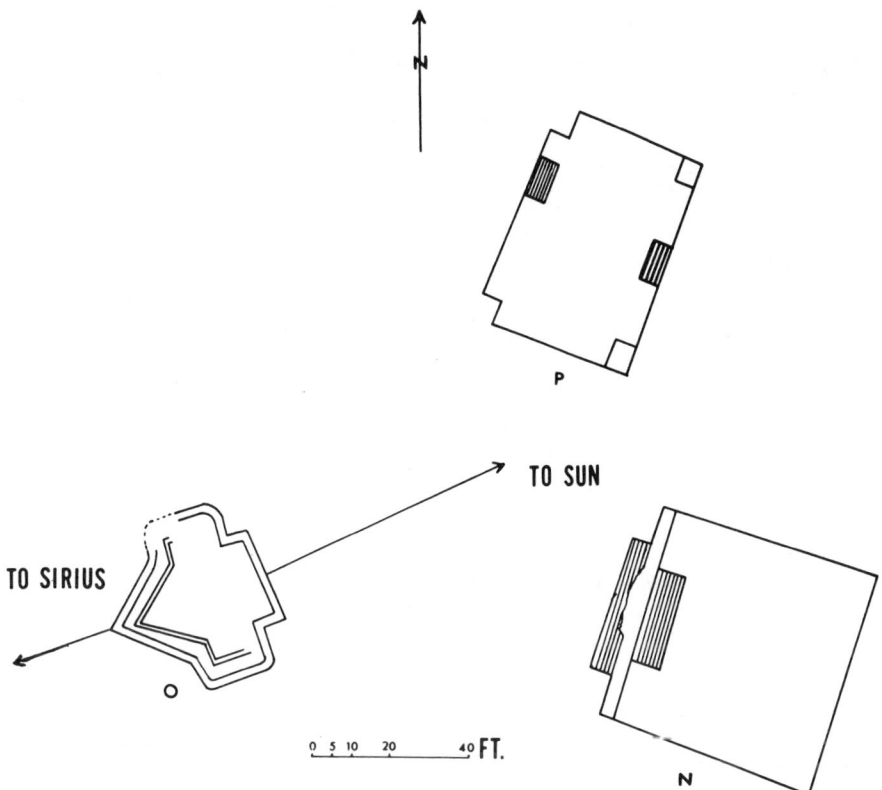

3   *Caballito Blanco*

$42°36'$ relative to the front of P and $48°36'$ relative to the front of N. In both J and O we note deliberate and sizeable departures from symmetry. Perpendiculars to the north-eastern (front) faces of both buildings deviate considerably from the angular bisector of the arrow point. Table 3 summarizes the basic alignments taken with the transit at both sites. The azimuths quoted are measured eastward from true north as usual. Apparent horizon elevations were determined also with the transit, and were utilized in conjunction with Aveni's (1972) tables to determine the astronomical rise/set phenomena occurring close to the directions of the alignments in 275 B.C., when the structure is presumed to have been built.

The perpendicular to the front face of Building O is found to point almost exactly to the rising position (first gleam) of the sun on the day of the summer solstice. Recalling that the corresponding alignment on Building J at Monte Alban is directed toward the rising position of Capella, the heliacal rising of which announced a solar zenith passage around 275 B.C., we find that both front face alignments appear to bear some relation to the sun.

For the point of Building O a close astronomical match-up occurs with Sirius, brightest star in the sky, which would have set only 11 minutes of arc off the arrow point at the time the building was erected. Again this star turns out to be of particular significance not only because of its brilliance, but also because the occasions of its heliacal rising and setting could have been employed for the purpose of dividing the year into nearly equal parts. At Caballito, Sirius made its first annual appearance in the pre-dawn sky close to the first day of summer (93 days after the vernal equinox), and the first day of winter was the last occasion on which the star would have been seen to rise in the east after sunset.

Does the difference between the orientations of the inner and outer points of Building O coincide with precessional changes affecting the setting azimuth of Sirius? The azimuth difference of the bisectors of the points amounts to nearly $+3^\circ$ (in the sense outer point minus inner point). The setting position of Sirius was altered between 500 B.C. and A.D. 0 in the same direction, but by only 6'.

From the results quoted, it is difficult to determine whether Building O was actually intended to serve an astronomical function or whether it is a mere imitation of Building J at Monte Alban. Indeed non-functional imitations of the "solar observatory" at Uaxactun (Ricketson 1928) appear at sites in the vicinity of that ancient city; however, in the present case two rather special astronomical events relate to Building O which strongly suggest to the author that it served a real astronomical function. It is worth noting that in the case of both J and O a solar relation may be connected to the front of the buildings while their arrow points seem to have been directed toward bright stars.

Table 3

Alignments at Monte Alban & Caballito Blanco Together with Possible Astronomical Matches

	Azimuth	Horizon Elevation	Astronomical Event (275 B.C.)
Monte Alban			
Perpendicular to front face of Building J	47°30'±30'	2°57'	Capella rise, 48°58'. Heliacal rise within two days of solar zenith passage.
Direction of "Pointer"	221°10'±30'	2°35'	5 of 25 brightest stars set 217°53'±4°
Caballito Blanco			
Perpendicular to front face of Building O	66°18'±30'	4°22'	Sunrise, first gleam, summer solstice, 66°23'
Direction of "Pointer"	251°18'±30'	1°04'	Sirius set, 251°29'. Heliacal rise 93 days after vernal equinox. Last visible rise at sunset 269 days after vernal equinox.

The Caracol at Chichen Itza

Perhaps the most famous "observatory" in all of ancient America, this partially ruined tower of northern Yucatan has already been examined for astronomical possibilities on at least two previous occasions (Ricketson 1928, Ruppert 1935). Ricketson's description of the building attests to its peculiar nature:

The Caracol is a circular tower with four outer doors facing the cardinal points of the compass. Within is a circular corridor from which four more doors, facing midway between the cardinal points, lead into another circular corridor. The inner circular corridor surrounds a masonry core, inside which a small, spiral staircase leads to the upper part of the building...Near the top of the structure is a flat area from which open three rectangular horizontal shafts...The largest of these, Window 1, faces west and until recently was the only one known. Windows 2 and 3...face southwest and south, respectively.

Fig. 4 is a cross-section of the top of the tower adopted from Ruppert's publication. Perhaps because the open windows accommodate a wide field of view, ($5°$ - $15°$), Ricketson suggested that the Maya architects who built them sighted astronomical phenomena along the diagonal rather than the midline of a window; he concluded that the extreme setting positions of the moon figured prominently in a diagonal sighting scheme. Ricketson's determinations of the azimuth of the diagonals of the windows, along with similar readings taken by Ruppert's group with the theodolite and our own results are summarized in Table 4.

Ricketson's corrected compass readings on Window 1 (inside left to outside right jamb, IL-OR in Table 4) and on Window 2 (IL-OR) match closely the lunar setting positions (LG = last gleam) at extreme declinations, but Ruppert's more accurate observations seem to work against this correlation. Our observations, consisting of 5 separate transit measurements per alignment taken on two different days, disagree slightly with Ruppert's results

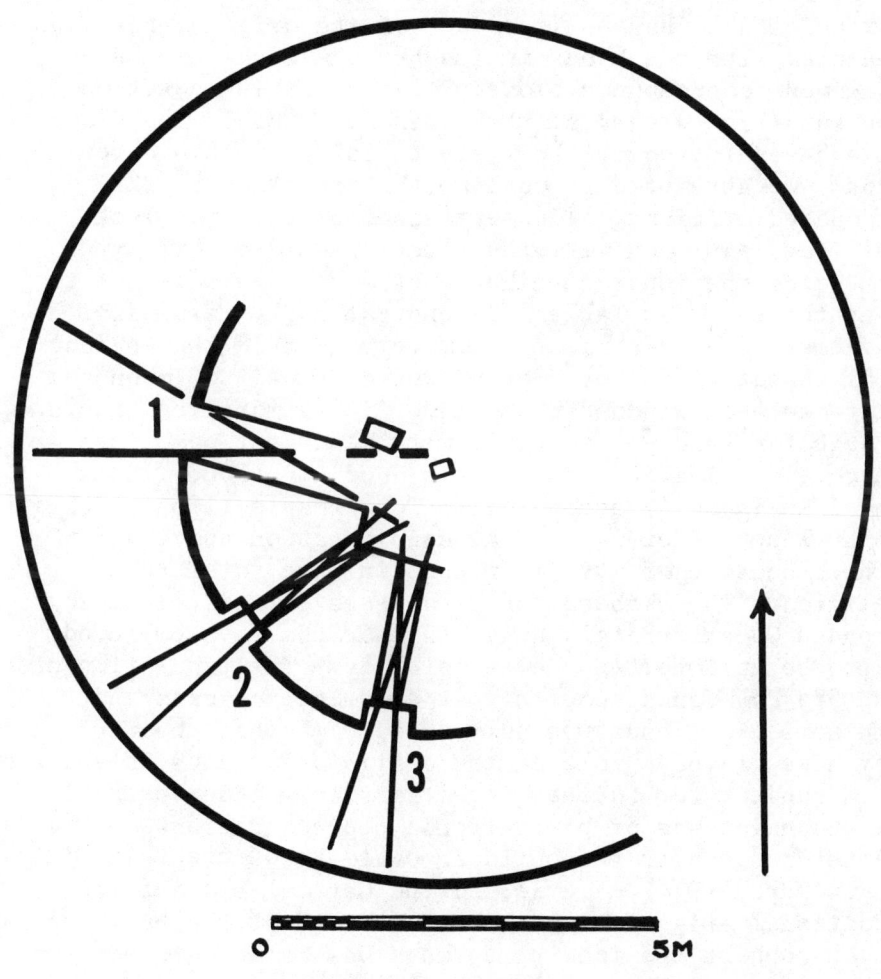

4 *Caracol Tower. (After Ruppert)*

but, nevertheless, seem to confirm most of the discordances he found with the earlier results of Ricketson. One wonders whether settling of the building might have something to do with the slight disagreement between our results and those of Ruppert.

We have sought every conceivable astronomical match-up which might be consistent with our measurements. The IR-OL alignment on Window 1 is particularly difficult since the end block on the right inner side of the window appears to have slipped out of its original position. Measured with the block in its present location the alignment corresponds closely to the setting position of the sun (last gleam) at the equinoxes. Could the block have been altered deliberately to point in this direction? We attempted to outline the direction of the "proposed original" alignment assuming the end block fell into position with the others. We did this by extending the inner straight portion of the wall out to meet the eye. As Table 4 demonstrates the resulting alignment deviates by approximately $2°$ from the equinoctial sunset position. For Ricketson's pair of moon set alignments on Windows 1 and 2 we find a more satisfactory, though far from perfect, fit with Venus set positions at extreme declinations attained around A.D. 1000 (lines 3 and 6 in Table 4) according to Tuckerman's (1964) tables. This planet deserves special consideration above all others because of its importance in Maya folklore and religion. The elaboration of a Venus calendar in the Dresden Codex strongly suggests that the Maya believed it to be an important celestial body. The tabulation of the 584-day Venus year in that document suggests that the motion of Venus was duly noted. Whether the Caracol may have played a role in the design of a Venus calendar is a subject for further investigation. A demonstration of the existence of hieroglyphic representations of the planet on or near the building would be of great interest.

In March 1974 we revisited the Caracol and obtained additional alignments on other portions of the building which support the idea that Venus was being watched from the building. The results of the recent investigation are discussed in detail in a paper presented at the XLI International Congress of Americanists meeting held at Mexico City 2-7 September 1974. They will appear in print in the conference proceedings.

Table 4
Azimuths of Alignments Taken through Caracol Windows

Window	Alignment	Ricketson (1925)	Ruppert (1935)	Present Study (1973)	Possible Astronomical Event Assuming Construction Date of A.D. 1000
1	IR-OL (present position)	270°	270°	270°57'	Sunset equinoxes LG (270°25')
1	IR-OL (proposed original)	--	--	272°27'±20'	
1	IL-OR	301°	299°45'	298°53'±10'	Moon set max northerly dec. LG (300°50') Venus set max northerly dec. (299°20')
1	Midline	--	284°53'	284°55' 285°40'	Sunset 28 Apr., 16 Aug. Sunset 1 May, 14 Aug.
2	IR-OL	--	228°	229°05'±05'	Moon set max southerly dec. (239°59')
2	IL-OR	239°	241°15'	242°11'±10'	Venus set max southerly dec. (241°11')
2	Midline	--	234°38'	235°38'	
3	IR-OL	180°	182°45'	182°13'±30'	Astronomical south; mag. south
3	IR-OL (alt.)		185°	184°46'	
3	IL-OR	--	198°	198°27'±10'	
3	Midline		190°	190°20'	
3	Midline (alt.)		191°30'	191°36'	

No obvious astronomical correlations have been found which pertain to the remaining Caracol window alignments. The midline of Window 1 points to the 28 April sunset, for which no significance seems apparent except that the Pleiades would have made their last appearance near the western horizon after sunset within a week of this date. The IR-OL alignment on Window 3 does not point to astronomical south as Ricketson suggested but rather about $2°$ west of south. Thus it deviates from astronomical south by approximately the same amount that Window 1 (IR-OL, "proposed original") misses true west. Magnetic south is a remote possibility which might be suggested. Unfortunately our knowledge of which way the compass needle pointed at Chichen Itza around A.D. 1000 is totally lacking. The apparent deviation of alignments from the cardinal directions in the clockwise (east of north) sense is again especially noteworthy in view of the results stated in the first section of this paper. Our investigation, currently in progress, of the orientations in the base of the tower reveals the same tendency. (The northern outer doorway opens slightly east of astronomical north.)

Andrews (1968) reports the discovery of a cylindrical tower in Puerto Rico, Campeche, which bears a distinct resemblance to the Caracol (Fig. 5). At least it possesses apertures, the location of which resembles those in the standing portion of the Caracol; but the building is totally different otherwise. It consists of a solid core with small openings, some of which could have been used as horizon sight tubes. The author plans to visit the site in the near future in order to examine its astronomical possibilities.

Uxmal and the Palace of the Governor

Most of the structures at Uxmal, one of the largest sites in northern Yucatan, are oriented approximately $9°$ off the cardinal directions, again in a clockwise sense. Spinden (1948) attaches considerable importance to the arrangement of the buildings there, particularly to the axis of the House of the Magician, ($9°17'$ north of west) which, he believes, duplicates the astronomical

Possible Astronomical Orientations

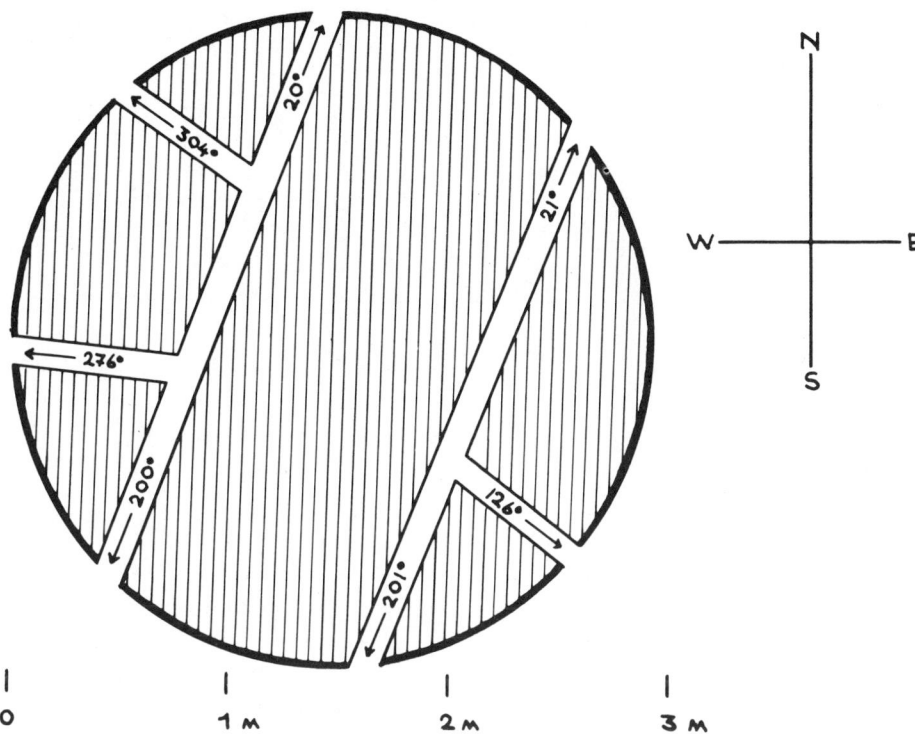

5 *Sketch of aerial cross-sectional view of cylindrical tower at Puerto Rico, Campeche, Mex.*

base line at Copan (9°14' north of west). The latter was believed to fix the start of the agricultural year. A notable deviation from the 9° axiality occurs in the Governor's Palace, which, erected on an elevated artificial platform, is skewed about 20° relative to the common axis. Hartung (private communication) pointed out the existence of a small mound on the relatively featureless distant horizon facing the front of the Palace. Viewing the mound with binoculars from the upper facade of the Palace, we found that it was part of a ruin about 6 km distant.

Securing transportation to the location sighted, we found ourselves at Nohpat, a fairly large field of ruins mentioned only infrequently in the literature. Stephens (1843) describes the ruins and tells of the appearance of the Governor's Palace as seen from the largest structure there:

...with the ruins of Nohpat at our feet, we looked out upon a great desolate plain, studded with overgrown mounds, of which we took the bearings and names as known to the Indians; toward the west by north, startling by the grandeur of the buildings and their height above the plain, with no decay visible, and at this distance seeming perfect as a living city, were the ruins of Uxmal. Fronting us was the great Casa del Gobernador, apparently so near that we almost looked into its open doors, and could have distinguished a man moving on the terrace; and yet for the first two weeks of our residence at Uxmal, no part of it was visible from the terrace or buildings there.

Nohpat appears to lie almost exactly along a line running out of the principal doorway of the Governor's Palace and perpendicular to its front face. The line passes over the base of a half-fallen stone column, which Stephens calls the picote or great stone, then over the center of a platform surmounted by a double-headed jaguar, and on to meet the horizon. The alignment from the Governor's Palace to Nohpat points almost exactly to the azimuth of Venus rise when the planet attained its maximum southerly declination around A.D. 750. A variation in the assumed building date of 200 years will result in a fit of the alignment to within 30 minutes of arc of the Venus rise position. Thus Venus would have risen directly over Nohpat as viewed from the Palace doorway. Atmospheric scintillation and its bright apparent magnitude would have rendered it an impressive spectacle above the flat Yucatecan horizon. The relevant data are supplied in Table 5 while Figs. 6 and 7 illustrate the situation.

Further evidence of astronomical considerations in the placement of buildings at Uxmal appears if we start at the House of the Magician (also called the House of the Dwarf). A straight line taken from the western doorway of that building and passed through the center of the ballcourt, continues on through the center of the northern plaza of the south group and finishes almost precisely at the principal doorway of the west group, deviating by less than one meter off its 500 meter course. This line lies within 1° of the Venus set position at its maximum southerly extreme around A.D.

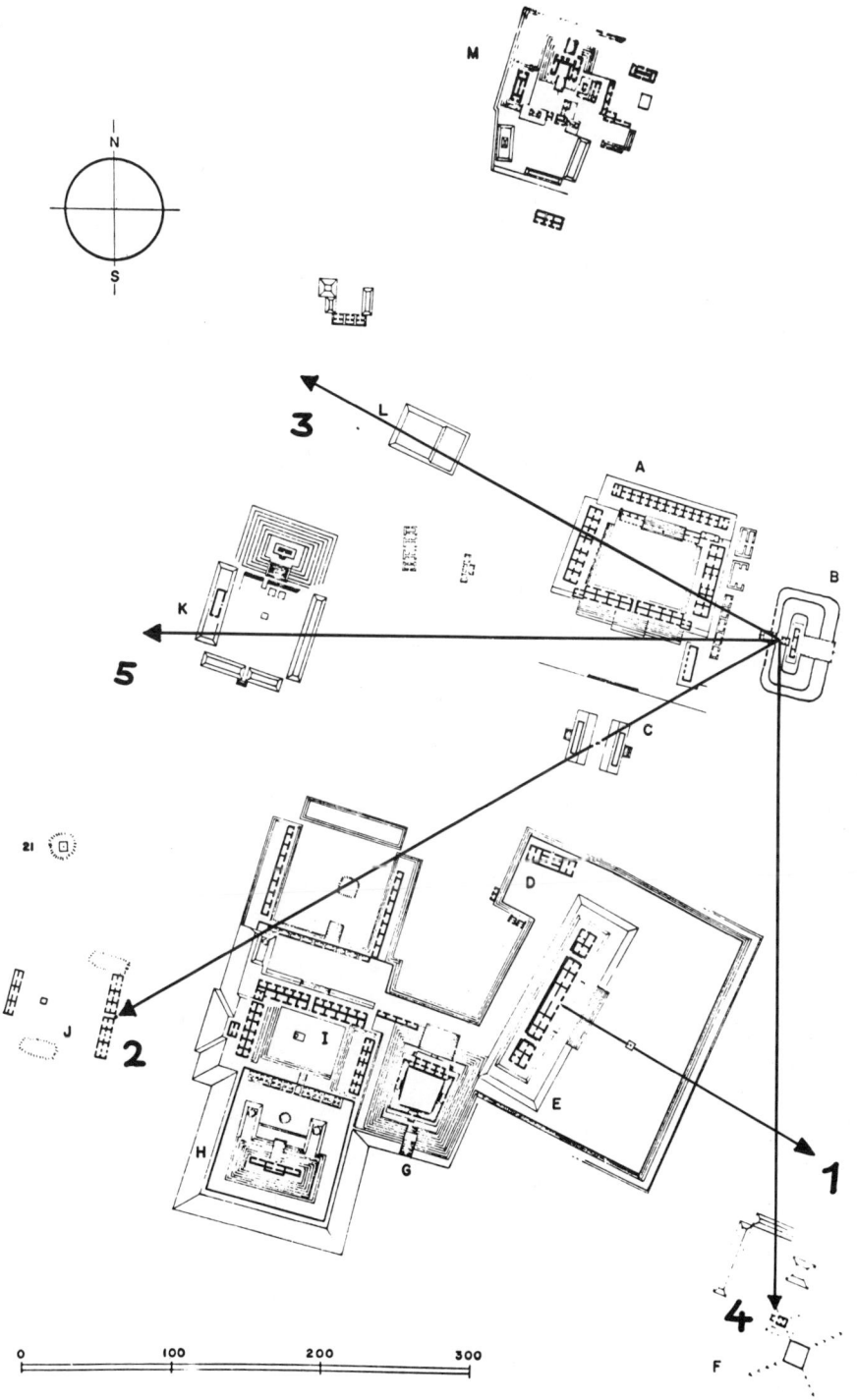

6 *Ground plan of Uxmal showing directions of possible astronomical significance.*

7 View of the southeastern horizon in front of the Governor's Palace at Uxmal taken with a telephoto lens. The mound on the horizon is Nohpat. The transit is in the foreground.

750; the extreme southern moon set position also gives a satisfactory fit.

Other suggested astronomical features of the Uxmal plan include (a) the nearly due westerly location of the Cemetery Group plaza relative to the House of the Magician, (b) the placement of the House of the Old Woman almost due south of the House of the Magician, and (c) the close correspondence between the last gleam sunset position on the first day of summer and a straight line passing from the western doorway of the House of the Magician over the geometric center of the Nunnery Quadrangle (see Table 5 and Fig. 6 for details). Utilizing transit readings taken at Uxmal of the placement of buildings within the site and of the positions of other neighboring sites relative to Uxmal, the author is currently examining the question of whether a grander astronomical plan might prevail.

Flannery (1972) has suggested that some Maya centers display a simple geometric relationship to others nearby. Secondary centers seem to be nearly equally spaced in a lattice-like arrangement about a primary center or capital. Marcus (1973) finds that the plan may have been even more elaborate. She suggests an "overall organization of the entire Maya lowlands" which resulted from the Maya's four-directional view of the universe; the existence of four regional capitals is suggested, each having its own lattice structure. Thus, cosmology may have played a major role in the large scale territorial organization of the Maya. It would be interesting to determine whether the locations of Maya cities were precisely determined according to astronomical phenomena occurring along the horizon. Careful measurement, together with a knowledge of significant astronomical events observed locally, should aid in deciding the question.

Conclusions

The aim of the present paper has been to apply a systematic, quantitative approach which might aid in understanding some of the underlying assumptions about architecture in ancient Mesoamerica and, in particular,

Table 5

Possible Astronomical Alignments at Uxmal

Alignment	Azimuth	Possible Astronomical Event Assuming Construction Date of AD 750
Palace of the Governor		
Perpendicular to principal doorway	118°05'	Venus rise, max. southerly dec.(118°03')
Center of doorway to Nohpat	118°13'	
House of the Magician — Uppermost Western Doorway		
To central doorway of West Group	240°20'	Venus set, max. southerly dec.(241°15')
To center of ballcourt	240°13'	Moon set, max. southerly dec.(239°02')
To center of Nunnery Quadrangle	296°50'	Sunset, max. southerly dec.(295°32')
To center of plaza, Cemetery Group	271°55'	West
To House of Old Woman	180°20'	South

the role that astronomical considerations may have played in the design of ceremonial centers and cities. It has been established quite clearly that the principal axes of many cities deviate in an easterly direction from astronomical north and that a 17° group exists which seems to originate in Central Mexico. The Pleiades star group is a possible celestial reference point for this alignment. Though it is still too early to decide, there seems to be no time dependent factor involved in the east of north preference.

Individual buildings at certain sites have been examined in order to determine whether their peculiar shapes or unusual orientations might be attributable to astronomical considerations. Astronomical phenomena of possible importance have been found to occur along that portion of the horizon coinciding with some of the significant alignments measured on each of these structures. The question of which astronomical objects actually served as orientation points can never be decided on an absolute basis and it would be premature to apply statistical considerations to the argument until more data become available. Venus and the Pleiades are two asterisms which frequently are best matches for some of the site alignments and both are prominently mentioned in native American folklore and legend. But what of Capella and Sirius which also fit some of the alignments? We know that both these objects could have served to mark special dates in the calendar, yet the ethnohistoric sources do not seem to support their importance. In this regard, more attention ought to be paid to Baity's (1969) notion of compiling an Astra Motif Index including surviving oral traditions and current folk practices as well as ritual motifs in ceramic and rock art which might have a bearing on orientation work.

We are a long way from determining the importance of astronomical considerations in the design and execution of Mesoamerican architecture. Objective evidence of the type presented in this paper certainly lies at the foundation of the investigation, but it is by far the easiest segment of input to obtain and it remains quite useless until it has been fully interpreted in the light of other facts, hopefully to be provided by a wider community of scholars involved in the interdisciplinary field of archaeoastronomy.

Acknowledgments

This research has been supported by the Anthropology Section of the National Science Foundation (Grant No. GS3-6262), the Osco Fund and The Society of Sigma Xi. The author is indebted to Michael Coe, Horst Hartung (who drew Figs. 2, 5, 6 & 7) and Sharon L. Gibbs for numerous helpful discussions which have influenced greatly the direction of his research. Robert Linsley has admirably carried out the plane table mapping and he and a team of Colgate University students have assisted in securing the transit measurements. Thanks are due particularly to Colgate students Leonard Weslowski, Lois Vermilya, Jeffrey Swallen, Stephen Nightingale, Joann Calderone and Barbara Toner for their valuable assistance in the reduction and interpretation of the data.

9

A Scheme of Probable Astronomical Projections in Mesoamerican Architecture

Horst Hartung

University of Guadalajara

Guadalajara, Mexico

(Translated by S. L. Gibbs from the original Spanish version.)

Astronomical knowledge is reflected in the architecture and in the emplacement of pre-Columbian buildings and can be recognized by analyzing possible manifestations in the interiors and exteriors of buildings, in points of observation, in orientations, and in astronomical directions using a rigorous criterion which does not confuse this class of lines and relationships with those arising from a visual-functional composition.

Existing documents concerning the extraordinary astronomical knowledge of the ancient Mesoamerican people, especially the Maya, permit us to suppose that it is manifested more than once in their architecture, particularly in constructions devoted to the observation of the heavenly bodies. Undoubtedly, astronomy played an integral part in their cosmic vision, and it is supposed that its influence was not limited solely to isolated constructions, but that it extended to the entire arrangement of the ceremonial centers (Hartung 1968, 1971, and 1972).

Frequent representation of temples with astronomical symbols in the codices indicates that not only certain very special buildings served for astronomical observations but that a large number of constructions were subject to conceptions of this kind or at least were following this intention in their form and orientation. Modern astronomers, having on hand the basic data, can calculate with electronic computers the positions of stars at any place and time with surprising accuracy. Nevertheless, the correct utilization of this method presents the difficulty of achieving a suitable balance between effective and judicious application in archaeoology, particularly in the archaeology of Mesoamerica.

The object of this paper is to establish some logical archaeological-architectural bases for recognizing items of astronomical interest in archaeological remains and to define points and constructions which may have served for observations of this type; that is, to try to find evidence which indicates a consciously formed astronomical direction and constructions which are specifically dedicated to the observation of the stars (observatories) without necessarily containing indications in their constructions.

In what lasting form did ancient astronomers mark directions (Fig. 1)? The simplest were lines traced with paint on stone or, more likely, on the stucco which generally covered the stone. These marks could have been made on horizontal surfaces, but also on vertical and inclined surfaces. As far as the author knows, no marks of this kind have been found; but one can include in this class the so-called astronomical circles (small stones placed one after the other in the stucco of a floor forming circles and lines), such as those of Teotihuacan and Uaxactun, sites located very distant from each other (Millon 1969, p. 113; Smith 1950, Fig. 15a; Hartung 1971, Figs. 8-C and 8-D).

The astronomical circle chiseled in the western part of Teotihuacan more appropriately belongs to the group of reliefs on stone blocks (Millon 1969, p. 113). Another circle, recently found at Cerro Gordo, is more representative. A surface (preferably vertical) can form a suitable direction. The edge of a corner (window, door, etc.) can define the approximate direction of an astronomical observation if it is viewed from a fixed point. Greater precision and facility of observation is attained if the observer's line of sight coincides with the vertical edges of two opposite corners (or more corners and/or points), as is the case in the interpretation of the use of the windows and doors of the Caracol of Chichen Itza (Ricketson 1928, pp. 218-222). There also exist other special arrangements for observation, as is supposed to be the case with tubes both horizontal (in the circular tower of Puerto Rico, Campeche (Andrews 1968, pp. 7-13)) and vertical (in the passage of the stairway of Mound P at Monte Alban). In addition to observational construction per se, points of reference can be found which, linked with a point of observation, mark determined directions. Examples can be seen in Fig. 2:

a) sculptured elements such as stelae, altars, rock reliefs, etc. Example: Piedras Negras--from the central doorway of Temple J-4 the center of the Ball Court (K-6) is due east and the Altar 2 is due south; both are at the same distance from the point of view at the central doorway (Hartung 1971, p. 36, plan 3).

b) outstanding architectural elements of a building: its door (or central doorway), its corners, or a characteristic detail. Example: Group E at Uaxactun--from an observation point on Pyramid E-VII-sub, the sun rises behind the north front corner of Temple E-1 at summer solstice and behind the south front corner of Temple E-3 at winter solstice (A detailed study was conducted by Ricketson and Ricketson 1937).

c) elements of the distant landscape, either artificial or natural. Example: a line perpendicular to the front of the Governor's Palace at Uxmal leads to a small, but notable elevation in the rather level horizon (Hartung 1971, p. 51, note 338, Fig. 15-C), which was identified by Aveni in 1973 as Nohpat, a ceremonial center 5 km distant (For an astronomical interpretation see contribution by Aveni, this volume.).

Illustrations in the codices suggest, and reliefs seem to emphasize that astronomical observations were made from the interior of a temple or from its doorways (Figs. 3 and 4).
The interior of a building can furnish one or various places for specific observations using the different practices mentioned above. They are most feasible in a

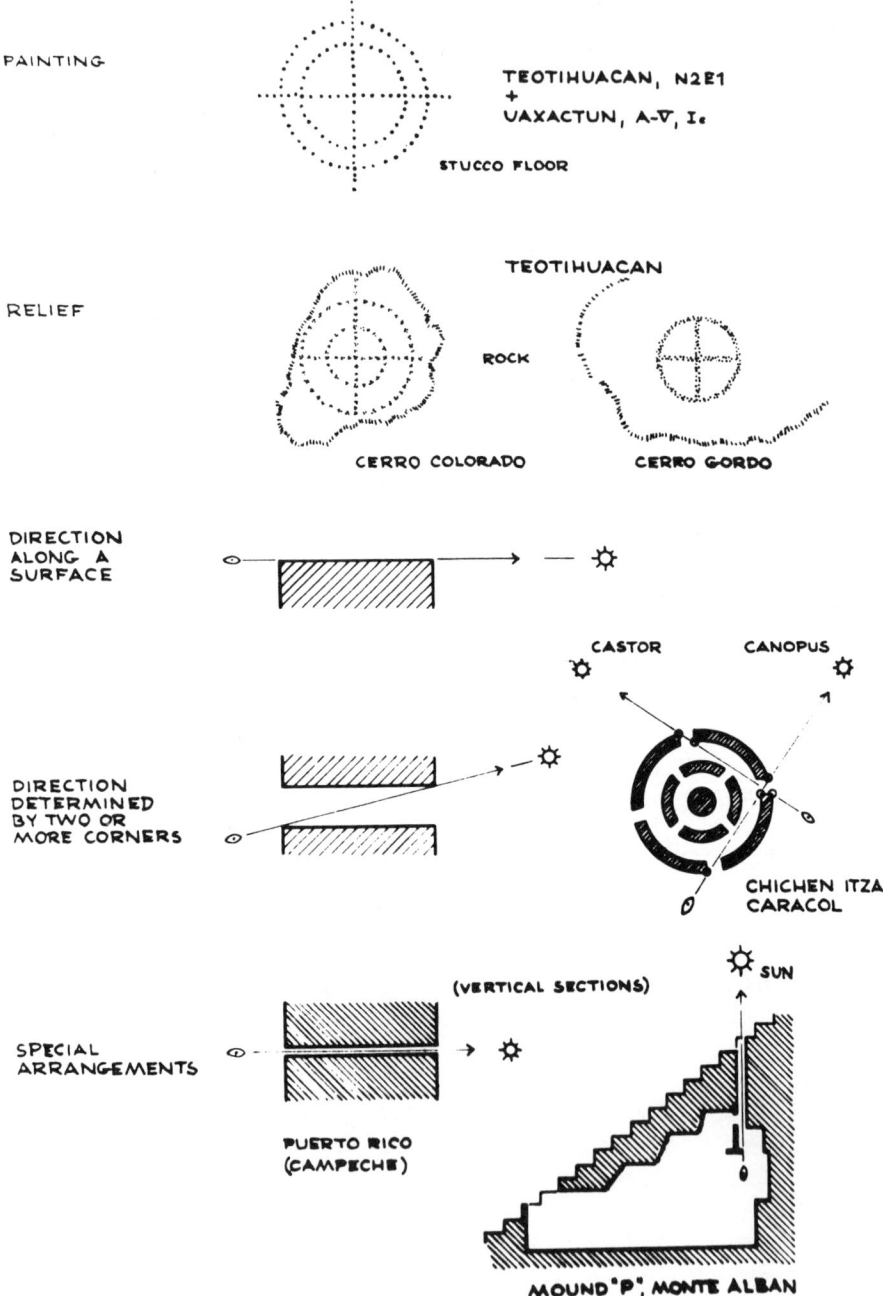

1 *Astronomical marks and visual lines to heavenly bodies. (After Hartung, 1971). All drawings are by the author.*

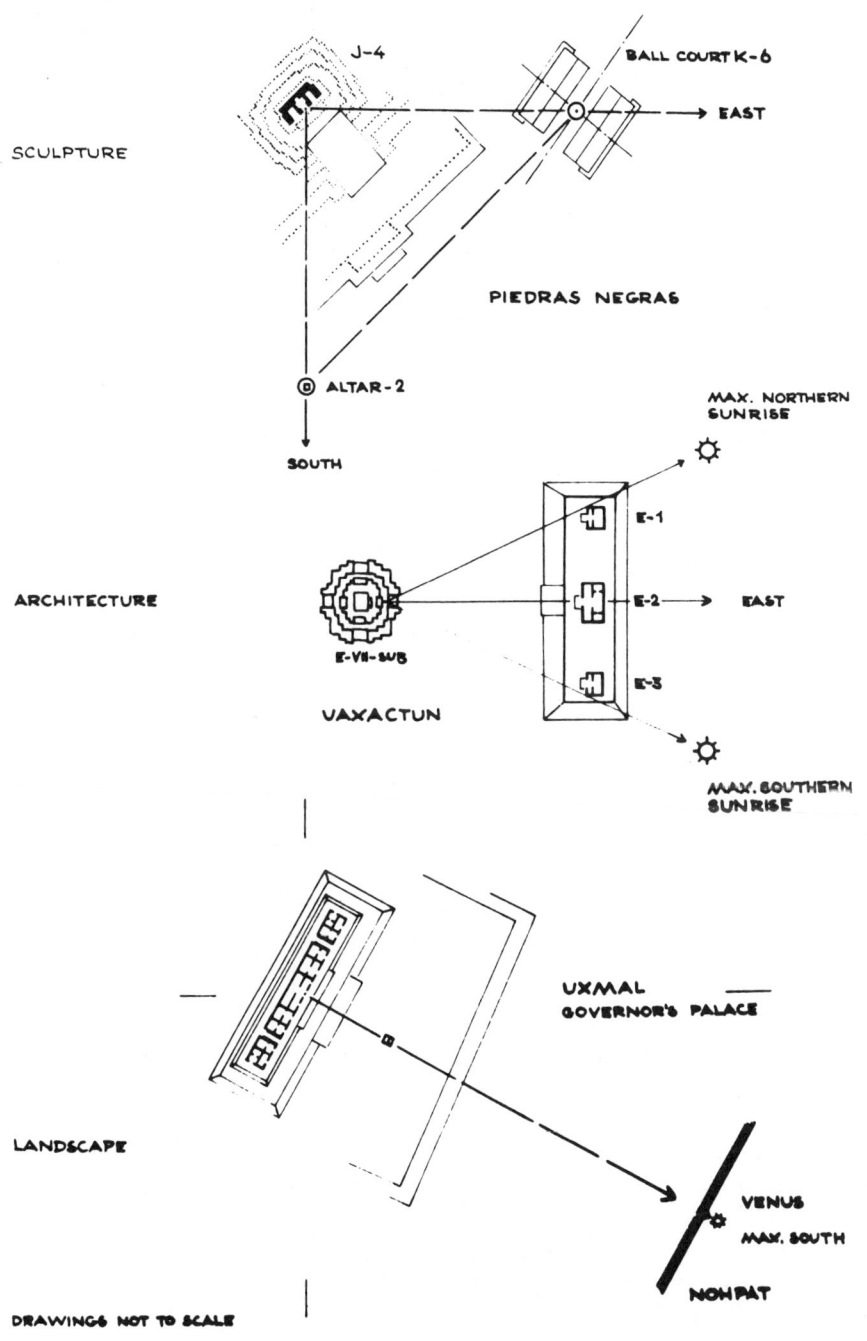

2 *Astronomical directions traceable in Maya ceremonial centers. (After Hartung 1971).*

3 *The head of a Mixtec astronomer is visible behind the doorway of a temple. He looks out through a pair of crossed sticks. Codex Bodley, 32-IV.*

4 *The entrance of a temple is marked by crossed sticks (?), similar to the picture of the Codex Bodley. The stone-slab (J-4) with reliefs was integrated in the west side of Mound J, Monte Alban.*

special room, but also have various possibilities from the interior of a temple using the edges and jambs of its doorway; although it appears that, in general, observations were made from the interior through the center of the doorway, that is, along the transverse axis of the temple (Fig. 5).

The doorway of a building (the central doorway if there is more than one) - the crucial place being between the interior and the exterior - served as much as a point of departure in the recognition of the transverse axis as a point of departure for lines of sight in other directions, provided that these latter could be confirmed by related points and elements of reference outside the temple (stelae, altars, etc.).

Farther outside, at the exterior, although with less possibility, other points of observation might be the landings of stairways, the projections at the middle or end of stairways, and even isolated elements such as platforms and altars.

When the line of sight from the interior of a building through the center of the doorway coincides with the transverse axis of the construction, it may be that the facade faces a determined astronomical orientation, that is, the facade is perpendicular to the line of sight (Fig. 6). Also the frontline of the facade can in itself point to an astronomical direction, especially when the building forms part of a group of constructions arranged around a square (quadrangle).

This situation can also be present even when the building is not situated near the determining building but is removed some distance. This disposition-relation becomes significant if the date of the constructions and/or the architectural styles of the buildings are similar. This arrangement (facing and/or aligning with) can be taken as the determinant of the so-called "astronomical orientation" of a construction. Within a group there can exist various systems of astronomical orientation, simultaneously or spaced in time.

Mound J at Monte Alban is a special case, where it is the pointed part of the building which appears to indicate a probable astronomical direction, apart from the astronomical reference of the stairway-side (Aveni and Linsley 1972) (Figs. 7 and 8).

Astronomical directions could have determined points in the plan of a ceremonial center or of a city; they could have been the source of the precise location in certain constructions of important points and/or the source of the building's orientation. The decisive points of several buildings in the northern part of Chichen Itza probably were fixed from the Caracol (Hartung 1971, Figs. 6-A, 6-B; 1972). As in this case, the central points of the ball courts seem to have a special importance in most of the Maya ceremonial centers (Hartung 1968).

In addition, it is possible to trace lines of relationship between certain constructions which agree with

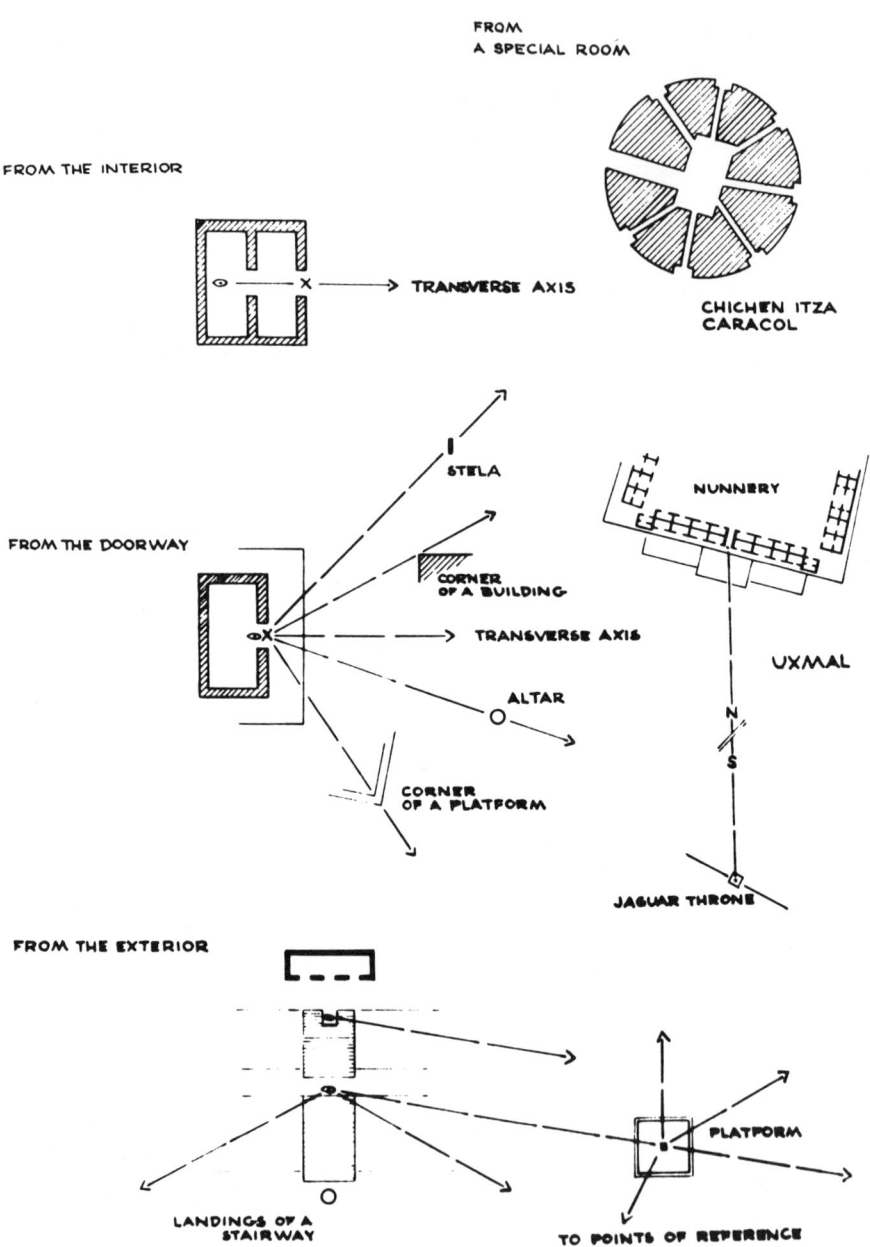

5 *Points of possible astronomical observation in ceremonial centers. (In part after Hartung, 1971).*

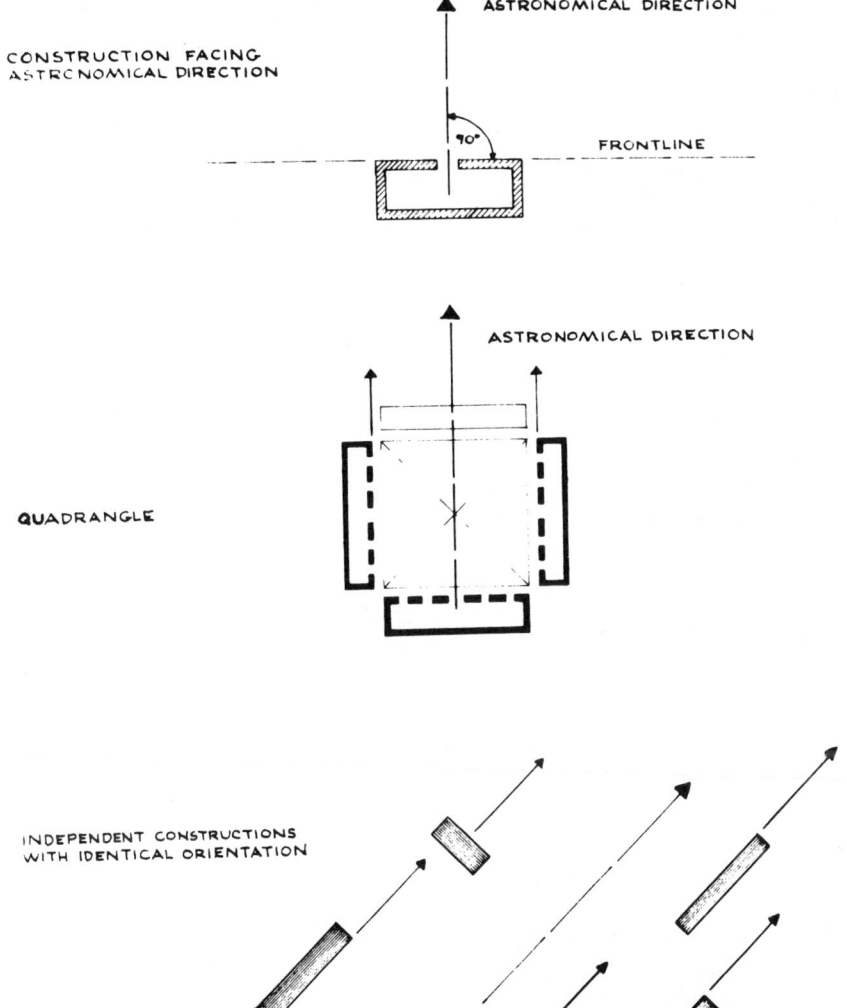

6 *Astronomical orientations of constructions: basic example, quadrangle and isolated examples.*

directions of astronomical origin (frequently the cardinal directions), even though these same constructions themselves do not show the effects of astronomical orientation (Fig. 9). The most interesting example is represented by Temple A-XVIII in Uaxactun, which is aligned due south of Temple B-XII and is related to the well-known Group E by a line to the winter solstice (Hartung 1971, Figs. 7-A, 7-B; 1972).

Groups of buildings of similar disposition exist in the Maya zone which can be derived from an original example (type site), such as Group E at Uaxactun, which is considered to be the most ancient Maya observatory. While in Group E the directions were defined with sufficient accuracy, such is not the base with later similar examples in nearby ceremonial centers which indicate quite different directions even though they retain the same general scheme. In these centers the groups take on a symbolic and traditional use over a practical use.

From the above discussion it is possible to conclude that astronomical directions were at times reflected in the orientation of buildings and/or in the emplacement of these buildings, preferably taking the central doorway as a reference point; but this need not lead to the supposition that all buildings had an astronomical function. On the contrary, many followed certain orientations in the composition of the ceremonial center solely because of formalism or order.

In the area of urban design, ancient Mesoamerican architects probably worked on the basis of relations between the surrounding space, the constructions, and the additional sculptural elements; these relationships can be found and fixed principally by visual lines.

Constructions of a similar or complementary function (i.e. the buildings with equal motifs in Yaxchilan; the special relationship between a temple, a ball court, and a sweat bath in Piedras Negras) can constitute an important axis or relevant direction in a very irregular plan of a ceremonial center, although these lines need not have any relationship to the stars (Hartung 1971). To date, the intricate system of lines of relationship at Piedras Negras has not been analyzed with respect to possible linkage with astronomical events (Fig. 10).

There are various factors, not always clearly explicable, which lead one to suppose that some lines are

7 *Monte Alban. View from the South Platform: Mound J in the center, to the right Mound P with its large stairway, in the upper part of which a small passageway enters (see Fig. 1) with a vertical shaft at its end, which probably was used to observe the zenith passage of the sun.*

8 *The pointed part of Mound J, Monte Alban, possibly indicating an astronomical event at the time of its construction.*

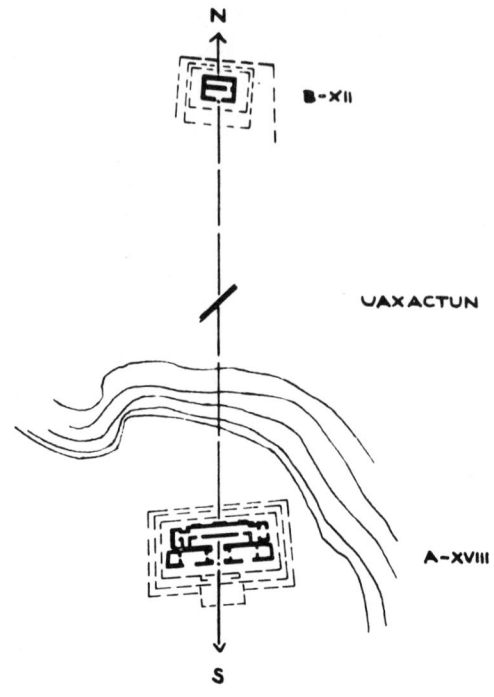
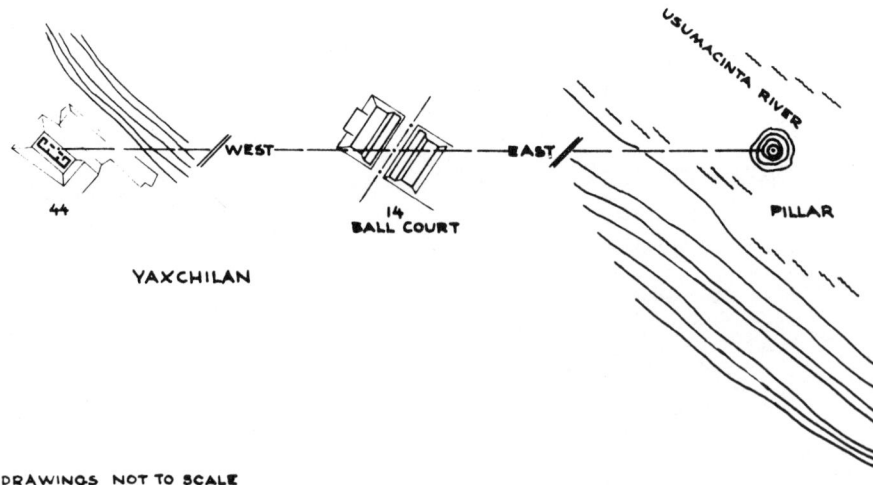

9 *Astronomical directions which can be traced between structures of different orientation (After Hartung, 1971, 1972).*

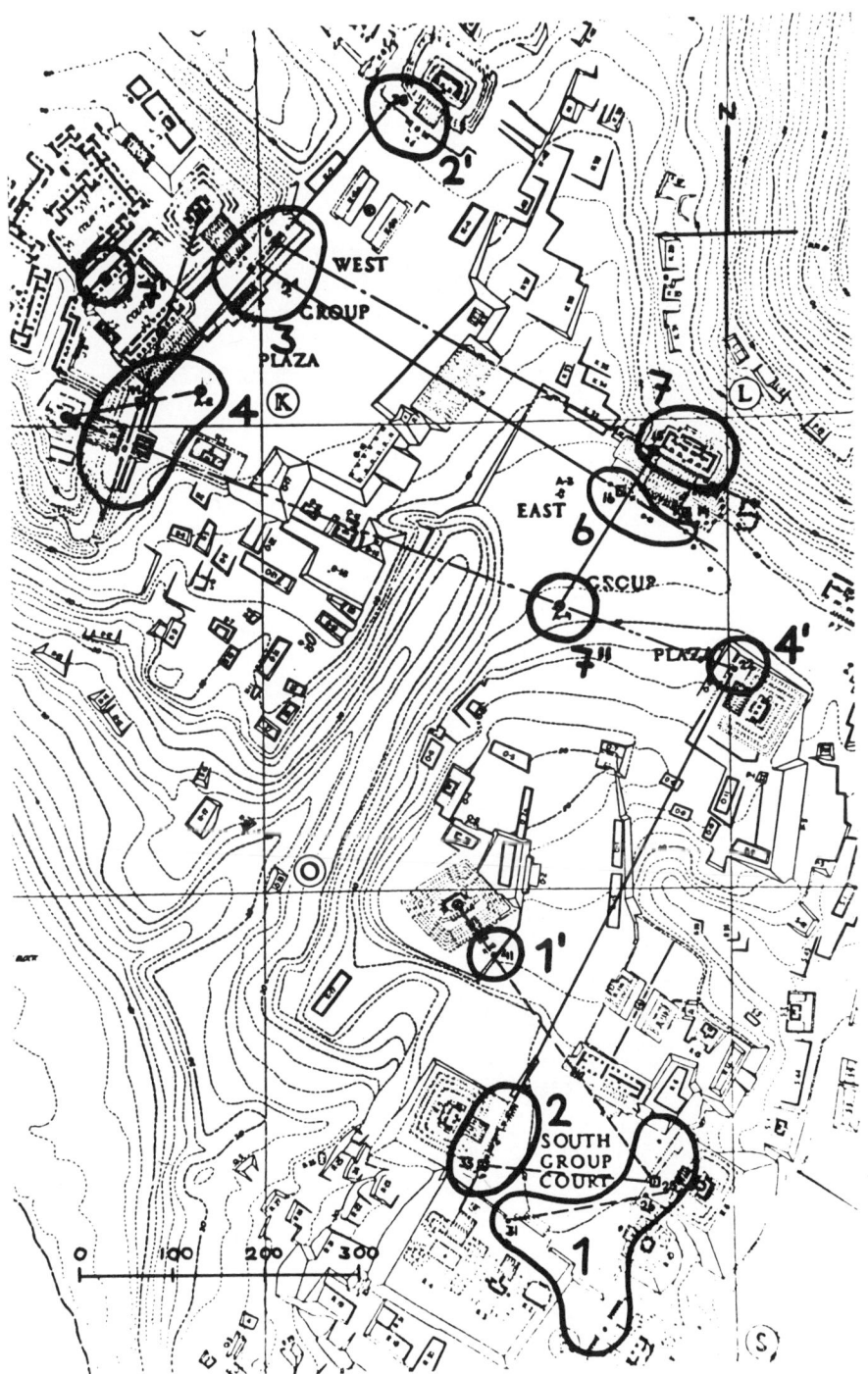

10 *Piedras Negras. In the central part of the map lines were drawn which relate stelae marked in groups, corresponding to the sequences of the rulers of Piedras Negras. (After Hartung 1971). This is only part of the intricate system of apparent lines of relation which remain to be investigated for possible astronomical orientation.*

astronomical and others are not. Generally astronomical lines tend to extend to the horizon while those which have an architectural-urbanistic relation remain inside the city or ceremonial center.

So, there exist lines (and sometimes dominant axes) in the structure of pre-Columbian Mesoamerican ceremonial centers which cannot be related to astronomical events, but which do not, because of this, cease to be of unquestionable value and importance in the architectural-urbanistic concept.

10

The Nature and Nurture of Archaeoastronomical Studies

Jonathan E. Reyman

Department of Sociology-Anthropology

Illinois State University

Normal, Illinois

During the past several years, I have had occasion to read or hear approximately one hundred reports relating to archaeoastronomical research; I have also had the opportunity to review four grant proposals in which archaeoastronomical techniques were to be employed. Most of these papers have been interesting, provocative, and potentially productive in terms of increasing our knowledge and understanding of prehistoric astronomical practices and the associated ceremonies and architecture. Often, however, perhaps even in the majority of cases, the results have not been very convincing. This is why I have used the term "potentially productive".

This lack of success, as I see it, results from the investigators' shortcomings in four distinct areas: (1) an inadequate conceptual scheme or theoretical approach; (2) an insufficient control of the relevant ethnohistoric, ethnographic, and/or archaeological data, particularly the last; (3) the failure to formulate specific field problems, hypotheses, and test implications; and (4) the lack of a consistent, systematic procedure for conducting fieldwork, coupled with the all too frequent use of unsuitable field equipment.
Due to limited space, I cannot offer a detailed solution to the fourth problem, but a paper should be forthcoming on this topic (Reyman and Sanders n.d.). Nevertheless two points need to be made regarding fieldwork.

First, one should follow the general procedure outlined by Hawkins (1968, pp. 48-50) and the precise field techniques discussed by Thom (1967, 1971). In terms of Hawkins' (1968) paper, however, it must be noted that, where written records do exist and where they indicate that a specific architectural or cultural feature was aligned to a particular star (see, e.g. Stephen 1936, p. 890), the date(s) for the alignment may help one to determine more accurately the date of construction. This would be most useful when one can document a sequence of rebuilding and realignment in accordance with changes through time in a star's position. Second, as discussed previously (Reyman 1971, pp. 40-41), one must use the best surveying equipment available, that is, a theodolite (cf. Thom 1971, pp. 119-222). It has been my experience (Reyman 1971, pp. 40-41) that other instruments are not sufficiently accurate and should not be used if space permits one to set up a theodolite. If desired, sextant readings taken from the sun (Smiley 1965) can be used to check the accuracy of azimuth orientations made with the theodolite. I shall now focus on the three remaining problems.

    The frequent lack of an adequate conceptual scheme is due to the investigators' failure to ask three basic questions: (1) given the environmental context and latitude of the site, what bio-culturally adaptive advantages may be derived from making and recording celestial observations? This is not to say that all astronomy was related to environmental adaptation. For example, the ethnographic data for the southwestern Pueblos strongly suggest that priestly power is tied closely to accurate astronomical predictions (see below). However, it seems that, on the first level of analysis, celestial observation must have been directed toward an increased understanding of seasonal changes, the better to cope with problems of subsistence, whether agricultural, hunting and gathering, fishing, etc. It was success in predicting these patterns of change in terms of regulating subsistence activities which led to the original acquisition of power, status, and prestige by part-time religious specialists (and ultimately to the expansion of this power); the initial establishment of priestly power was linked with and derived from the successful food production based on their predictions. Problems arise precisely when increasingly accurate predictions are required or demanded for astronomical events which are either unrelated or which rarely coincide. When their predictions and "control" fail, difficulties occur, and their positions may be in jeopardy.

However, the Zunis seem never to have been able to decide on the relative merits of solar and lunar calendar, and the desire to have observation of the solstice occur at the full moon disarranges the calculations and naturally leads to dissention (sic) among the various priests. (Bunzel 1932, p. 534).

It is at this point, when the system has been significantly modified or diverted from its original function--astronomical observation for the purpose of subsistence--when it is least adaptive ecologically, that the greatest threats arise to its continued existence.

(2) Which celestial features are best suited for the purpose determined in response to the previous question? (3) What are the best means of "permanently" recording these observations? This last question is specifically addressed by Hoyle (1966) with regard to Stonehenge, and all three must be considered in conjunction with planning any archaeoastronomical project and prior to undertaking the actual fieldwork.

An insufficient knowledge and control of the pertinent ethnohistoric, ethnographic, and archaeological data have often plagued investigators. The last area has been most pronounced with regard to fieldwork, specifically that, first, many buildings have been stabilized and "reconstructed" so that present alignments may not be the original ones, and second, certain structures were frequently rebuilt by the original inhabitants and also by later occupants, again with possible changes in orientation, both in terms of the building, per se, and in its relationship to the site. Indeed, if there is one predominant characteristic of human settlements, it is that we are constantly changing their appearance. Some fieldworkers seem either to be unaware of these two situations or else have chosen to ignore them. This is not permissible because such disregard probably invalidates the results of archaeoastronomical research; alignments found under such poorly controlled circumstances are not acceptable. One need only look at Proskouriakoff's (1946) drawings for the various stages of Structure A-V at Uaxactun to realize how radically the complex was altered during the occupation of the site. The same is true for the Palacio tower at Palenque and numerous other Maya ceremonial centers, most of which have been stabilized and rebuilt by archaeologists (cf. Maudslay 1889-1902), the Great Sanctuary at Chetro Ketl (Vivian and Reiter 1960, Figs. 14 and 17), Stonehenge (Atkinson 1960), and every other major site throughout the world. A good indication of the potential significance of this problem for archaeoastronomy is contained in the following statements regarding Fewkes's (1917, 1917a) excavation of Far View House of Mesa Verde. For Room 17 it is noted that, "the original character of this room was lost in the rebuilding" (Burgh 1934). Furthermore, "No records of the extent of the rebuilding done by Dr. Fewkes exist. Therefore, it was necessary to consider all damaged and fallen walls as of prehistoric construction" (Burgh 1934, p. 8).

The lack of control which too frequently has been exercised over these three classes of data may help us to explain why many of these projects have not included a <u>specific</u> formulation of a problem within a cultural-ecological context, the hypotheses necessary to study the problem thoroughly and systematically, nor the test implications which need to be met for the hypotheses to be validated. The fieldwork often appears to have been conducted without the use of a consistent, ordered methodology. The search for alignments, at times, seems to reflect a haphazard, almost random "groping," and the accompanying explanations have tended to be after-the-fact (see Aveni and Linsley 1972). In short, archaeoastronomers have all too rarely used anything approaching the scientific method.

Yet even when fieldwork <u>has</u> been conducted with care and precision, the resultant explanations are sometimes unacceptable, generally because the basic hypotheses are inadequate. For example, Thom (1971) has argued that a number of Megalithic sites in south Argyllshire were lunar observatories for use in navigation. There is no question that these sites contain lunar alignments of certain stones, and several may have been used for navigating the Sound of Jura. Nevertheless, Thom does not explain <u>how</u> these structures were used for navigation, why so many were necessary, nor how these located at some distance from the sea could have served this purpose; and what function was served by the solar alignments which are also found at certain sites? (Reyman 1972). Moreover, as Thom suggests (1971, p. 11), the study of lunar motion was carried far beyond the degree of knowledge required for navigation, <u>per</u> <u>se</u>. Clearly, then, his basic navigational hypothesis is inadequate in terms of many of the existing data, although it may account for some of them. These sites would seem to represent a broad cultural complex, built over a period of at least several centuries and probably the work of various groups, which was built to serve multiple functions.

Although there are areas of the world for which there is a paucity of ethnographic and ethnohistoric materials relating to astronomical practices, this is not the case for much of the New World, particularly the Southwestern United States, Mexico, and Peru. Coupled with these data are vast amounts of archaeological materials including much architectural information. Thus the problem which faces the scholar is twofold: First, one must become thoroughly acquainted with this literature, and second, one must critically evaluate the information before using it. Many archaeoastronomers have been somewhat lax in these matters.

It took years for the native astronomers to determine the alignments and then to construct the buildings and other features so as to incorporate these orientations; one cannot expect to control these data with only cursory preparation. On the basis of my own research, I estimate that, for a large site such as Teotihuacan, <u>at</u> <u>least</u> six months of steady reading and study are necessary <u>just</u> <u>to</u> <u>become</u> <u>acquainted</u> <u>with</u> the archaeological literature which directly or indirectly pertains to astronomy and astronomical practices; when there are also numerous ethnographic sources, still more time is required to cover them. This is especially true for the southwestern Pueblos where several dozen to hundreds of sources exist for a specific community; there are several dozen for Picuris and more than 500 for either the Hopi or Zuni.

A visit to the site is also mandatory because site maps are notoriously inaccurate (Reyman 1971, pp. 23-25). This problem is probably most pronounced in the Maya area and similar regions where the jungles or dense upland forests often make it impossible for surveyors to plot precisely the positions of buildings relative to one another. Thus the overall site orientation, as shown on site maps, is often incorrect, although aerial photography and large scale clearing projects such as have been undertaken at Tikal and Chichen Itza are reducing the errors. Archaeoastronomers, working from site maps, have frequently assumed orientations to be correct when, in fact, they were not (e.g., Fuson 1969); many of the maps in Marquina's (1964) otherwise seminal work, Arquitectura Prehispanica, contain inaccuracies in building orientations (Reyman 1971, p. 25). Yet even when maps are extremely accurate, as is the recent one made by the Teotihuacan mapping project, they are still unsuitable for archaeoastronomical purposes because they lack critical data such as the heights of the observer's horizons along the axis of orientation (Reyman 1971, p. 24); these heights are crucial because they affect declination. Furthermore, incorrect constructions, like that noted above for Far View House, are recognizable only when one is thoroughly familiar with the site and the pertinent literature and is discerning in ones evaluation and use of data. With these problems in mind, let me take up the matter of sources for formulation of archaeoastronomical hypotheses and their test implications.

There are several basic sources for the New World, and I shall confine discussion to this area. In my opinion, the two most useful general references are (1) The Catalogue of the Library of the Peabody Museum of Archaeology and Ethnology, Harvard University and (2) The Human Relations Area Files. The former is organized by author and subject, with the subject division cross-indexed in numerous ways including topic, area, site, and ethnic group. The HRAF is the most complete single compendium of ethnohistoric and ethnographic materials. Archaeoastronomers would do well to begin their projects by thoroughly researching these two sources for relevant data.

Spanish clerical accounts are abundant and extremely useful, for example, the Jesuit Relations, Landa's

Relacion de las Cosas de Yucatan (Tozzer 1941), and Sahagun's (1956) Historia General de las Cosas de Nueva Espana, also known as The Florentine Codex. A number of the native codices are available, such as, Fejervary-Mayer (Seler 1902; Burland 1950), and Dresden (Gates 1932); other ethnohistoric works such as The Book of Chilam Balam of Chumayel (Roys 1933) are also widely distributed. These form a rich corpus of material from which hypotheses can be formulated. Major areal syntheses are similarly useful, particularly Marquina's (1964) Arquitectura Prehispanica, Pollock's (1936) Round Structures of Aboriginal Middle America, and Parsons' (1939) Pueblo Indian Religion. Such works are indispensible but, like all materials, they must be used critically. Baity's (1973) article and bibliography are very useful for archaeoastronomical studies, per se. With the exception of this last reference, all of the above sources plus hundreds more are currently listed in the Peabody Catalogue and most are at least partially abstracted into the HRAF.

I noted earlier that it took years for the native astronomers to determine the alignments to which ceremonial structures were oriented. This is well-illustrated by the following account from Zuni:

The man who went to the Sun was made Pekwin. The Sun told him, "When you get home you will be Pekwin and I will be your father. Make meal offerings to me. Come to the edge of the town every morning and pray to me. Every evening go to the shrine at Matsaka and pray. At the end of the year when I come to the south, watch me closely; and in the middle of the year in the same month, when I reach the farthest point on the right hand, watch me closely." "All right." He came home and learned for three years, and he was made Pekwin. The first year at the last month of the year he watched the Sun closely, but his calculations (for winter solstice) were early by thirteen days. Next year he was early by twenty days. He studied again. The next year his calculations were two days late. In eight years he was able to time the turning of the sun exactly. The people made prayersticks and held ceremonies in the

the winter and in the summer, at just the time of the turning of the sun. (Benedict 1935, Vol. 2, pp. 66-67)

This is probably the basis for the orientation of the suntower at Zuni, the use of which was described by Bandelier:

The year begins about the winter solstice. The suntower is used as follows. In winter, they look over a notch in the western wall, over the pillar to the east. When the sun rises over a certain point, there in a line with the pillar, then it is midwinter or the beginning of the year. In summer they look from a pillar in the summer gardens to the sun tower (sic). Sign of the year according to the Zuni. (Lange and Riley 1970, pp. 69-70; see also Cushing 1941, pp. 128-129)

Similar materials exist for the other southwestern Pueblos, as well as for many other areas of the New World. These resources must be used in the formulation of research programs, specific hypotheses, and their test implications.

This brings me to another point: Determination of alignments is not and should not be the endpoint of archaeoastronomical studies. Indeed, this cannot be if archaeoastronomy is to be more than just an interesting technique. As noted above, these alignments formed an integral part of some adaptive strategy (also see below), and it is to the understanding of that process that archeaoastronomical studies must be addressed. Therefore, using the American Southwest as an example, I shall briefly outline how some of the above materials can be incorporated into a coherent anthropological research project.

From an environmental standpoint, the Southwest was and still is a marginal agricultural area, particularly in the Anasazi region. Even today, as the past five years at the Velarde region north of Santa Fe, New Mexico demonstrate, late spring and/or early fall frosts can significantly reduce crop production. Nevertheless, the prehistoric Anasazi lived in communities which, to a large extent, were established and maintained on an agricultural subsistence base. How was this possible?

It is immediately apparent that understanding seasonality and anticipating seasonal changes will have a profound affect on one's ability to maintain a successful agricultural system. Therefore, one purpose of celestial observation is to predict accurately the length of the growing season and seasonal changes; the group that watches and records the movements of astronomical phenomena has a selective advantage in terms of its survival relative to other groups. In conjunction with this purpose, the sun, moon, and certain star groups (constellations) are useful, the last serving as a nighttime check on the sun's path as it appears to move through them from solstice to solstice and back again. The best way of recording these observations is by first using horizon calendars (or shadows in the case of the sun), and later by incorporating fixed observation points and orientations into architectural structures (see Ricketson 1928). The final development, achieved by the Maya, Aztec, and probably the Inca, was a "written" calendar, on paper or in stone, which also included predictions of specific astronomical events like eclipses. However, direct celestial observations never seem to have been eliminated.

From existing ethnographic data, it is clear that solar observation is the most important type of skywatching; <u>every</u> Pueblo has institutionalized solar observation in the office of Sunwatcher (Ellis and Hammack 1968, p. 31; Parsons 1939, pp. 122-123; Parsons in Stephen 1936, pp. 3-4; Stephen 1936, pp. 30, 62 and 1075). Given the relative constancy of the environment through time, we can hypothesize that one function of prehistoric Southwestern towers was for solar observation. Among the test implications are: these towers were built with a clear view of the horizons, and cross-jamb or parallel-jamb orientations of windows or other wall openings align to significant points in the solar year, such as, solstice sunrises. After carefully checking the excavation reports for possible errors made during stabilization, one can test the hypothesis in the field using the towers at Mesa Verde and elsewhere. Alternatively, towers can be initially excavated with this hypothesis as the research problem. The hypothesis of stellar alignment would have the same test implications except in terms of stars, but one must also check for the possibility that the rebuilding and reorientation of the feature represent a solution to the problem of the stars' movements through time due to precession and proper motion.

Of course, there are other adaptive strategies which may have been operative, such as the use of alignments and astronomical predictions made from them for the consolidation and expansion of priestly power. Thom (1971) has dealt specifically with this problem, and priestly responsibility for celestial observations and "control" of astronomical events is much in evidence in the Southwest (see Bunzel 1932, p. 534; Lange 1959, p. 321; Parsons 1939, pp. 122-123 and 508-510; Simmons 1942, p. 59; Parsons in Stephen 1936, p. Xl; Stephen 1936, pp. 389-390; White 1942, pp. 205-206).

Systematic studies can be conducted even when one has only the site plan with which to work and there are no extant ethnographic materials. The consistent orientation of buildings within a site at least suggests the possibility that a celestial referent was used. Thus the overall site plan may be another source for the formulation of specific hypotheses. Most of the archaeoastronomical work at Teotihuacan seems to have been predicated on the definite biaxial regularity of the city. Similarly, the peculiar orientation and/or shape of a particular structure relative to other buildings at the site also suggests the possibility of an astronomical alignment, as with Mound J at Monte Alban (Aveni and Linsley 1972). As Dow (1967, p. 326) states, however, geographical reference points and chance associations must also be considered. His argument is especially important where the range of possible orientations is strictly limited as, for instance, within the cave sites at Mesa Verde (Reyman 1971, pp. 190, 194-195). Nevertheless, even when one is working from a site plan, the adaptive aspects of the potential alignments must be considered.

In summary, we must (1) ask what selective advantage accrues to those who watch the sky and record astronomical events? What celestial features are most useful for the purpose? and what are the best methods for permanently preserving such information? (2) use the relevant ethnographic, ethnohistoric, and archaeological data to support hypotheses generated from our observations, to formulate general research problems, specific hypotheses for testing, and the test implications for these hypotheses; (3) employ an ordered methodology and consistent field techniques to collect the data needed to evaluate the hypotheses. Explanation must result from the systematic testing and evaluation of problem-oriented hypotheses, not as an after-the-fact development from the haphazard discovery of "alignments". I have tried to suggest, albeit very briefly, both a general program and a few sources for the formulation of archaeoastronomical hypotheses. If implemented, this program should provide greater consistency and comparability of results, as well as more convincing conclusions than have heretofore been the case.

Finally, it should be obvious that the nature and nurture of archaeoastronomical studies require close cooperation between anthropologists and astronomers. We need astronomers to tell us how the sky works and why, and what the sky was like at the time the sites which we study were occupied. Astronomers need anthropologists to provide the archaeological and ethnographic data, and to remind them that people built and aligned these structures for very specific cultural, adaptive reasons. Until such cooperation is forthcoming and constant, the results of archaeoastronomical studies will continue to be something less than convincing.

Acknowledgments

Thanks are made to Edward B. Jelks and Robert Dirks for their critical reading of the manuscript, their comments and suggestions. However, the author alone is responsible for all statements contained herein. Research for this paper was supported, in part, by grants from the National Science Foundation (GS-2829 and GS-40410).

11

Effigy Mounds and Stellar Representation: A Comparison of Old World and New World Alignment Schemes

Thaddeus M. Cowan

Department of Psychology

Kansas State University

Point to Point Alignments

The evidence suggests that the proto-astronomers in the Old World were primarily concerned with significant solar and lunar events as they appeared on the horizon. A sighting line for such an event required at least two points to define it. For the megalithic astronomer, the observer's station, marked by a menhir, served as one point. The second point was probably some prominent feature on the horizon, at least in the early stages of the art (Thom 1970). The alignment between observer and any single solar or lunar event was point to point, and the relationship between observer location and all event locations was a one-to-many mapping.

At some later time someone thought of bringing the distant sighting points closer to the observer. The erection of circular or quasi-circular megalithic structures not only accomplished this, but it also freed the location of the observation point. This eventually led to refinements by which the movements of the sun and moon across the horizon could be tracked more precisely. The edge-on projection of menhirs standing in a circle formed a sinusoidal frequency pattern on the horizon, the spaces being wide in the middle but closely packed toward the ends. This served as a good model for the movement of the moon as it swung from standstill to standstill. Yet even with this modification the relationship between observer and sighting point, as well as between observer and astronomical event, was one-to-many.

There is at least one site in which a nearly one-to-one mapping of sighting line points was observed. At Temple Wood the relationship between multiple observer stations and multiple horizon markers was very close to being isomorphic; for virtually each horizon marker there was a unique observation post as well as a unique astronomical event.

In the abstract, a point to point isomorphic mapping can assume the characteristic of simultaneity, and such a mapping of points is often called a "template" alignment. Since it is likely that there was only one megalithic observer viewing a unitary event, it would be absurd to think that a template type of alignment was used by Neolithic man in Western Europe. Does it make sense to speak of template-type alignments at all in archaeoastronomy? I think it does, and what is more I propose that the early North American Indian used such alignments in connection with the constellations.

The Effigy Mounds

The great earthworks of Ohio and the effigy mounds of Wisconsin are thought to be the products of the Hopewell culture, and there is some evidence that the link may extend as far back as the late Adena people. This period, known as Early to Middle Woodland, spans the first millennium A.D., and carbon dating places the Effigy Mound Cult within this era (McKusick 1964). The mounds assume various shapes, but the bear and bird effigies seem to be the most common. In fact, these often appear together.

The proposition that these effigies represent contellations is derived from two simple observations: First they face skyward making them difficult to recognize from the ground level. They were often constructed on hilltops and ridges, although they may have been placed there for more mundane and pragmatic reasons: valleys tend to flood during heavy rains. Secondly, they are large. (One bird effigy measures 190 meters from wing tip to wing tip.) If an effigy mound was a copy of some constellation then its size would be determined by the perceived size of that constellation. The Indian of course had no inkling of the great distances to the stars, and therefore imposed his own subjective distance measure. It would be reasonable to suppose that these subjective distances were some function of his perceived distance to the horizon which gave him some experiential standard for infinity.

In making my case I will draw from three sources. First, there are observations made by others which might be reviewed briefly; these provide only indirect support. Secondly, it will be necessary to discover how the Mound Builders regarded the stars or how they interpreted their patterns. Present-day Indian folklore might provide a lead here. Finally, the template alignments of effigy to constellation will be considered.

Prior Speculations

There have been a number of suggestions that at some point the mound builders turned skyward for their religious inspiration. Hyde (1962, p. 17), for example, comments:

At the close of the Archaic period and in the early Woodland times that followed, the Indians of the woodland country east of the Mississippi began to develop an incipient civilization. Many of the Indians were no longer content to worship earthbound animal gods alone, and they turned their eyes to the heavens seeking new gods.

Add to this, if you wish, that Pidgeon's De-coo-dah, the last of the mound builders (however mythical he might be), told that his ancestors regarded the mounds as symbols of heavenly bodies (Silverberg 1968, p. 252).

One of the more interesting comments regarding the problem here was made by Nuttall (1901, p. 274) almost 75 years ago. In her discussion of the swastika as a primitive symbol she observed:

The swastika or cross, the most ancient of primitive symbols, was primarily a graphic representation of the annual rotation of the Septentriones (the seven stars of Ursa Minor) around Polaris. It thus constituted not only an image of the most impressive of celestial phenomena but also a year-symbol....In Mexico and the Ohio valley it is linked with the serpent, to the symbol of which reference should be made.

The connection between the famous Serpent Mound in Ohio and the pole star is teasingly close. The Serpent Mound is not only the largest of the effigies, it is also one of the most aesthetic in its artistry (Fig. 1).

Indian Lore

The North American Indians both knew and named the stars. The way in which they regarded the constellations, however, was unique. Many of their stories seemed to relate characters and figures to individual stars and not their patterns. The star-husband stories are cases in point (Thompson 1929).

Star clusters were often noted where individual stars in close proximity were seen as a group but without loss of the identity of each individual star. One legend has it that the constellation Ursa Minor represents a small child (Polaris) leading a lost hunting party home (Scott 1942). Among the Paviotso the stars in the belt of Orion were three sheep in a line following the Pleiades (Curtis 1926, Vol. XV). Similarly, the northeast Yavapai called Orion's belt Muu (mountain sheep) according to Gifford (1936). However, Gifford notes that the western Yavapai's constellation of Muu comprised four stars in a row which clearly is not Orion's belt. We might gather from this that there was a tendency to perceive the stars in linear arrangements. These perceived clusters of stars have a loose Gestalt-like meaning for the observer; their patterns are representative, but unlike the myths of the western world, each point in the star group retains its identity. Each star is not seen as an individual playing a smaller part in a larger scenario.

Effigy Mounds

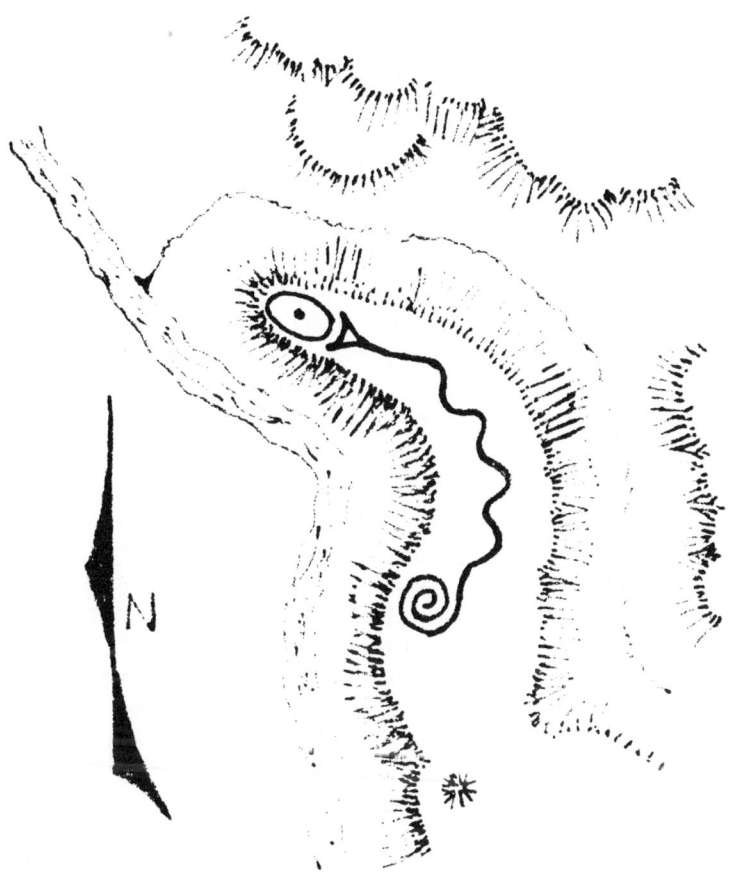

1 *The Serpent Mound located in Adams County, Ohio. The effigy is some 400 meters long following its coils. (Copied from Squire and Davis 1848.)*

There are legends that treat the stars as genuine patterns. These include paw-print legends of which I discovered two. The Pomo name for the Milky Way means "bear foot" and they say that a bear walked across it (Loeb 1926). (It was a common belief that the Milky Way was a path taken by the dead.) To the Maricopa there is an unnamed star group (the Pleiades?) that represents a hand print (Curtis 1926, Vol. II). Also the four stars outlining the bowl of the Big Dipper suggested a bear to the Fox Indian (Marriott and Rachlin 1968). In these instances the individuality of each star is lost, and each point becomes a subordinate part of a more meaningful whole.

This treatment of the stars as individuals, groups, and patterns seems to be ubiquitous among the North American Indian, at least as far as I have been able to discern, and I have looked at legends ranging from the Eskimo to the Yavapai to the Algonquin. It is not impossible that the ancient Hopewells or Adenans regarded the constellations in much the same way.

The North American Indians had no system of writing nor did they have an established oral tradition like the pre-semasiographic Middle Easterner. There are exceptions such as the Walum Olum of the Delaware Indians (Hyde 1962), but these are rare and are primarily historical in content; they do not contain mythology. Therefore, there is no direct historical evidence that the mound builders perceptually arranged certain stars in groups, like the linear clusters of the Yavapai, or patterns, like the hand print constellation of the Maricopa. From the present hypothesis, however, we might expect the star configurations to be reflected somehow in the shapes of the mounds. For example, the numerous chained conical mounds just might represent linear star arrangements. These are strings of mounds, commonly in groups of three, connected by less obvious earthen fill. As for the paw-prints there are three earthworks which are highly suggestive of such prints but which have not been so identified, surprisingly. These are shown in Figs. 2, 3, and 4. The last (the Ross County Stoneworks) appeared to Peet as two intertwined snakes with five tails, an observation about which Fowke had a wry comment or two. Quipped Fowke:

Peet possesses a peculiar faculty for seeing snakes. But he is justified in exploiting this discovery; in fact, he should give it more prominence than he has done, for there is probably not another work in the world where two snakes are represented as the proud possessors of five tails--or five sets of rattles, whichever it is (Fowke 1902, p. 297).

In summary, Indian lore suggests a variety of ways the stars can be regarded (individual stars, groups of

Effigy Mounds

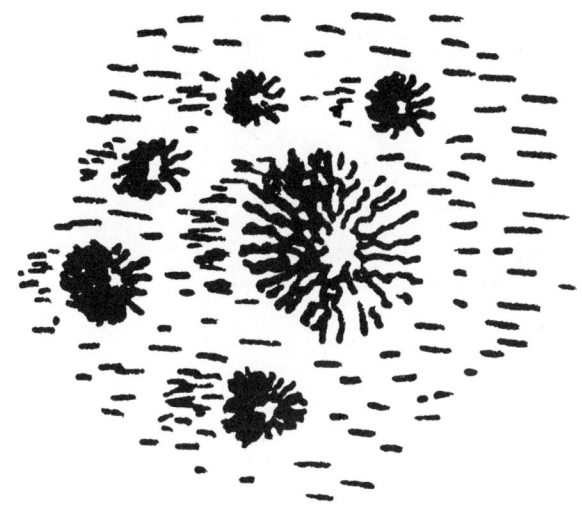

2 *A mound group resembling a paw-print located in Hamilton County, Ohio. The largest mound is about 8.23 meters high. (Copied from Fowke 1902.)*

individual stars, patterns). Similarly, the mounds might be seen as following the same course (conicals, chains, effigies). It is not suggested here that the mounds necessarily represent the legends associated with the constellations. What is implied is that the stars were regarded in characteristic ways, and these ways were reflected in both the construction of the mounds as well as the legends. Therefore, these characteristics might serve as a basis for mound shape classification.

Classification of Mounds

In the separation below, the mounds are typed by their shape and presumed characterization of stellar pattern. I will use the term "cluster" to mean any grouping in which the points (stars or conical mounds) retain their identity. The term "pattern" will be taken to mean a grouping in which the individual stars or conical mounds have lost their individual identity and may only represent

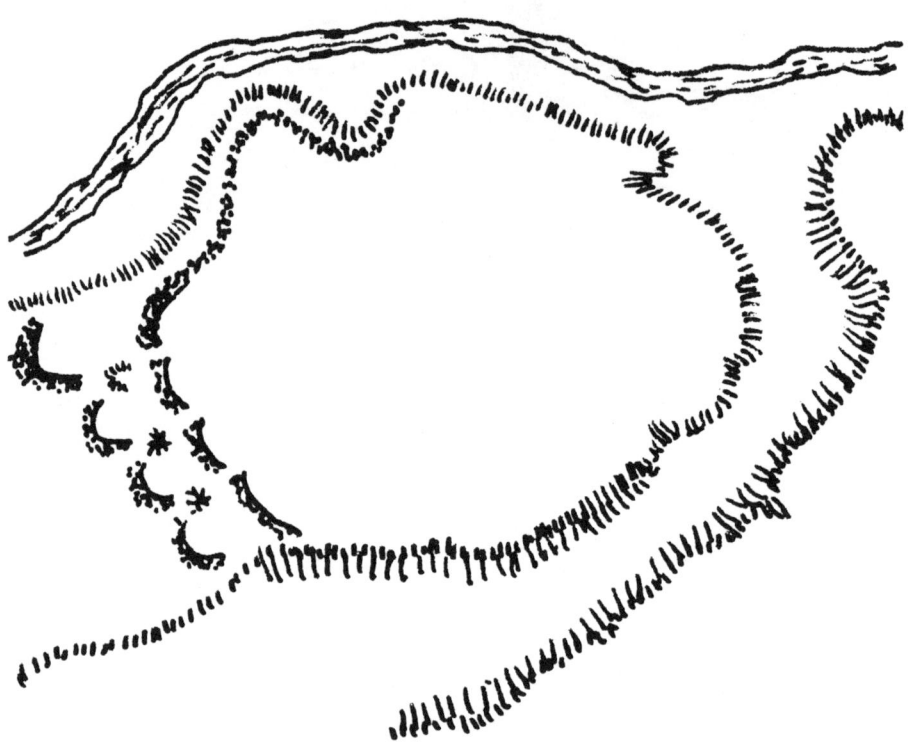

3 *An earthwork resembling a paw-print located in Green County, Ohio. The central area is twelve acres in size. (Copied from Fowke 1902.)*

critical points in the perception of the figure's form (Attneave 1954). A summary of this classification scheme with examples is given in Table 1.

   Type I. The conical mounds represent individual stars. No clusters or patterns are evident. One might guess that, since these mounds were often burial mounds, each represented a star present at the time of death. It might have served to immortalize the man or men there interred.

Effigy Mounds 225

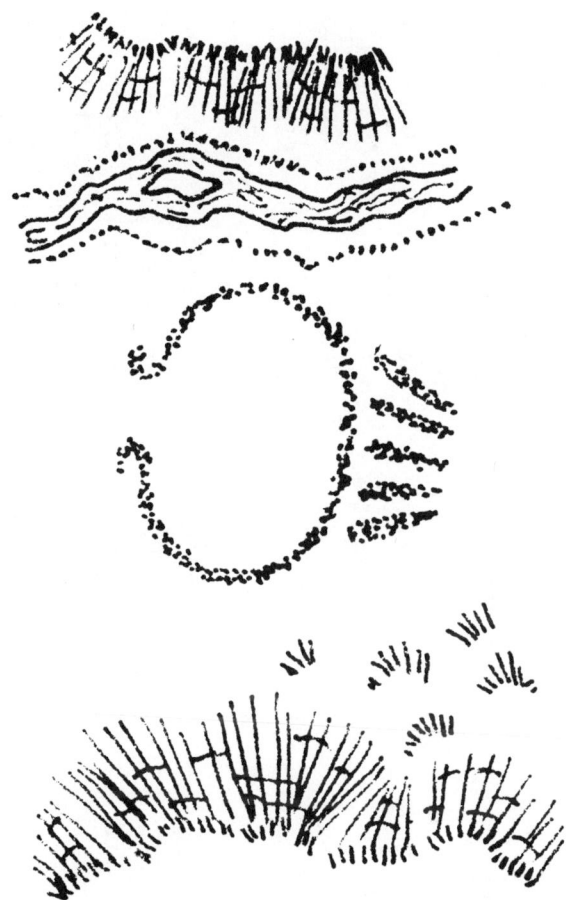

4 *A stonework resembling a paw-print located in Ross County, Ohio. The width at its widest measure is a little more than 190 meters. (Copied from Fowke 1902.)*

<u>Type II (a)</u>. <u>The conical mounds represent individual stars. They are placed in clusters suggested by obvious star groups.</u> These "obvious" groups might include the belt of Orion or the Pleiades, for example. The mound representations include any of the numerous linear arrangements of disconnected conicals or the many conical clusters.

Table 1
Mound Classification by Shape and Constellation with Correlated Legend

	Constellation's Salient Feature		Representative Legend	Representative Constellation	Representative Mound Shape
	Primary	Secondary			
Type I	Star Only	---	Star Husband Stories	Castor or Pollux	Single Conical
Type II a	Star	Cluster Shape	Mountain Sheep (Yavapai)	Orion's Belt	Rows of Conicals
b	Star and Cluster Shape	---	Mountain Sheep	Orion's Belt	Chained Conicals
c	Cluster Shape	---	Path of the Dead	Milky Way	Linear Mounds
Type III a	Star	Pattern	Hand Print (Maricopa)	Pleiades(?)	Paw Print
b	Star and Pattern	---	Hunters and Bear (Fox)	Ursa Major Cygnus	Tremper Effigy, Eagle Mound
c	Pattern Only	---	Hunters and Bear	Ursa Major Cygnus	Bear and Bird Mounds

Effigy Mounds

Type II (b). The conical mounds represent individual stars and the cluster characteristic is emphasized by earthen fill between the conicals. The chained conicals alluded to earlier would belong to this group.

Type II (c). Individual star representation is lost, but the cluster shape is retained. The linear mounds would fit this category. There are numerous linear mounds; they are perhaps copies of the brightest stars (magnitude 2 and brighter) which seem to come in pairs. Also, the Milky Way, regarded as a path for the dead, is a cluster whose shape might be represented by the linear mounds.

Type III (a). The conical mounds represent individual stars. They are made to form patterns suggested by obvious star groups. There are few mound groups in this class. The paw-print mounds might be included here.

Type III (b). The conical mounds represent individual stars and the pattern characteristic is emphasized by earthen fill between the conicals. I found three earthworks that belong to this class (Figs. 5, 6, and 7). One is the Eagle Mound at the Newark Mound Group in Ohio. Squire and Davis (1848) noted that this bird effigy was really a composite of four mounds. Careful excavation of the Tremper Effigy revealed a set of connected mounds (Shetrone 1930). The anthropomorphic effigy found in the Turner Group in Hamiton County, Ohio (Willoughby 1922), is also very much in this class. Its primary mounds are ringed with stones.

Type III (c). Individual star representation is lost, but the pattern shape is retained. Most of the effigy mounds in the Wisconsin region are of this type.

It is tempting to assign a temporal sequence to these types. Unfortunately, the field data will not allow it. The Serpent Mound in Ohio and the Bird effigies at the Poverty Point site in Louisiana belong to Adena complexes, and some of the most recent mounds are conical.

Template Alignments

A general approach to the alignment problem might be to record the effigy orientations, then pose questions such as the following: (1) Are there more alignments in

5 *The Eagle Mound from the circular enclosure at the Newark Group in Ohio. Its greatest length is a little more than 47 meters. (Copied from Fowke 1902.)*

one direction than others? (2) Are the more frequent effigy orientations correlated with important constellation positions (positions during seasonal changes)? The extremely varied orientations of the mounds (Hyde 1962, p. 74) quickly discouraged me from pursuing this approach.

In lieu of this general attack on the problem, specific effigies might be singled out for examination. The possible association between Serpent Mound in Ohio and the constellation Ursa Minor was alluded to earlier. Recall Nuttall's (1901) observation that in the Ohio Valley region the snake served as a symbol for the Septentriones. One of the more unusual characteristics of the Serpent Mound (Fig. 1) is the egg that, seemingly, is about to be swallowed. Was this egg meant to represent Polaris? If it was, and if the serpent represents Ursa Minor as Nuttall suggests, then the bend in the "handle" of the constellation is not congruent in the same plane with the bend of the Serpent Mound. On the other hand, suppose the end of the tail is seen as anchored to Polaris. Then not only is the serpent

Effigy Mounds

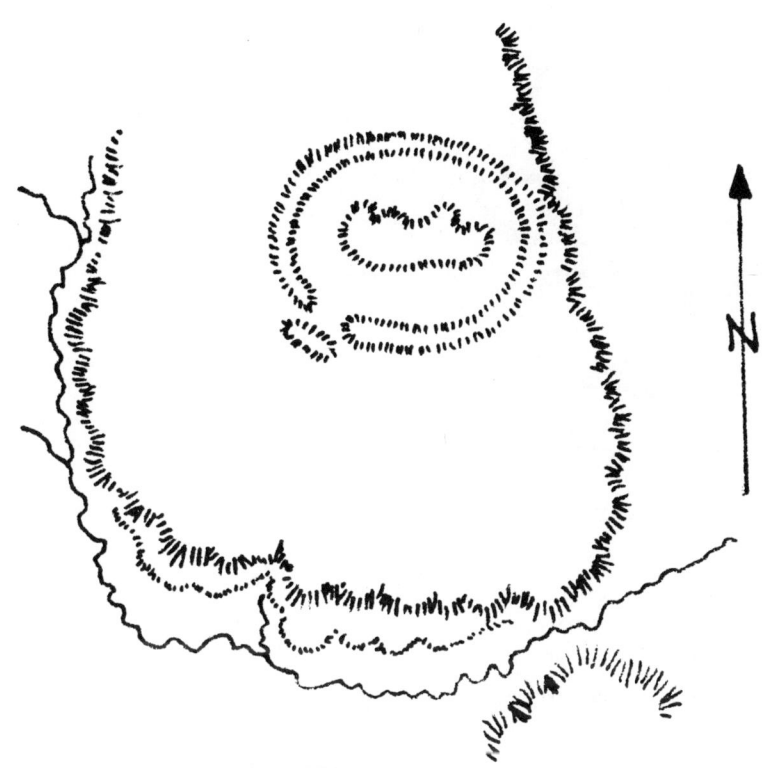

6   *The Tremper Effigy. The oval is about 61 meters at its greatest measure. (Copied from Fowke 1902.)*

congruent with Ursa Minor, but the movement of the serpent, as indicated by the coil of the tail, is in keeping with the movement of the constellation around Polaris. (The term "Septentriones" can refer to either Ursa Major or Ursa Minor. In either case the argument stands. The direction of the handle of Ursa Major is such that the bend of the star path from the end of the handle to Polaris would be in compliance with the bend in the mound serpent's body only if the tail was coiled around the Pole Star.)

Where does this leave the status of the egg? According to Dr. Elizabeth Baity (personal communication), whose field of expertise is Asian proto-history, there is an oriental legend of a serpent swallowing the moon associated with lunar eclipses. While it is not the

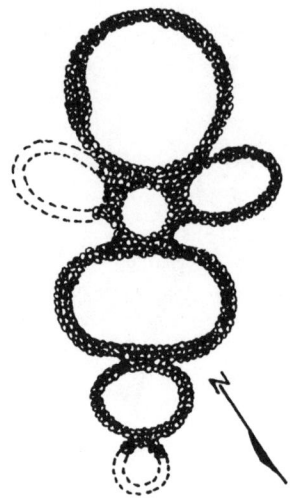

7 The anthropomorph at the Turner Earthworks in Hamilton County, Ohio. The large circular mound is 30.5 meters in diameter. Each mound is ringed by small stones shown here. (Copied from Willoughby 1922.)

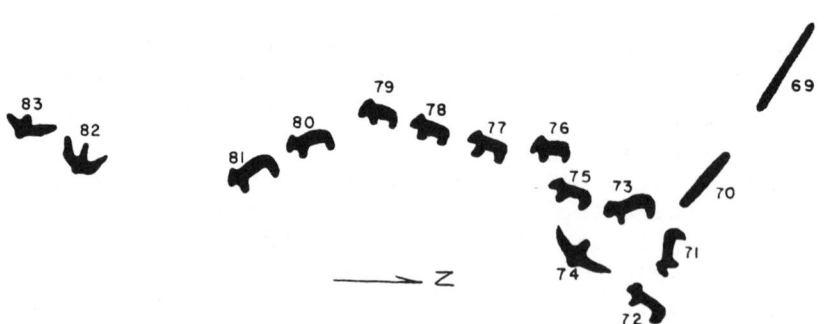

8 The Marching Bear Group from the Effigy Mounds National Monument in McGregor, Iowa. Bird Effigy 74 has a wing span between 50 and 55 meters. (Copied from Parsons 1962.)

purpose here to suggest the monogenesis of legends, the implication that the Serpent Mound was a pictographic description of such an event is most tantalizing.

A second effigy group which lends itself nicely to an alignment analysis is the Marching Bear Group at the Effigy Mounds National Monument on the Iowa-Wisonsin border (Fig. 8). The most evident feature of this group is the line of the march which forms an arc from northeast to southeast. The line is nearly symmetrical around an east-west line, and it is suggestive of the movement of Ursa Major around Polaris observed over sequential evenings during the summer months. Ursa Major is located at the top of its arc in the early evening during midspring (April) and reaches the bottom position in the early evening during late summer (August-September). This is a time of year conducive to mound building, incidentally. It is certainly better than the winter months when the group lies frozen and the warmth of a fire is more enticing than star-gazing.

If the Marching Bears are regarded as a template representation of the march of Ursa Major around Polaris then two problems are encountered. First the matching effigies' path is a vertical transpose of the constellation's line of movement. The effigies march from northeast to southeast while Ursa Major moves from southeast to northeast. Yet the placing of the effigy march on the ground in this way was quite natural and expected. This is not as contradictory as it sounds. Suppose someone draws an arrow on the board pointing downward. You are asked to reproduce it. You draw an arrow which points to the bottom of the paper in front of you. The two arrows are vertical transposes of one another; if the arrow on the board were geometrically projected directly onto the paper in front of you, it would be pointing to the top of the page, and not the bottom.

Might it be that the path of the Marching Bears is curved only because it follows the line of some natural terrestrial formation? According to Thomas A. Munson (private communication), Superintendent of the Effigy Mounds National Monument, the Marching Bear mounds could have been built in a straight line but they were not. Even ignoring this fact, however, the discovery that lines of the mounds follow some natural formation (like the serpent mound for example) should not give us too much concern. It is possible that the natural formation was selected specifically because of its alignment with certain constellations or their paths of movement. This problem might be resolved by investigating other natural features in the vicinity which could have been selected but were ignored. This approach has been suggested elsewhere (Cowan 1971), and it is an exercise that will not be pursued here.

There is another apparent misorientation of the bear effigies and the Pole Star; all of the bears' feet are pointing away from Polaris because the three stars in the "handle" are regarded as the bear's tail. To the Fox Indians, however, these three stars were three hunters chasing the bear which was represented by the remaining four stars (Marriott & Rachlin 1968). It would make good sense to perceive the bear's feet pointing toward Polaris under these circumstances since this would place them in the same plane as the Indians giving chase.

There are three bird effigies in this group which are aligned in an easterly to southeasterly direction, the same direction toward which the bears are headed at the end of their march. In August and September Ursa Major reaches the bottom of his summer swing through the sky, and the constellation Cygnus is located directly overhead pointing in the same general direction as the bear in the manner that the mound alignments suggest. Both Ursa Major and Cygnus are pointing toward the east to northeast corner of the sky which of course means that their vertical transpose would head them toward the east to southeast. Also note that Bird Mounds 81 and 82 are in close proximity to Bear Mound 81. This is suggestive of the appearance of Cygnus in this orientation when Ursa Major reaches the bottom of his arc in the sky.

Bear Mound 71 poses a special problem since it seems to be out of line with the other bears. But it suggests the following speculations, however liberal, concerning the temporal sequence of the construction of this mound group. After the construction of the linear mounds, Bear 71 was made. The orientation of this mound suggests that Ursa Major was at the bottom of his arc at the time of construction. Thus Bird Mound 74 was the next to be built. At this point the idea of constructing a line of effigy bears to mimic the path taken by Ursa Major emerged. Then Bear Mounds 72-81 were layed down in order. (It might be stated here that Mound 81 was constructed more recently than Mound 71 according to soil analysis (Parsons 1962). This is weak evidence for the constellation hypothesis, however, since this temporal ordering might be expected for any such line of marching effigies.) Since Mound 81 appears headed toward the end of the arc, the final mounds, Bird Mounds 82 and 83, were constructed.

Summary and Conclusions

The analysis of Old World sighting schemes suggested that a template matching technique seemed to be a step away from the schemes that were in use at the time. Template matching was not used because it is completely inappropriate for aligning solar and lunar events as they appear on the horizon. The North American Indian may have used a (transposed) template alignment procedure. Among other things, this helps to establish their independence from any hypothetical influences from Western Europe.

The evidence in support of the proposition that effigy mounds are images of the constellations can be summarized as follows. In general: (1) The effigies are large and face skyward. (2) There have been independent suggestions that at the time the Mound Builders began, they turned to the stars in their quest for religious inspiration. (3) Indian legends dealing with the constellations seem to reflect the same perceptual schema found in the patterns of mound structure. This schema can serve as the basis of a well-patterned system of mound classification.

In particular: (4) The swastika or cross, a symbol of the stellar rotation around Polaris, was associated with the serpent in the Ohio Valley region where the Great Serpent Mound is located. (5) The shape, orientation, and movement of Ursa Minor (or Ursa Major), and the depicted act of swallowing the egg might be indicative of an eclipse. (6) At least one mound in this region was made in the shape of a cross and oriented northward.

The analysis of the Marching Bear Effigies indicates: (7) The effigies' marching line is oriented in an expected way with the march of Ursa Major around Polaris. (8) The orientation of each bear is what would be expected from Indian legend. (9) The path of the bear effigies follows the summer path of Ursa Major. (10) The direction of the Bird Effigies relative to the end of the Bears' march is in keeping with the direction of Cygnus relative to Ursa Major at the end of his arc. (11) The distance between the bird mounds and the bear mounds representing the bottom of the arc suggests the appearance of Cygnus at the time when Ursa Major reaches its bottom most point in the sky. (12) Finally, what little dating evidence there is coincides with the predicted temporal order of the effigies' construction.

This evidence is hardly definitive, yet it has enough substance to suggest that the proposition put forth here merits further investigation. Nothing has been said of other earthwork effigies, the Great White Horse of southern Britain or the enigmatic lines and effigy figures located on the plains of Nazca in Peru. They are no less baffling than the effigy mounds of North America. The implication here that even these are in some way related to animal figures seen in the stars deserves consideration.

In my searches I discovered an Algonquin poem in Leland's (1898) book which is relevant to the subject here. I think it would be an appropriate note on which to close this chapter on Birds, Bears, and Stars:

The Song of the Stars

We are the stars which sing,
We sing with our light;
We are the birds of fire,
We fly over the sky.
Our light is a voice;
We make a road for spirits,
For the spirits to pass over.
Among us are three hunters
Who chase a bear;
There never was a time
When they were not hunting.
We look down on the mountains.
This is the song of the stars.

12

The Use of Eclipse Data to Determine the Maya Correlation Number

Nancy Kelly Owen

Solana Beach, California

On pages 51 to 58 of the Dresden Codex, as reproduced by Gates (1932), appear a series of numbers and several dates which should afford a means of correlating the Maya and Julian dating systems. Five of the dates are unambiguously written.

Date I	9.16.4.10.8	12 Lamat	1 Muan	Codex p. 52, col. 4
Date II	9.16.4.11.3	1 Akbal	16 Muan	Codex p. 52/red, col. 4
Date III	9.16.4.11.18	3 Eznab	11 Pax	Codex p. 52, col. 3
Date IV	10.9.10.6.8	12 Lamat	9 Cumhu	Codex p. 51/red, col. 1
Date V	8.16.4.10.0	5 Ahau	3 Tzec	Codex p. 51, col. 1

An additional date, 9.19.8.7.8 7 Lamat --?-- (Codex p. 52/red, col. 3) is inconsistently recorded, and has not been interpreted at this time. It is reasonable to suspect that these dates refer to observed historical eclipses because they appear at the beginning of an eclipse table. (Three of them are just 15 days apart.) Also there is evidence that the dates accompanying the Venus table (Codex p. 24) and some other astronomical tables relating to planetary periods refer to historical events (Makemson 1946; Owen 1972). The series of numbers was early recognized by Meinshausen (1913), Willson (1924), and Teeple (1930) to be an ephemeris, or table of eclipse syzygies, and proof of this was presented by Makemson (1946), who compared eclipse intervals between A.D. 495 and 1037. This ephemeris consists of a series of 69 intervals: 52 of 177 days, 8 of 178 days, and 9 of 148 days, making up one complete Maya Eclipse Cycle (1 MEC) of 11,960 days. The 69th and all of the 148-day intervals are followed by a conspicuous design, or picture, clearly intended to call attention to these "pictured intervals". Both solar and lunar eclipse glyphs appear within, or adjoining, these pictures.

During the golden period of Maya astronomy, there occurred three rare and spectacular sequences of eclipses which were all observable in Central America.

1) In the year A.D. 495, two partial solar eclipses occurred 30 days apart with a lunar eclipse between them.
2) In a 98-year period, A.D. 489 to 587, there was a clustering of 25 lunar eclipses, visible in Chiapas, at the 10 pictured intervals provided that an appropriate correlation number is used in the dating.
3) In a 61-year period, A.D. 692 to 750, there was a clustering of 18 solar eclipses visible in Chiapas.

Makemson arrived at her correlation number, 489,138, by making the basic assumption that Dates I, II, and III, at 15-day intervals, represented the two partial solar eclipses and the bracketed lunar eclipse of the year A.D. 495. This particular lunar eclipse occurred on A.D. 495, 25 May, or Julian day number 1,902,001. Identification of this Julian date with 9.16.4.11.3, the Maya day count 1,412,863, leads to Makemson's correlation number of 489,138. In her opinion, this number fulfilled several necessary conditions, both astronomical and historical, but she stopped short of detailed consideration of the ephemeris, and therefore made no use of the spectacular sequences 2) and 3). For reasons discussed later, Owen (1973) proposed the correlation number 487,410 derived by identification of the solar eclipse of A.D. 490, 31 August, Julian day number 1,900,273, with 9.16.4.11.3, the Maya day count 1,412,863. This particular solar eclipse was total at high noon over Chiapas, was of long duration, and visible throughout Mexico and Yucatan. Furthermore, it was immediately preceded and followed by visible lunar eclipses on

9.16.4.10.8 and 9.16.4.11.18. As expected, both of these correlation numbers precisely relate historical eclipses with the 69 intervals, which constitute in fact an ephemeris, but in at least two respects this correspondence appears to be unique and not shared by datings based on other correlation numbers. This will be made clear following presentation of the tables.

In keeping with the assumption that Dates I, II, and III mark historical eclipses, a useful comparison of correlation numbers may be made. The correlations of Makemson, Owen, and Smiley place these dates near the beginning of lunar eclipses on the pictured intervals, and place Date IV at the end of a spectacular clustering of solar eclipses visible in Chiapas in the eighth century. The other correlations place these dates several centuries later, where the distribution of eclipses is more nearly average. In Table 1, the pictured intervals are related to lunar eclipses on the basis of various correlation numbers throughout 3 MEC. The intervals in each MEC are identified consecutively by the letters A to J in column 1. The numbers in column 2 are the day intervals for the immediately succeeding picture, and are composed of shorter intervals on which eclipses, not necessarily visible in Chiapas, are expected to occur. In the remaining columns, headed by a particular correlation number and the names of the author who proposed it, appear the corresponding dates of lunar eclipses which occurred on the indicated intervals. The numbers in parentheses are the identifying numbers used by Oppolzer (1968), whose Canon of Eclipses is the source of the dates. Those eclipses not included in Oppolzer's treatise were checked against the eclipse tables of Schram (1908) and the new moon tables of Goldstine (1973). With some of the correlation numbers, certain eclipses did not fall exactly on the predicted date, but occurred a day early (-1) or late (+1), as indicated. An asterisk indicates that the eclipse was visible in Chiapas. The Owen correlation number identifies Date I with A.D. 490, August 16 and places the lunar eclipse of this date directly on the pictured interval A. Succeeding pictured intervals are occupied by lunar eclipses as shown in Table 1. Exactly the same result is produced by the Makemson number, but the series must be begun at picture B with Date II representing the lunar eclipse of

Table 1

Lunar Eclipses at the Ten Pictured Intervals Dated According to Various Correlation Numbers

	Interval		489,138	487,410	482,699
	Days		Makemson(1946)	Owen(1973)	Smiley(1960)
A	1743		490 Aug 16	490 Aug 16	*473 Jan 30(2588)
B	1033		*495 May 25(2626)	*495 May 25(2626)	*477 Nov 6(2597)
C	1211		498 Mar 23(2630)	498 Mar 23(2630)	*480 Sep 5(2601)
D	1742		*501 Jul 16	*501 Jul 16	*483 Dec 30(2607)
E	1034		*506 Apr 23(2643)	*506 Apr 23(2643)	488 Oct 6(2615)
F	1211		509 Feb 20(2647)	509 Feb 20(2647)	491 Aug 5(2619)
G	1564		*512 Jun 15(2653)	*512 Jun 15(2653)	494 Nov 28(2625)
H	1211		516 Sep 26(2660)	516 Sep 26(2660)	*499 Mar 13(2632)
I	709		520 Jan 20(2665)	520 Jan 20(2665)	*502 Jul 6(2637)
J	502		*521 Dec 29(2669)	*521 Dec 29(2669)	504 Jun 14(-1)
A	1743		*523 May 15(2670)	*523 May 15(2670)	505 Oct 28(2642)
B	1033		*528 Feb 21(2678)	*528 Feb 21(2670)	510 Aug 5(2650)
C	1211		*530 Dec 20(2682)	*530 Dec 20(2682)	*513 Jun 4(2655)
D	1742		*534 Apr 14(2687)	*534 Apr 14(2687)	516 Sep 26(2660)
E	1034		*539 Jan 20(2695)	*539 Jan 20(2695)	521 Jul 5(2668)
F	1211		*541 Nov 19(2699)	*541 Nov 19(2699)	524 May 3(2672)

Continued

Interval		489,138	487,410	482,699
	Days	Makemson(1946)	Owen(1973)	Smiley(1960)
G	1564	*545 Mar 14(2704)	*545 Mar 14(2704)	*527 Aug 27(2677)
H	1211	*549 Jun 25(2711)	*549 Jun 25(2711)	*531 Dec 10(2684)
I	709	*552 Oct 18(2716)	*552 Oct 18(2716)	*535 Apr 4(2689)
J	502	*554 Sep 27(2720)	*554 Sep 27(2720)	537 Mar 13(-1)
A	1743	*556 Feb 11(2721)	*556 Feb 11(2721)	538 Jul 27(2694)
B	1033	*560 Nov 19(2729)	*560 Nov 19(2729)	543 May 4(2702)
C	1211	*563 Sep 18(2733)	*563 Sep 18(2733)	546 Mar 3(2706)
D	1742	*567 Jan 11(2738)	*567 Jan 11(2738)	549 Jun 25(2711)
E	1034	*571 Oct 19(2747)	*571 Oct 19(2747)	554 Apr 3(2719)
F	1211	*574 Aug 18(2751)	*574 Aug 18(2751)	557 Jan 30(2723)
G	1564	*577 Dec 11(2756)	*577 Dec 11(2756)	*560 May 25(2728)
H	1211	*582 Mar 25(2763)	*582 Mar 25(2763)	564 Sep 6(2735)
I	709	*585 Jul 17(2768)	*585 Jul 17(2768)	567 Dec 31(2740)
J	502	*587 Jun 26(2772)	*587 Jun 26(2772)	569 Dec 9

Continued

	553,279 Kelley (1970)	584,284 Goodman-Thompson (1950)	674,265 Hochleitner (1970)
A	*666 Mar 26(2899)	*751 Feb 15(3029)	997 May 24(3403)
B	671 Jan 2(-1)	755 Nov 23(3037)	1001 Sep 5(3410)
C	*673 Oct 31(2911)	758 Sep 21(3042)	1004 Dec 29(3416)
D	*677 Feb 23(2916)	*762 Jan 15(3047)	*1008 Apr 23(3421)
E	681 Dec 1	766 Oct 22(3055)	*1013 Jan 29(3429)
F	*684 Sep 29(2927)	*769 Aug 22(3059)	1015 Nov 28(3433)
G	*688 Jan 23(2932)	*772 Dec 15(3064)	1019 Mar 23(3438)
H	692 May 5(+1)	777 Mar 28(3072)	*1023 Jul 5(3446)
I	695 Aug 29(2944)	*780 Jul 21(3077)	1026 Oct 28(3451)
J	697 Aug 8	782 Jun 29(3081)	*1028 Oct 6(3455)
A	698 Dec 22 (2949)	783 Nov 13	*1030 Feb 20(3456)
B	703 Sep 30(-1)	788 Aug 21(-1)	1034 Nov 28(3464)
C	*706 Jul 30(2961)	791 Jun 20(3093)	*1037 Sep 27(3468)
D	*709 Nov 22(2966)	794 Oct 13(3098)	*1041 Jan 20(3473)
E	714 Sep 1(-1)	799 Jul 21(3106)	1045 Oct 28(3481)
F	717 Jun 28(2978)	*802 May 21(3109)	*1048 Aug 26(3485)
G	*720 Oct 21(2983)	805 Sep 12(3113)	*1051 Dec 20(3490)
H	725 Feb 1(+1)	809 Dec 25(3120)	*1056 Apr 2(3497)
I	728 May 27(2995)	*813 Apr 19(3125)	1059 Jul 27(3502)
J	730 May 6(+1)	815 Mar 29	1061 Jul 5
A	731 Sep 20(3000)	816 Aug 14(-3)	1062 Nov 19(3507)
B	736 Jun 28	821 May 20(3136)	*1067 Aug 27(3515)

Continued

	553,279 Kelley (1970)	584,284 Goodman-Thompson (1950)	674,265 Hochleitner (1970)
C	7 9 Apr 27(+1)	824 Mar 18(3140)	*1070 Jun 26(3519)
D	742 Aug 20(3017)	827 Jul 12(3145)	1073 Oct 18(3524)
E	747 May 29	832 Apr 18(3153)	1078 Jul 27(3532)
F	750 Mar 26(+1)	837 Jan 24(+1)	*1081 May 25(3536)
G	*753 Jul 20(3034)	839 Nov 24(3164)	*1084 Sep 16(3541)
H	757 Nov 1	843 Mar 19(3176)	1088 Dec 30(3548)
I	761 Feb 24(+1)	*847 Jul 2(3176)	1092 Apr 24(3553)
J	762 Jul 10(3048)	850 Oct 24(3181)	1094 Apr 3

A.D. 495 May 25. The Smiley and Goodman-Thompson numbers allow entry into the pictured intervals at B through solar eclipses on Date III adjusted by 15 days to change over to the lunar series. The Kelley and Hochleitner numbers do not lead to any eclipses on Dates I, II, and III, but through Date III they have been made to enter the pictured series at A in their proper centuries by arbitrarily subtracting 2260 and 2290 days, respectively. Two observations readily follow from inspection of Table 1. First, the Makemson and Owen numbers lead to the same eclipse dates, which were predictable because their derivations are based on two eclipses separated by the interval between A and B, 1743 days. Second, and of more importance, is the fact that eclipse dates based on these two correlation numbers represent a series of visible eclipses throughout 2 MEC without a break.

Since Date IV is exactly 8 MEC later than Date I, it might be expected to mark the end of a time period. Accordingly, all visible solar eclipses occurring in the 2 MEC preceding Date IV are recorded in Table 2. The last entry under each group of eclipses is Date IV expressed in Julian terms by the indicated correlation number. With the Owen number a visible lunar eclipse fell on this date; the others are solar (not visible), and for the Smiley and Hochleitner numbers a new moon occurred without record of an eclipse. As before, the first column shows the identifying number or letter of the eclipse interval and the parentheses contain the eclipse number assigned by Oppolzer. In the lower part of the table appear two pairs of eclipses separated by only six months (Hochleitner 1227-8; Goodman-Thompson 995-6), but two such occurrences within the long period 877 to 1253 are not unexpected. In the upper part of the table the three longer groups, which overlap over much of their range, display an extraordinary number and clustering of solar eclipses between A.D. 689 to 711. Visible solar eclipses only six months apart occur three times in the Makemson and Smiley and four times in the Owen columns; three eclipses occurring within the two years A.D. 710 to 711 are common to all three. At the rate that eclipses were occurring in the 65 years preceding A.D. 752, a Maya priest-astronomer had the rare privilege of witnessing almost a score of solar eclipses within his lifetime without leaving home. This is such a rare situation that it could have happened only once during the period of Maya supremacy.

The difference between Dates IV and V is 1.13.5.14.8 or 239,688 days; this is 20 MEC plus 488 days, the interval between the J and A pictures minus the minimum period between solar and lunar eclipses (239,200+502-14). Therefore Date V falls on the J picture of the preceding epoch of 20 MEC and 239,688 may represent the period between the A picture in a lunar series (752 July 31, Date IV) and the J picture in a solar series (96 May 10, Date V + 2), thereby exactly enclosing a 20 MEC epoch if a 2 day correction is applied for the accumulation of fractional days that far back in time. None of the correlation numbers lead to eclipses falling precisely on both Dates IV and V, but all, except those of Kelly and Hochleitner, place Date V one to four days before a solar eclipse. The Owen correlation number is used in this paragraph because it places Date V on 96 May 8, two days before a long total solar eclipse occurring in turn just five days before a superior conjunction of Venus. Although this eclipse was far south of Chiapas (midpoint at Quito), a visible Venus near conjunction at midday would be noteworthy. A correction of two days to allow coincidence of this eclipse with the J picture on 96 May 10 allows the beginning of long periods in which lunar and solar eclipses fall on the pictured intervals without a break. The A picture beginning the 1st MEC of the new epoch 502 days later than 96 May 10 is 30 days ahead of a solar eclipse, but thereafter throughout 16 MEC solar eclipses are in phase with the pictures. Lunar eclipses are in phase from the beginning of the 9th MEC until the beginning of the 13th MEC in the succeeding epoch, about 656 years. Thus, the phase relationships overlap between the 8th and 17 MEC, and Dates I and II fall at the midpoint of this range. This suggests that the Maya used the pictured intervals as a weak solar-lunar ephemeris of rather long periods (1½ to 4½ years) that emphasized visibility in Chiapas. Possibly, the solar series was used up to A at the beginning of the 13th MEC (Date II) and then by subtraction of 15 days to reach A on the lunar series (Date I); thus lunar eclipses would be in phase with the pictures for 20 MEC. In any event, it seems abundantly clear that the pictures on pages 51 to 58 of the Codex are not random decorations, but are essential elements of the Maya ephemeris.

To make a definite decision regarding the one proper correlation between the Maya and Julian calendars, there is a wide variety of information which should be considered and much data, the accuracy and relevance of which, should be established. Some correlations are derived with heavy emphasis on heliacal risings of Venus, dates on stelae, etc., as well as eclipses. The present discussion, limited to numbers and dates on pages 51 to 58 of the Dresden Codex, presents strong evidence that Dates I to V are observed eclipse dates, that the pictured intervals are important elements of the Maya ephemeris, and that the Makemson, Owen, and Smiley correlation numbers probably place these dates in the proper centuries. Based on these facts alone, the Owen number 487,410 receives the strongest support from the data because the other numbers do not lead to the very long series of eclipses in phase with the pictured intervals and Dates IV and V. This may be due to the fact that the series had to be entered at picture B. Furthermore, there may be considerable significance in the fact that with the number 487,410, Date V marks a solar eclipse very near a conjunction of Venus and Dates II and IV, which are colored red in the Codex, serve to define this correlation, to determine the cross-over from solar to lunar pictured series at the 13th MEC, and to define the end of a 20 MEC epoch.

13

The Solar Eclipse Warning Table in the Dresden Codex

Charles H. Smiley

Department of Astronomy

Brown University

In the Dresden Codex, there are two sections which concern solar eclipses. One is the so-called Venus Table (pp. 24, 46-50) in which there are three horizontal sequences of dates, each with the same day numbers and day glyphs, but with different month numbers and month glyphs. The lubs are 1 Ahau 13 Mac for the top row, 1 Ahau 18 Kayab for the middle row, and 1 Ahau 3 Xul for the bottom row. Each sequence extends over 104 tuns or two calendar rounds. The shortest interval forward from a date 1 Ahau 3 Xul to the next date 1 Ahau 18 Kayab is 9,360 days, which is an approximation to a solar eclipse interval. The shortest interval forward from a date 1 Ahau 18 Kayab to the next date, 1 Ahau 13 Mac is 11,960 days, a very good approximation to a solar eclipse interval. It seems likely, therefore, that the table was designed to help predict solar eclipses using the relatively uniform motion of Venus among the stars as a basis. If this be the case, then it is probable that solar eclipses were expected on days named 1 Ahau. The Smiley correlation is based on the assumption that the date, 9.9.9.16.0, 1 Ahau 18 Kayab was one on which a solar eclipse was visible in Central America, and Venus was just reappearing four days after an inferior conjunction with the sun, namely A.D. 344, December 21 of the Julian Calendar. The annular eclipse of the sun on A.D. 1973, December 24 which was visible in Central America will be similar in many respects to that of A.D. 344, December 21.

The second section in the Dresden Codex which concerns

solar eclipses is found on pages 52 to 58. It has commonly been called the Lunar Table; we shall show that it is a Solar Eclipse Warning Table. Consider the central line of dates running across the top of pages 53 to 58 and the central line across the bottom of pages 51 to 58. The Lub or starting date may be taken as 12 Lamat on page 52 and it is followed by 7 Chicchan, 2 Ik, etc. First a vertical list of the 70 dates in the table should be made, a day-number and a day-name for each (Table 1). From this list, a list of intervals between consecutive dates should be made (Table 2).

Three intervals will be found in this sequence: 177 days, 178 days, and 148 days. Remembering that the day number runs from one to thirteen and then repeats and that the twenty day names are used in order and then repeated, we can build an auxiliary table which will help us to identify the intervals.

$$177 = 13 \times 13 + 8 = 14 \times 13 - 5 = 8 \times 20 + 17 = 9 \times 20 - 3$$
$$178 = 13 \times 13 + 9 = 14 \times 13 - 4 = 8 \times 20 + 18 = 9 \times 20 - 2$$
$$148 = 11 \times 13 + 5 = 12 \times 13 - 8 = 7 \times 20 + 8 = 8 \times 20 - 12$$

This means that after a 177-day interval, the day number will be increased by 8 or decreased by 5, while the day name will be moved down 17 places or up 3 in the regular day-by-day sequence, Imix, Ik, Akbal, Kan, etc. But a table can be set up with the day names reordered as they would be after 177-day intervals: Imix, Eznab, Men, Eb, etc. If these names are listed vertically, a 177-day interval corresponds to one step down, a 148-day interval to four steps down, and a 178-day interval to six steps up (Table III). Thus if we start with a date, 3 Ix, and use successive intervals of 177, 148, and 178 days, we have 11 Chuen, 3 Cauac, and 12 Caban.

With this auxiliary table, we are prepared to determine the intervals between successive dates in the Solar Eclipse Warning Table. For each date, we have two determinations of the interval, one from the day number and one from the day name. Hopefully they will agree. For date 19, no day name is available but the day number is 9. The day number has increased by 8, so the interval is 177 days. For date 20, there is no day number but the day name is Cimi, so the interval is 148 days. For date 12, the day name is obviously in error and is easily seen

Table 1

Solar Eclipse Warning Table

12	Lamat	+177	8	Men	+177	3	Imix	+177
7	Chicchan	+177	3	Eb	+177	11	Eznab	+177
2	Ik	+148	11	Muluc	+148	6	Men	+148
7	Oc	+177	3	Caban	+177	11	Akbal	+177
2	Manik	+177	11	Ix	+177	6	Ahau	+177
10	Kan	+177	6	Chuen	+178	1	Caban	+178
5	Imix	+178	2	Muluc	+177	10	Men	+177
1	Cauac	+177	10	Cimi	+177	5	Eb	+177
9	Cib	+177	5	Akbal	+177	13	Muluc	+177
4	Ben	+177	13	Ahau	+177	8	Cimi	+177
12	Oc	+177	8	Caban	+177	3	Akbal	+177
7	Manik	+177	3	Ix	+177	11	Ahau	+148
2	Kan	+148	11	Chuen	+148	3	Lamat	+177
7	Eb	+178	3	Cauac	+178	11	Chicchan	+178
3	Oc	+177	12	Caban	+177	7	Akbal	+177
11	Manik	+177	7	Ix	+177	2	Ahau	+177
6	Kan	+177	2	Chuen	+177	10	Caban	+177
1	Imix	+177	10	Lamat	+177	5	Ix	+177
9	Eznab	+148	5	Chicchan	+148	13	Chuen	+148
1	Cimi	+177	10	Ben	+177	5	Cauac	+177
9	Akbal	+177	5	Oc	+177	13	Cib	+177
4	Ahau	+177	13	Manik	+177	8	Ben	+177
12	Caban	+178	8	Kan	+177	3	Oc	+178
						12	Lamat	

Table 2

Julian Calendar Dates with Associated Solar Eclipses

477 Sept 22	177	Oct. 22		
478 Mar. 18	177	Apr. 18		
478 Sept 11	148	Oct. 11		
479 Feb. 6	177		Apr. 8	
479 Aug. 2	177		Oct. 1	
480 Jan. 26	177		Mar. 27	
480 July 11	178	Aug. 11	Sept 9	
481 Jan. 15	177	Feb. 14		
481 July 11	177	Aug. 11	Sept 9	
482 Jan. 4	177	Feb. 4		
482 June 30	177	July 31		
482 Dec. 24	177	Jan. 24		
483 June 19	148	July 20		
483 Nov. 14	178		Jan. 14	
484 May 10	177		July 8	
484 Nov. 3	177	Dec. 4		
485 Apr. 29	177	May 29		
485 Oct. 23	177	Nov. 23		
486 Apr. 18	148	May 19		
486 Sept 13	177		Nov. 12	
487 Mar. 9	177		May 9	
487 Sept 2	177		Nov. 1	
488 Feb. 26	178	Mar. 29	Apr. 27	
488 Aug. 22	177	Sept 21	Oct. 20	
489 Feb. 15	177	Mar. 18		
489 Aug. 11	148	Sept 11		

Continued

490	Jan. 6	177		Mar. 7
490	July 2	177		Aug. 31
490	Dec. 26	178		Feb. 24
491	June 22	177		Aug. 21
491	Dec. 16	177	Jan. 15	Feb. 14
492	June 10	177	July 10	
492	Dec. 4	177	Jan. 4	
493	May 30	177	June 29	
493	Nov. 23	177	Dec. 24	
494	May 19	148	June 19	
494	Oct. 14	178		Dec. 13
495	Apr. 10	177	May 10	June 8
495	Oct. 4	177	Nov. 3	
496	Mar. 29	177	Apr. 29	
496	Sept 22	177	Oct. 22	
497	Mar. 18	148	Apr. 18	
497	Aug. 13	177		Oct. 11
498	Feb. 6	177		Apr. 7
498	Aug. 2	177		Oct. 1
499	Jan. 26	177	Feb. 26	Mar. 27
499	July 22	177	Aug. 22	Sept 21
500	Jan. 14	177	Feb. 15	
500	July 10	148	Aug. 10	
500	Dec. 5	177		Feb. 3
501	May 31	177		July 31
501	Nov. 24	178		Jan. 24
502	May 21	177		July 20

Continued

502	Nov. 14	177	Dec. 15	
503	May 10	177	June 10	
503	Nov. 3	177	Dec. 4	
504	Apr. 28	177	May 29	
504	Oct. 22	148	Nov. 22	
505	Mar. 19	177		May 19
505	Sept 12	178		Nov. 11
506	Mar. 9	177	Apr. 9	May 8
506	Sept 2	177	Oct. 2	Nov. 1
507	Feb. 26	177	Mar. 29	
507	Aug. 22	177	Sept 22	
508	Feb. 15	148	Mar. 17	
508	July 12	177		Sept 11
509	Jan. 5	177		Mar. 6
509	July 1	177		Aug. 31
509	Dec. 25	178	Jan. 25	Feb. 24
510	June 21		July 21	

to be Manik. The last date must agree with the Lub, so we move down to the bottom row of dates, 12 Lamat.

Now with the list of 69 successive intervals complete, we are ready to step forward in the Christian calendar from the date corresponding to the Mayan date, 9.16.4.10.8, 12 Lamat 1 Muan. By the Smiley correlation, this is A.D. 477 Sept. 22 in the Julian Calendar. One can use a table of Julian Day Numbers or one can build a second small auxiliary table (Table 4), providing for steps of 177 or 148 days. 177 days forward from May 16 will lead to Nov. 9; the same interval from June 2 will take us to Nov. 26. Similarly, 148 days forward from March 11 will take us to August 6, and 148 days after May 3 comes Sept. 28. For 178 days, simply add one day to the answer given by the 177-day table and for an interval with a Leap Day (February 29) in it, subtract one day from the answer given by the auxiliary table.

It is hoped that the reader will actually carry out the various operations indicated, even though the results are given here. In Table 2, one will find to the right of each date one or two dates on which solar eclipses occurred on the earth. Checking these dates by the use of Oppolzer's <u>Canon der Finsternisse</u> (1962), it is found that in the interval covered, A.D. 477 to 510, not a single solar eclipse occurred anywhere on the earth without warning, nor was there a single false warning. The list of dates in the Solar Eclipse Warning Table are the dates of new moons such that at the next new moon, or the following one, or at both, solar eclipses occurred.

On checking carefully, one finds that this same method of warning would have worked without change from A.D. 206 June 23 to A.D. 647 March 12, warning over that interval of 441 years, of 1034 consecutive solar eclipses, all that were visible <u>anywhere</u> in the world, and not once warning when there wasn't a solar eclipse somewhere at one or both of the next two new moons. If one looks to find how long the method would have warned of solar eclipses visible in Central America, the interval runs from A.D. 50 May 9 to A.D. 866 Oct. 29 or 267 consecutive solar eclipses visible in this area.

With slight changes, the same method would work in the twentieth century. The reader can verify this by starting with the date A.D. 1900 April 29 and using the

Table 3

Auxiliary Table—Maya Calendar Solar Eclipse Warning Table

```
                Ahau
                Caban
                Ix
                Chuen
                Lamat
                Chicchan
                Ik
                Cauac
                Cib
                Ben
                Oc
                Manik
                Kan
                Imix
                Eznab
                Men
                Eb
                Muluc
                Cimi
                Akbal
```

177 day interval, down one and +8 or −5
148 day interval, down four and +5 or −8
178 day interval, down six and +9 or −4

intervals as they follow the date A.D. 500 July 10 in the original table, that is, 148, 177, 177, 178, etc. Oppolzer's Canon der Finsternisse will provide the dates of the solar eclipses in the 20th century and allow the reader to verify that the method works. Using the Canon of Solar Eclipses of Meeus (Meeus et.al. 1966), one can show that with similar slight changes, the method will work for all of the 25th century. When someone uses a big computer to run out the precise dates of the solar eclipses of the 30th century, the Solar Eclipse Warning Method of the Maya will undoubtedly still work.

Why did the Maya give up this method? Perhaps because some young Maya astronomer made the mistake of assuming

Table 4

Auxiliary Table—Christian Calendar Solar Eclipse Warning Table

	177 days		148 days	
Jan.	July −4	June +26	June −3	May +28
Feb.	Aug. −4	July +27	July −2	June +28
Mar.	Sept −7	Aug. +24	Aug. −5	July +26
Apr.	Oct. −6	Sept +24	Sept −5	Aug. +26
May	Nov. −7	Oct. +24	Oct. −5	Sept +25
June	Dec. −6	Nov. +24	Nov. −5	Oct. +26
July	Jan. −7	Dec. +24	Dec. −5	Nov. +25
Aug.	Feb. −7	Jan. +24	Jan. −5	Dec. +26
Sept	Mar. −4	Feb. +24	Feb. −5	Jan. +26
Oct.	Apr. −5	Mar. +26	Mar. −3	Feb. +25
Nov.	May −4	Apr. +26	Apr. −3	Mar. +28
Dec.	June −5	May +26	May −3	Apr. +27

Leap Day in interval, subtract one day.
178 days in interval, add one day.

that the average of two Maya solar eclipse intervals, the Fox of 11960 days (Forty-six times the Maya sacred round of 260 days) and the Thix of 9360 days (Thirty-six times 260 days), would also be a solar eclipse interval (Smiley 1973). Unfortunately this interval of 10,660 days (Forty-one times 260 days), the Fone (pronounced "phony"), is not a solar eclipse interval. Perhaps the easiest way to show this is to note that 10,660 days is only 88 days more than the well-known solar eclipse interval, the Inex. Another way would be to consider the Diophantine equation in synodic periods: $6x + 5y = 361$. The evidence in support of this suggestion is found on pages 51 and 52 of the Dresden Codex. There in the first column on the upper part of page 51 is a date, 10.19.6.1.8, which is 166,140 days after the Lub of the Solar Eclipse Warning Table. But 166,140 is 13 Foxes plus a Fone of 10660 days. And in the fifth column in the upper part of page 52 is another date 9.19.7.7.8 which is 22,620 days after the Lub of the Warning Table. And 22,620 is one Fox plus one Fone.

Briefly the evidence seems to indicate that the Venus-Solar Prediction Table on pages 24, 46 to 50 of the Dresden Codex ended about 9.14.15.6.0, A.D. 448 Nov. 25 for the middle row and about 9.16.8.10.0, A.D. 481 Aug. 24 for the top row, and that the famous meeting of astronomers commemorated by Altar Q at Copan, Honduras occurred on A.D. 485 May 12. It seems likely that they decided to have the new Solar Eclipse Warning Table start on 9.16.4.10.8, A.D. 477 Sept. 22 in order to have the beginning near an autumnal equinox.

14

Planetary Data on Caracol Stela 3

David H. Kelley

Department of Archaeology

The University of Calgary

Calgary, Alberta, Canada

Among the monuments of Caracol, British Honduras, one of the more important is Stela 3. This monument is now on loan from the University of Pennsylvania to the Natural History Museum of Denver, Colorado. Although it has not been published, Linton Satterthwaite, who excavated at Caracol, kindly sent me a drawing of the stela made by William Coe, a good series of photographs, and his interpretations of the dates on the monument. The monument contains the well-known "star" glyph, long misread as "Venus," in different compounds. It also contains day names which are frequent in astronomical contexts, and dates found with similar glyphs at Naranjo. There, Thompson long ago recognized apparent astronomical intervals relating to Venus and Jupiter although there seemed no reason for this in his correlation. In a preliminary computer run, in a program written for me by Professor K. Ann Kerr of The Departments of Physics and Chemistry of the University of Calgary, the dates from this stela revealed a further striking series of planetary intervals. These showed both internal and external relationships of importance.

The values used for the average length of the synodic intervals of the planets were taken from Charles P. Bowditch (1910, p. 335) and included a value of 377.750 days for the revolution of Saturn. I have subsequently found out that the figure used by Escalona Ramos (1940, p. 28) is 378.09142 and that used by Stahlman and

Gingerich (1963, p. xv) is 378.09208. This difference of more than a quarter of a day adds up rapidly, but does not significantly change the relationships of dates over short intervals on a single monument, as in the case of Caracol Stela 3. Thus, some of the external relationships for Saturn suggested by that program may be in error, but the Saturn intervals recognized on Stela 3 are probably intentional.

The ideal way to study the astronomical import of an inscription is to convert a Maya date into its Julian equivalent and see what was happening at that time. Unfortunately, after examining more than 50 proposed correlations of the Maya calendar with our own, I have been unable to convince myself that there is any substantial probability that any of them is correct. I have, therefore, continued to work with average intervals to try to determine what the Mayas were saying rather than examining events to see what they might have been talking about.

The most striking interval on Caracol Stela 3 goes from the opening date 9.6.12.4.16 (1,343,616) 5 Cib 14 Uo to 9.9.13.4.4 (1,365,564) 9 Kan 2 Zec, a period of 21,948 days. Fifty-five synodic revolutions of Jupiter, calculated at an average of 398.88421 days, are 21,938.63155 days, about nine days less than the interval. Fifty-eight synodic revolutions of Saturn at 378.09208 days make 21,929.34064 days, about 18 days less than the interval. Sixty tropical years of 365.2421 days equal 21,914.526 days, about 33 days less than the interval. Finally, five sidereal periods of Jupiter, calculated at 4,332.8486 days give 21,664.2430 days, about 284 days less than the interval. (See Table 1 for possibly relevant data referring to Jupiter and Saturn over this interval.)

Implicit in these figures are the facts that Jupiter-Saturn conjunctions occur about once every twenty years and that every third Jupiter-Saturn conjunction recurs in about the same area of the sky. In a rather limited check of Jupiter-Saturn conjunctions, I found intervals as small as 21,590 days and as large as 21,890 days, the two planets sometimes remaining together in near conjunction for up to 240 days.

The fact that the Jupiter sidereal period is about 4,333 days means that Jupiter moves through approximately one twelfth of the sky, or one zodiacal sign, each year. In Asia, this early led to a 12-year Jupiter cycle which, in turn, was combined with other factors in a way to produce a 60-year Jupiter sidereal cycle. If it were desired to combine the Jupiter sidereal cycle with the Jupiter synodic period, 54 synodic periods average 124 days less than 5 sidereal periods, while 55 synodic periods average 275 days more. The latter approximation keeps an even multiple of the 12-year cycle and hence, apparently, was preferred. The fact that it also approximates the Saturn synodic period would have been helpful in calculations involving that planet.

The nature of the Maya's interest in planetary movements was very different from that of modern scholars and it seems likely that both the day names and the months in calculations were picked partly for astronomical reasons and partly for religious reasons. Thus, the month Uo was under the patronage of the old black god of the number seven. I have recently argued (Kelley 1972) that this deity is the god of the planet Saturn, so this month would be particularly appropriate for ceremonies and calculations involving that planet. Besides the 14 Uo date already mentioned, Caracol Stela 3 includes the date 9.7.14.10.8 3 Lamat 16 Uo. The interval between the two dates is 22 vague years of 365 days plus 2 days. The date 9.9.13.4.4 9 Kan 2 Zec, which ends the 60-year interval from 5 Cib 14 Uo, shows a similar relationship to 9.10.4.7.0 8 Ahau 3 Zec, from which it is separated by 11 vague years plus one day, or exactly half of the previously mentioned interval.

The previously mentioned date 9.7.14.10.8 3 Lamat 16 Uo is 1320 days after 9.7.10.16.8 9 Lamat 16 Ch'en. The interval is a close approximation of 3 1/2 synodic periods of Saturn (3 1/2 x 378.09208 = 1323.32228 days). This date pattern has a remarkable parallel at Tikal, where Lintel 3 of Temple IV (Satterthwaite 1961, p. 56) gives the date 9.15.12.2.3 12 Akbal 16 Ch'en and Lintel 2 of Temple IV gives 9.15.15.14.0 3 Ahau 13 Uo, some 1,317 days later. That the association of a Saturn interval with comparable dates at widely separated sites is no accident is verified by the fact that the interval from 9.7.10.16.8 9 Lamat 16 Ch'en at Caracol to 9.15.15.14.0 3 Ahau 13 Uo at Tikal is 59,352 days. 157 average synodic revolutions of Saturn at 378.09208 are

Table 1

Multiples of the Tropical Year with the Sidereal and Synodic Periods of Jupiter and Saturn

Saturn sidereal	2	x	10,764.44	=	21,528.88
Jupiter synodic	54	x	398.88421	=	21,539.8+
Tropical year	59	x	365.2421	=	21,549.+
Saturn synodic	57	x	378.09208	=	21,551.2+
Jupiter sidereal	5	x	4,332.8486	=	21,664.2+
Tropical year	60	x	365.2421	=	21,914.5+
Saturn synodic	58	x	378.09208	=	21,929.3+
Jupiter synodic	55	x	398.88421	=	21,938.6+

59,360.45656 days. The interval is also exactly 162 1/2 tropical years (162.5 x 365.2421 = 59,351.84125 days). There are no recognizable direct parallels in the immediately accompanying glyphs at the two sites, but both include star glyphs and other indications of astronomical interest.

It is perhaps worth noting that Lintel 2 of Temple IV at Tikal also includes the date 9.15.12.11.12 6 Eb 0 Pop. Caracol Stela 3 includes an interval of 7,018 days, apparently counted from 9.10.4.7.0 8 Ahau 3 Zec. If counted backwards from that date, one reaches 9.9.4.16.2 10 Ik 0 Pop. This date is within a day of the same position in the tropical year as the date 9.15.15.14.0 3 Ahau 13 Uo, already mentioned as 162 1/2 tropical years after 9.7.10.16.8 9 Lamat 16 Ch'en. The interval is 47,117 days (129 x 365.2421 = 47,116.2309 days). This type of intricate cross-relationship between dates at Caracol and Tikal is far from understood, but strongly suggests a common body of elaborate and precise astronomical data throughout the Maya area.

The Jupiter and Venus intervals, which originally attracted Thompson's attention at Naranjo, are repeated on Caracol Stela 3 with other intervals which are also multiples of average synodic periods of Jupiter. Table 2 shows the Jupiter and Venus intervals as found on Caracol Stela 3. Average parameters used for the synodic intervals of the planets are: Jupiter (398.88421 days), Venus (583.92166 days), Saturn (378.09208 days).

Table 2A

Dates from Caracol Stela 3

Interval	Long Count	Calendar Round
	9.7.14.10.8 (1,351,648)	3 Lamat 16 Uo
+12752		
	9.9.10.0.0 (1,364,400)	2 Ahau 13 Pop
+ 3203 15955		
	9.9.18.16.3 (1,367,603)	7 Akbal 16 Muan
+ 397 16352		
	9.10.0.0.0 (1,368,000)	1 Ahau 8 Kayab
+ 1580		
	9.10.4.7.0 (1,369,580)	8 Ahau 3 Zec
	9.9.5.13.18 (1,362,868)	4 Lamat 6 Pax
+ 1532		
	9.9.10.0.0 (1,364,400)	2 Ahau 13 Pop
+ 1505 3037		
	9.9.14.3.5 (1,365,905)	12 Chicchan 18 Zip

Table 2B

Related Synodic Periods

Interval	Jupiter	Venus	Saturn
12752	32x=12764.29472		
3203	8x=3191.07368		
15955	40x=15955.36840		
16352		28x=16349.80648	
397	1x=398.88421		
1580	4x=1595.53684		
1532			4x=1512.36832
1505			4x=1512.36832
3037			8x=3024.73664

These data alone are enough to suggest that the dates shown in Table 2 refer to some important positions of Jupiter. In spite of the fact that the nature of the relationships is not always clearly specifiable, the new evidence on Jupiter, Saturn, and Venus puts a considerable number of new restraints upon any suggested correlation. Further constraints are placed on data on Mercury and on the tropical year which cannot be considered here.

15

A New Astronomical Interpretation of the Four Ballcourt Panels at Tajin, Mexico

Carmen Cook de Leonard

Centro de Investigaciones

Antropologicas de Mexico

Mexico City, Mexico

The contents of the Tajin panels are described in detail by Jose Garcia Payon (1959), and before that by Ellen S. Spinden (1933). They describe the development of the ceremonial Ball Game as follows: (1) Selection of the Candidate, (2) Consecration, (3) The Game, (4) Death of the Losing Player.

This author is not completely in accordance with the happenings as described, and shall point out differences in this paper. The scenes take place on Earth, with the Lower Regions below, symbolized by scrolls, and the Sky above, represented by three astronomical bodies: Sun, Moon, and Venus. The Moon is identified by Garcia Payon as the body which appears on one extreme, and always above a skeleton emerging from a pot, partly submerged in water. The Moon is shown twice from the front and twice in profile. The other two bodies appear interlinked, but interchanging their position from left to right. The interlinking I interpret as conjunctions. One of the bodies has a cross in the center, which I take to be the Sun, as the Maya and other groups represent the 'Day' or 'Sun' in this manner. Even today the Huave Indians (Cook and Leonard 1949) represent the Sun with this type of slanted cross. The other is a serpent-like creature, obviously Venus, the Feathered Serpent, but which Garcia Payon considered to be the Sun.

It is my contention that the interrelationship of these astronomical bodies indicates the time elapsing and the time at which each of the scenes is enacted. The position of the three bodies on the four panels change as follows (Fig. 1):

Celestial Bodies   *Possible Number Indications*

		Border	Moon	Venus
1. SE:	V - S - FM	6☉ +	3☉ -	(3+4)= -7☾
2. SW:	HM - S - V	8☉ +	1☉ -	-
3. NW:	V - S - HM	3☉ -	1☾ -	(2+3)= -7☾
4. NE:	HN - S - V	6☉ +	3☉	2

V=Venus; S=Sun; FM=Full Moon; HM=Half Moon; NM=New Moon.

The movement of Venus gives us the frame within which the ceremony develops:

```
250 d Evening Star + 8 days invisibility    258
236 d Morning Star                          236
 90 d Occultation behind Sun                 90    584 days¹
250 d Evening Star + 8 days invisibility
      to Emergence as Morning Star          258    842 days²
```

I start out with Venus as an Evening Star because in Panel 1 the planet is on the left side, and if one faces the sinking sun, Venus would be on the left when setting after the Sun.[3]

The Moon's phases accompany the ceremony and probably pinpoint the actual day of the culmination of the happenings:

Venus   *Moon*

```
1. SE 250+8    6 FM=177d* +3 FM=88.5** = 265.5-7d=258.5d
2. SW  236     8 HM  (+1)                      236
3. NW   90     3 HM  (-1)                         88.5(-5)
4. NE 250+8    6 NM=177d +3 NM=88.5 - (5+2)=-7  258.5
      842 days         T O T A L S             841.5 days
```

* 177 days = Ecliptic Cycle   ** Mercury Cycle (?)

Since the Moon has to be set back 7 days (the seven feathers over Venus as seen in Fig. 1 to conform with the Venus cycle of 250+8 days, the new counting of the Moon now starts with a Half Moon, or the Moon in the first

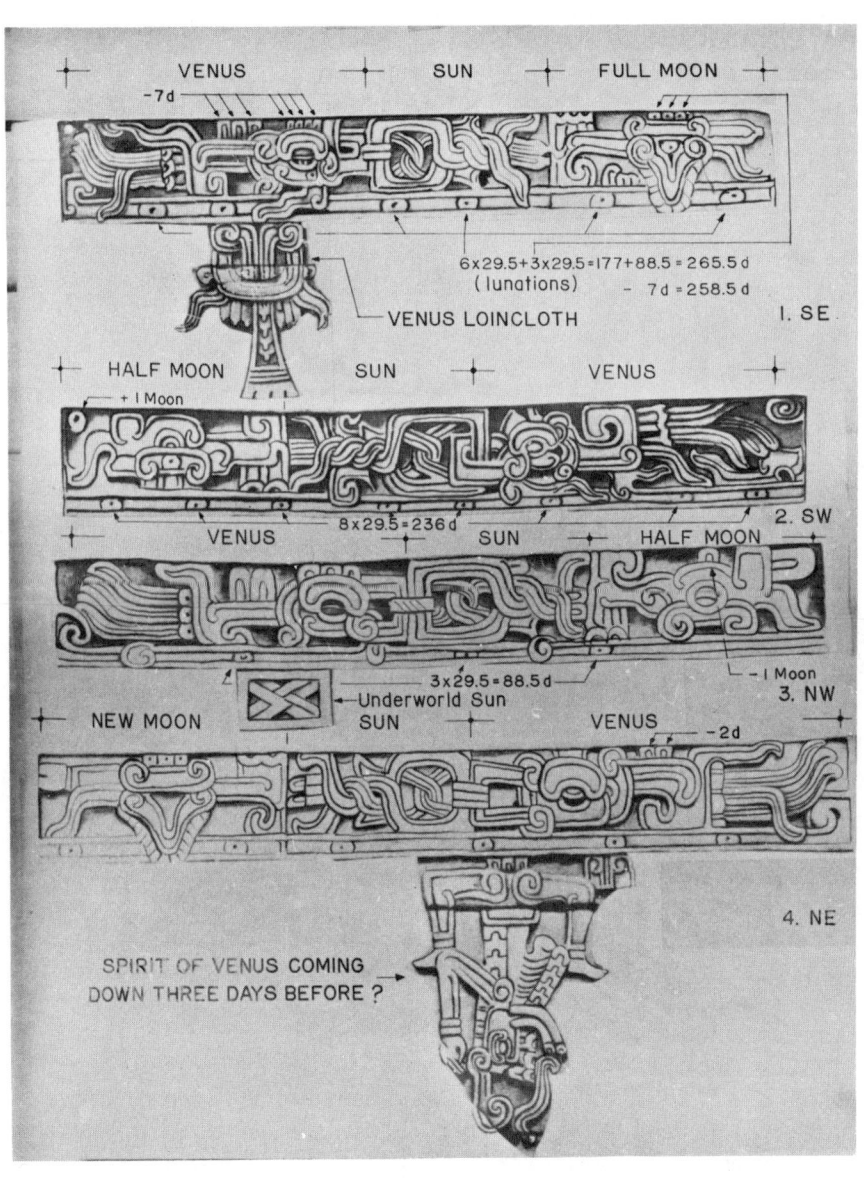

1 *An analysis of the upper skybands of the four Tajin Panels.*

quarter, which would be the actual meaning of the Moon in profile. The regressing Moon is looking towards the Sun. After 8+3 lunations, again 3+2+2 days = 7 days are taken out, or counted back to conform to the 258 days from Evening Star to emergence as Morning Star, and we count now New Moons. Although the first and last Moon are both shown en face, No. 1 Moon has a small circle in the center, which is lacking in No. 4. This would be the difference between a Full Moon and a New Moon in their representation in this culture.[4]

First Panel. Venus Evening Star. Candidate Selected and Vows Taken. (Fig. 2).

The candidate sits on a throne representing the open jaws of a serpent[5]. He holds three arrows pointed downwards renouncing sexual life, the arrow being a phallic symbol.[6] From the sky hangs what Garcia Payon interprets as a 'maxtlatl', or loin cloth, with which I agree. Its down-hanging band is decorated with the rattles of the snake, a symbol of Venus-Quetzalcoatl. This, I believe, to be symbolic of the coming down of Venus to Earth, the cloth being the one he will wear in his physical body. In accordance with the widespread Mexican legend, Venus walks on Earth when in its phase of Evening Star, taking the form of a man, as symbolized here by the male piece of clothing.

Witnessing and accepting his oath of purity, two deities stand and sit before him. The central figure could represent the Sun God, with a tongue above him protruding from the animal figure over him, and on the left the sitting figure has been interpreted by Spinden as feminine and Garcia Payon, accepting, calls her Xochiquetzal, Goddess of Love, without explaining her presence. Since the Moon is in the sky, she probably represents this goddess in her Moon form. The vow is made before her, although she at the same time puts temptation before him. She may, in some form, be involved in aiding the breaking of the vow and the sinning of Venus during his sojourn on Earth, all according to legend.

2 Panel 1. Venus descends to Earth and is initiated.

Second Panel. Venus Morning Star. Venus Sins. (Fig. 3).

The candidate rests on a bed and over him hovers or 'dances' a bird as Garcia Payon has it. The latter identifies the bird as the quetzalcoxcoxtli because of his crest or comb. He is the disguise of Xochipilli, the God of Love and Lust. Usually the bird holds some musical instrument in his hand, most often a rattle, but here two musicians on each side, one with a rattle, and the other with a drum, supply the rhythm to the singing of the bird, who is known to sing at dawn. Thus Venus has spent the night, or rather he spends 236 nights with music, song, and love.

Most ceremonial dances have no women in them even to this day, and the women are represented by men in female disguise. If Xochipilli here is to represent his wife or mother, Xochiquetzal, the disguise is not feminine, yet the act is, notwithstanding, producing a birth, which is represented above the scene.[7] A sky monster is being born from a rectangular object, where his left foot is still held. This is the conventional form of showing a birth in the codices. The object from which it is being born is similar to the division between the Moon and the other celestial bodies. I do not know of any legend of Quetzalcoatl in which there is a product of his sin, but here it seems to be represented on a sculpture dating from around the tenth century A.D.

Third Panel. Occultation behind the Sun. The Ball Game in the Underworld. (Fig. 4).

The scene is placed in a Ball Court, the walls of which are seen on both sides. From the sky hangs the symbol of the Sun over the figure that represents him. He has his arms folded over his chest. Venus faces him, and is asking him to accept the knife to kill him. Obviously they have played the game and Venus has lost, probably weakened by lustful nights for which he is submitting to voluntary death by the hand of the Sun. Astronomically Venus has disappeared behind the Sun which makes

3 Panel 2. Venus sins.

the latter the Conqueror. Between them are two twisted bands, representing the movement of the ball, but also Venus' movement back and forth around the Sun. Below is a Ball Court ring. These two latter forms might give the ball game sexual implications, further enhancing the quality of Venus' sin. On the left sits a man with a roll of paper in his hand, probably the judge. On the right kneels, also on top of the wall of the Ball Court, a human figure with a mask of a big-eared dog or maybe an opossum, probably representing the god Xolotl who might also be a symbol of the planet Mercury whose revolution around the Sun is probably twice depicted (88 days). He is also the leader of the dead to the other world.

Fourth Panel. Evening Star to Morning Star. Death by Sacrifice. (Fig. 5).

Venus lies on the curved stone of sacrifice held by an assistant. Before him, the Sun is pointing his knife over Venus' heart. On the right sits the judge with his scroll. Over Venus hangs his Spirit from which oozes the symbol of penitence (the symbol identified by Seler). Along his body and his backbone we again see the identification of the rattles of the snake. Garcia Payon believes the victim to be the Sun that dies daily leaving Venus victorious, but neither the astronomy nor the representations on the panels favor this interpretation.

There are various elements on the panels which can be added to the known legends of Quetzalcoatl as they survived in the sixteenth century, and likewise can be used to identify parallels to other sources. One is the time element. The idea of Venus taking on the form of man during the Evening Star period and sinning during this period is common to all versions, but the fact that this sin takes place during the time when Venus is Morning Star is new. In the legend this exalted position might be equal to his position as king.[8]

The ball game takes place during the obscuration of Venus behind the Sun which reminds us of the Divine Twins in the Popol Vuh who play the game in the Underworld. Very probably the game played in the Third Panel

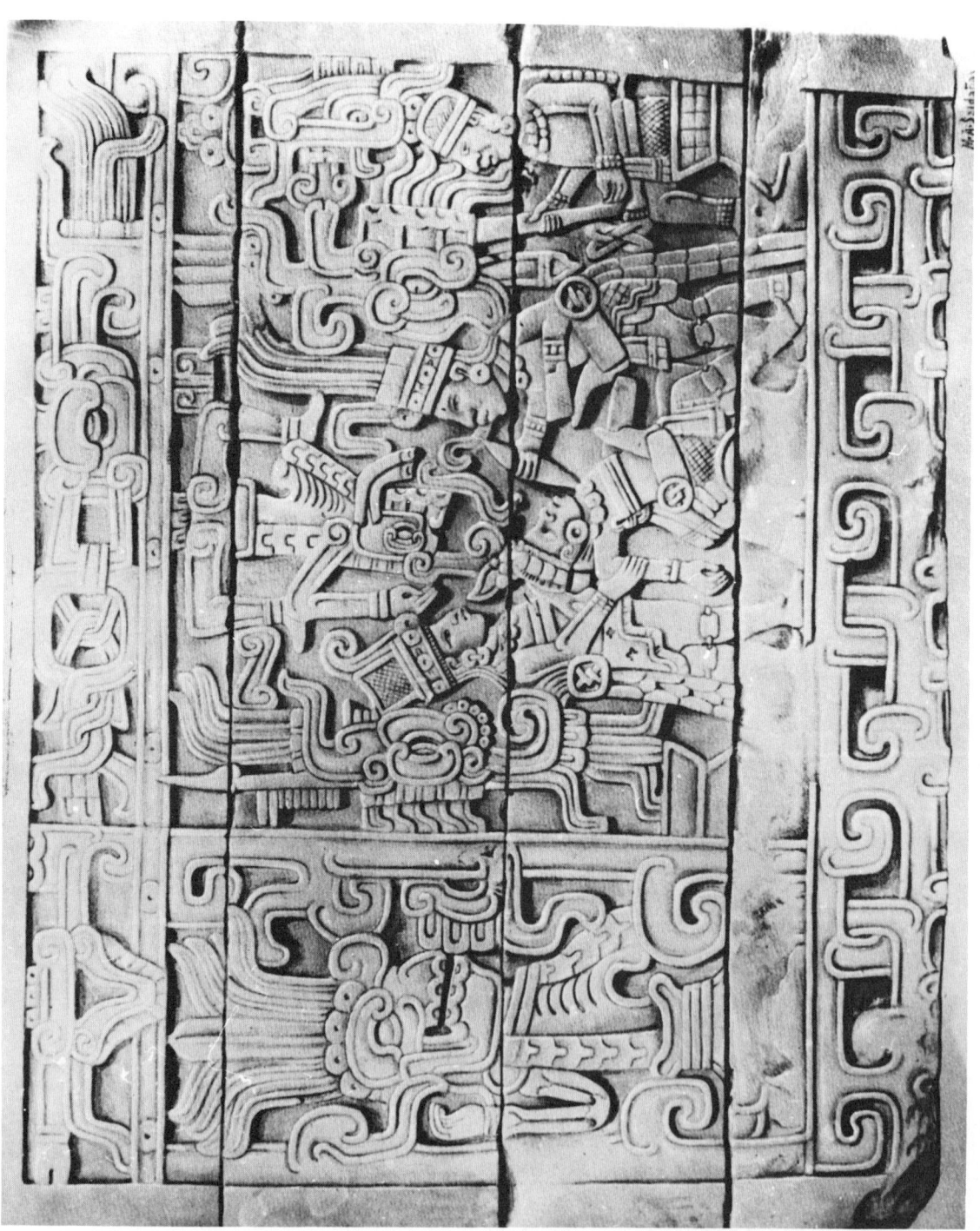

should be interpreted as played in the Underworld. As a matter of fact, the floor they step on in the first two panels is of a different consistency, being adorned with feathers which the second two lack.

The sign which Garcia Payon interprets as Mars is more likely the symbol of the Night Sun, the Sun in its travels through the Underworld during the night. Also I have not been able to locate any astronomical period which coincides with the Mars period, although I did make an effort to search for it.

After the game, astronomically, Venus first proceeds West until its greatest elongation to return East in an apparent retrograde movement as seen from the Earth. Death occurs in the "conjunction" with the Sun, after wanderings during 250 days. For eight days Venus remains invisible within the solar fire, to finally surge forth as the Morning Star.

The Astronomical Reckoning.

Although the disk counted as 'one' in all cultures of Mesoamerica that had writing, the feather as a unity to be deducted does not seem to have been mentioned before. At the moment I cannot say if this is a generalized symbol or just a regional element. Likewise the division of 4+3 feathers probably has a meaning.[9]

The exact dates on which events culminate probably are to be searched in other elements. For example, +1 lunar cycle on Panel 2 and -1 on Panel 3 might have some meaning, but since they cancel each other, the final result is not affected. The two small feathers over the larger one on Panel 1 beside the Moon also could have some interpretation, but since, for the moment, I have no basis for same, I cannot speculate on it. Parallels exist in several codices and a comparison might further enlighten us, not only as to astronomical and arithmetical aspects, but also as to the identification of the gods and other participants. This will have to be done in a separate study to be published soon.

The reason for the division of the lunations in Panels 1 and 4 in 6+3 could be to emphasize the cycle of eclipses of 177 days plus one revolution of Mercury, if

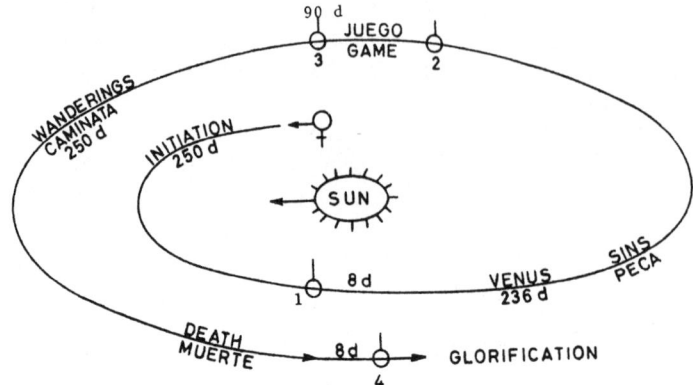

6 *The Venus Cycle represented on Panels 1 to 4 of the Ball Court of Tajin. The four 'conjunctions' are numbered corresponding to each of the panels.*

this were known to them. Revolutions are more difficult to observe.

Since Sun and Venus are always shown in conjunction, that is, interlinked, these points of contact seem to be of importance to fix events and maybe their culmination. Fig. 6 shows the points where contact probably is depicted, due to their interchanging positions. The conjunctions do not correspond to a modern idea, but rather to the moment when the planet or star becomes invisible because of its proximity to the Sun. The four "conjunctions" correspond to the four stages or steps that the God takes in space to fulfill his destiny. This, I believe, is contained in one of his names: "Nacxit"=4 feet or steps (<u>nahui</u>=4; <u>sitli</u>=step or foot).

In this study I have taken advantage of several combined elements for the interpretation of the Tajin Panels: tradition and legend, the pictorical representation, symbology, and astronomy.

Notes

1. ON THE SCHEMATIC DIVISION OF THE VENUS CYCLE. The schematic cycle of Venus resulting from the calculations made from the Tajin tablets is very similar to the known Venus cycle found in the Dresden Codex, which is 236-90-250-8. It will be noticed that the Tajin people made an effort to adjust the Moon cycle to the one of Venus. This adjustment might have resulted in the fixed scheme used by the Maya also. Whether this scheme originated with the Maya or in Tajin is not so easy to say, but probably the Gulf coast could have produced this calculation. The Tajin culture is known to have spread towards the Maya region, and contacts along the Gulf coast were surely constant. According to Teeple, the Dresden Codex may have its origins in the 11th century. Thompson (1972) places the Codex between the 13th and 15th centuries. The Tajin sculptures are earlier, but the codex, of course, in all probability is the result of previous knowledge. Seler believes the legend and the science of Venus were imported from the Highlands of Mexico, yet he finds a different division of the Venus cycle in the Borgia Codex (243-77-252-12), which is nearer to nature.

The difference in the number of days in the superior and inferior conjunctions is explained by the fact that when Venus is behind the Sun during the superior conjunction, Venus is progressing in the same direction as the Sun along the ecliptic, as seen from the Earth. Forstemann gives us the possible number of true days, considering a distance of $10°$ from the Sun to make Venus invisible, and reached 12 days for the inferior and 77-80 for the superior conjunction. But since Venus is seven times farther away from the Earth during the superior conjunction, the visibility is diminished and the 90 days may be justified. Forstemann also believes that a clearer sky on the Mexican coast and a more sudden fall of night might turn the twelve days into eight. Forstemann, though, confesses he cannot explain the difference between the other two periods (Morning and Evening Star), which should be equal, at 243 days each. The adjustment to the Moon cycle might then be the answer.

2. ON THE 842 DAY CYCLE. Neither the Dresden nor the Borgia emphasize an 842-day Venus cycle. The cycles shown there (in the Dresden) are five groups of thirteen Venus periods, each equal to 7578 days, which also equals 9x842=7578, with a difference of half a moon, or 14 days. Again 257 Moons + 105 days (the difference between 365 days and the 260-day almanac) is 7592 days. While Tajin tries to correlate the Venus cycle with the Moon, the Maya Dresden Codex and the Borgia of the Mexican Highlands seem to correlate Venus with the almanac of 260 days; both may be in some way implying the other not so obvious cycle.

3. ON THE STARTING POINT OF THE VENUS CYCLE. The Dresden begins, like the Borgia, with the emerging of Venus out of the Sun's rays passing from Evening to Morning Star, without the preliminary 250 plus eight days found in Tajin, in which Venus comes down to Earth and is transformed into man.

4. ON STARTING WITH FULL MOON. Thompson (1960, p. 230) discusses whether Indians of the Americas, and the Maya specifically, started their count with a New Moon or with a Full Moon; he decides in favor of the former. My decision to start the count for Tajin with a Full Moon seems to be confirmed by the circle within the en face Moon glyph on Panel 1, and because on this same panel Venus being Evening Star and on the West of the Sun, the Moon was on the opposite side of the Sun, therefore rising in the East. Again, on Panel 4 the Moon is West of the Sun just separating from conjunction, while Venus is on the right just emerging as Morning Star. Also, as mentioned, the two half Moons on Panels 2 and 3, are looking towards the Sun and are returning or regressing to the conjunction with the Sun. Venus as Morning Star forms a conjunction with the Moon, astronomically, which could be another way of interpreting Venus' sin of copulating with the wife of the Sun, the sin being adultery.

5. THE SERPENT THRONE. The idea of the candidate sitting within the fangs of a serpent, I believe, confirms the identification I made of the seat of the man to be sacrificed in sculpture Figure 1 of Chalcatzingo (1967) as a coiled serpent. Aside from this, last year I found

another confirmation in the oral tradition in Amatlan, Morelos, not far from Chalcatzingo: When a man asks the Gods for corn, he sits on a serpent, and a woman on a mat in the form of a milliped (ciempies), a symbol of the female organ. Even in the early 20th century, little girls were not allowed to sit on a bench, lest they lose their feminine qualities.

6. ARROW SYMBOLISM. Venus or gods representing Venus in the Dresden hold two arrows pointed downward in the same manner as on Panel 1 of Tajin. It may be significant that the first two, essentially Maya deities, do not hold them so, while the three seemingly imported from the Mexican Highlands, do so. The Borgia and the Dresden both have many examples of figures holding arrows in different positions, complicating the symbol so much as to make a discussion here beyond the scope and finality of this study. In the case of the Dresden, there are pictures below the arrow or spear-holding gods which have been wounded by a spear. Pages 46-50 of this codex have been interpreted as the central figure killing the lower figure. It may be, though, that there is more complexity than meets the eye, which will be discussed further in the coming publication. The arrow symbolizing the male sexual attack, for one other interpretation, is confirmed by the story of Mixcoatl darting Chimalman, which is told in the "Leyenda de los Soles," p. 124:

Then Mixcoatl went to conquer Huiznahua; he was met by the woman Chimalman, who put her shield on the floor, threw away her arrows and her atlatl, and stood naked, without a skirt or blouse. On seeing her, Mixcoatl shot his arrows at her: the first time the arrow went above her head; the second to one side and the arrow doubled up; the third she stopped it with her hand; and the fourth she removed it from between her legs. After shooting her four times, Mixcoatl turned away. The woman immediately fled, in order to hide in the cave of the big barranca. Again Mixcoatl came for new arrows, and looked for her, but did not find her. So he mishandled the woman of Huiznahuac, and they said: "Let's look for her". They told her: "Mixcoatl is searching for you, and because of you he is abusing your younger sisters". So she went back and again she met Mixcoatl, in the same manner, standing up and her shame uncovered; again she put down her shield and arrows, and Mixcoatl repeated the shooting and the arrow again went over her head, to one side, stopped by her hand, and between her legs. After this, he took her, he lay with the woman from Huiznahua, who was Chimalman, who became pregnant. When Ce Acatl was born, for four days his mother was much afflicted; so when he was born his mother died. Ce Acatl was raised by Quillaxtli, Cihuacoatl.

The story has been told translated in detail, because it applies not only to the symbolism of the arrow, as handled by the old savants, but also to the psychology of their goddesses, who must be demasculated before they can perform motherhood, here symbolized by her putting down her arrows - and other concepts, such as the number four and eight (see Note 8).

7. ON THE INTERCHANGING OF SEXES. Although there is one probable woman on Panel 1 of Tajin, and one certain one on page 49d of the Dresden, we find that men are chosen for the scenes related to Venus' sin, a sex act in spite of chastity vows. The male characteristics are emphasized. We, therefore, cannot speak of transvestism, and because this contact between two males bears fruit, we might coin the similar expression of transfunctionalism. Although in the codices the acts are performed by gods, it is very probable that the cycle of Venus was represented in ceremonies to the exclusion of women, and that the acts of the gods are rather a projection of the desires of human performers.

The Maya and the people of Tajin do not stand alone in transvestism and transfunctionalism. Bleibtreu (1970) discusses this universal phenomenon which reaches into our modern culture (Kinsey 1958). According to Kinsey the male-female transvestism ratio is 100:6, confirmed by Giese (1962) for Germany. In present native cultures I have tried to explain the problem by the too intimate sleeping together of the family with the children witnessing constantly the parent's act with different reactions: the majority craving for an early marriage and a few denying sex altogether. But especially the custom of the mother or older sister carrying the child until nearly five or six years old will result in a too close and dependent relationship to older females with whom they identify. The classic example is precisely Quetzalcoatl who was being cared for by his older sister with whom he finally commits an incestuous sexual act in a drunken orgy. (See Note 9 and Cook de Lenard 1967).

Bleibtreu quotes several authors who have observed the relationship of shamanism and transvestism, explained because of the wish to identify with a goddess as well as the instinctive procedure of the mothers who pledge their son to shamanism, dressing the boy with girl's clothes. This might explain Quetzalcoatl dressed as a girl in Borgia 57. This scene might identify with the story of the later King of Tula who, according to legend, is raised by Quillaxtli-Cihuacoatl. These goddesses, who are identical, are said to be the goddesses of war, representative of the mociuaquetzque, the "warriors who appear dressed as women"; they carry a shield, and are, in other words, a 'mother' who would have their son swear celibacy and rear him in their image. Such is probably the older sister of Quetzalcoatl, who at the same time is the older sister of the Mimixcoua, the stars, and he being Venus would relate to her.

Transvestism has been discussed and classified in its different aspects by several psychologists. The most convincing to me personally is that explanation given by Dr. Otto Fenichel (1953), which I have already applied to the interpretation of an Olmec sculpture (1967). He calls a woman transvestite a phallic woman, and certainly within this category falls also Chimalma, Quetzalcoatl's mother, who like Cihuacoatl carries weapons. A typical act of the goddess is grinding the bones of sacrifice for the creation of humanity, which she mixes with the blood of the penis of the gods.

The desire of such men, tending to imitate the duties of women, takes them likewise to desire giving birth and I believe my interpretation correct for page 48 of the Dresden (Leonard, Ms.) where in the parallel Borgia, Tonacatecuhtli lies pregnant with Venus-Quetzalcoatl in the Ballcourt of the Underworld. Another male god to give birth is God E in the Codex Madrid 74a. The Corn God E is a passive figure and appears as a victim on page 48 of the Dresden where he will be further discussed.

8. A BRIEF SUMMARY OF THE VENUS LEGEND. Prince Ce Acatl Topiltzin is born after his father dies and his mother dies at birth (see note 6). When nine years old, he asks for his father who was killed. He unburies his bones and places them in the temple of Quillatzin. He becomes a priest of the God Quetzalcoatl and vows chastity. His fame as an architect and inventor spreads, and he is called to Tula to become king. Here he builds a house surrounded by four walls which he never leaves. He sends for his sister with whom he goes down to the river every midnight to pray. His enemies plan his downfall. First they bring him a mirror in which he sees himself (his true self, in love with his sister). He is horrified and says he cannot show himself to his people as ugly as he is. So his enemies send him a mask builder who makes him a beautiful mask so he may show himself to his people. They then bring him a gift of vegetables and chile (to make him thirsty) and offer him pulque (an intoxicating beverage) which he finally tastes with his finger after resisting. He likes it and drinks four glasses, plus one more. Intoxicated he sends for his sister, who also drinks four plus one glasses. Next day he tragically cries over his incestuous orgy and decides that only death can wash away his sin. He has a stone coffin built, in which he lies for four days. He buries his treasure and with great lamentations he wanders towards the Gulf coast with his followers. There he puts on his mask and steps into the fire. His ashes turn into birds of all colors and his heart becomes the Morning Star.

9. VENUS NUMEROLOGY. The question of why the number 7 to be subtracted in Tajin Tablet 1 should be divided into 4+3 leads us to the several times number 4, or 7, or 8 is mentioned in relation to Venus. For this we find both astronomical and legendary information which permits us to conclude that four is the number of the planet Venus.

a. The Dresden Codex (Teeple 1937; Thompson 1972) makes a correction of the average Venus Synodical Revolution after 61 Venus revolutions (VR) to recover the lub 1 Ahau by subtracting 4 days, "but as the error at the end of 61 VR was nearly five days, an occasional 8-day correction

Ballcourt Panels at Tajin

was necessary at the end of the 57th VR. Corrections had to be four days or multiples thereof, as corrections had to recover the day 1 Ahau" (Thompson 1972).

While the Dresden is set on correlating the VR to the 260-day calendar, Tajin tries to fit the Venus cycle to the Moon cycle, and likewise must subtract 4+4 days to fit the schematic Venus cycle to the actual Moon movement, which can only be done by setting the beginning Moon count back a quarter phase twice during an 842 day period. Since we have 9x29.5=265.5-7, Tajin could just as easily have subtracted 4+4 instead of 4+3 (or to be exact 3.5).

b. The highest common factor to harmonize Venus with the Sacred Almanac of 260 days is four (Thompson 1960, p. 221): 146x260=37,960; 146x4=584 d (Venus cycle); 65x584=37,960; 65x4=260 d (Sacred Alamanac).

c. 584÷20 (days of calendar) has a remainder of 4, so that heliacal risings of Venus as Morning Star can occur only at intervals of 4 days in the official revolution of the planet, although observed risings could occur on any day because the actual revolution of Venus varies in length from 581 to 587 days (Thompson 1960, p. 221).

d. Tajin represents the Venus cycle in four tablets or four steps, which I believe confirms the meaning of <u>Nacxit</u>, one of Venus' names, as Four Steps or Four Feet, the four steps Venus takes in space. The name appears in the Chilam Balam of Tizimin (fol. 11, verso) and other sources.

e. The Maya symbol of Venus probably represents the revolution of Venus around the Sun, with the center dot either the Sun or the Underworld, as reckoned in the Dresden Codex, where the Underworld (p. 48) is the center of the cycle.

f. The schematic division into four parts of the Venus cycle in 236-90-250-8 days.

g. Page 46 of the Dresden Codex starts, according to Forstemann, with the moment at which Venus disappears in the rays of the rising sun (3 Cib). According to Seler it is possible that the end of the visibility is meant and not the beginning of invisibility; then the day 4 Caban = 4 Olin would correspond to the day in which Venus unites with the Sun, which is the symbol of the Sun, but also the day for which Quetzalcoatl is regent.

h. According to the Anales de Colhuacan (Lehmann 1938), Quetzalcoatl stays four days in Tollantzinco, and in Tule he makes his Fasting House fourfold. He drinks four plus one glasses of pulque, which is his downfall, as does his sister. When he knows he must leave his city and dies, he has a stone coffin made in which he lies for four days. When he dies the people say that for four days he is not seen, is invisible, and for four more days he is a skeleton. And eight days later he comes back (as Morning Star).

i. According to the "Leyenda de los Soles" (see note 6), his mother was much afflicted for four days which might correspond to what Sahagun (1938, VII, Ch. 3) says: "When the planet Venus appears once more on the horizon, it disappears four times first before it reappears in its full splendor, shining like the Moon. This probably means that it is only on the fifth day that Venus is far enough away from the Sun to be seen in all its shining glory."

16

An Astronomical Calendar in a Portion of the Madrid Codex

Marion Popenoe Hatch

Department of Anthropology

University of California

Berkeley, California

Introduction

Although astronomical inquiry into the decipherment of Maya texts has currently taken second place to the excellent and productive inroads made possible by the recent historical approach, there is no reason why the contributions of the latter should necessarily eliminate astronomy as a tool of scientific investigation into the archaeological past of Mesoamerica. Its pre-Conquest calendar was a masterpiece of defining and observing the passage of time; no calendar has yet been devised by man which was not dependent to some degree upon celestial observations. A great deal is known about the various day counts of the ancient Maya, but there is little in the ethnographic record (which comes to us through a filter of European world-view) to give a clear understanding of the structuring of the universe which provided the framework for the development of the Maya calendar. I hope to demonstrate that pages XII to XVIII of the Madrid Codex (also called Codex Tro-Cortesianus) lend themselves to an astronomical-calendrical interpretation, in which the patterning is too regular and the correlations too precise to be attributed to coincidence alone. What is presented here is exploratory in nature, and derived from the standpoint of an archaeologist who has acquired only secondarily a minimal knowledge of astronomy, but the paper is offered in the hope of eliciting some interest in the subject.

There is at the present time no certain date for the writing of the Madrid Codex. Thompson assigns to the Dresden Codex a date of ca. A.D. 1250 based on lunar data contained in the text; with the same information, Satterthwaite concludes that it dates no earlier than A.D. 1345 (Thompson 1972, p. 15). Such datable lunar material appears to be lacking in the Madrid Codex. I have estimated a date for it of about A.D. 1400 based on the following assumptions:

1. It is certainly of pre-Columbian date, and was very likely in use at the time of the Conquest, at which time it came into the hands of the Spanish. If this is so, one might estimate a century or so of its having been in existence. Longer use than that would seem improbable in view of obsolescence, deterioration, and wear.

2. On stylistic grounds, one might guess it to be later in date than the Dresden Codex because the Madrid Codex is less carefully executed and generally inferior to it in style, the difference being greater than the contrast between individual artists. One is therefore tempted to assign a greater time gap between it and Classic Maya style than that for the Dresden Codex.

3. The results of my analysis also support a date of about A.D. 1400, although a margin of plus or minus a century or two would not seriously affect the interpretation.

The above assumptions do not eliminate the possibility of an earlier date for the Madrid Codex. Since completing my analysis, a research project under the direction of Arthur G. Miller at Tancah, Quintana Roo, Mexico, has come to my attention. From the results of his study, Miller (1973) suggests a date of A.D. 1200 for the Codex. For my analysis, a slightly later date for the Codex fits better; also I find it hard to imagine a codex in continuous use for over 300 years in a tropical lowland climate. Since I have already devised the charts based on data for A.D. 1400 and now lack the time to change them, I am retaining them as originally designed. The difference between A.D. 1200 or 1400 is minimal, involving for most of the analysis a difference of but half of a degree

Astronomical Calendar

in right ascension, or less than a whole day. A date of
A.D. 1300 may be a satisfactory compromise. During the
course of the discussion, I will attempt to point out
any alterations which would be involved should the
earlier date for the Codex be proved correct.

Analysis

Fig. 3 shows a photograph of the text under discussion.
This is copied from pages XII to XVIII of the Gates' edition of the Madrid Codex (Gates 1911). The main subject
is the so-called Chicchan serpent, set in the context of
a day count. Each page is divided into two panels, an
upper smaller panel which I shall term Panel A, and the
lower one, larger in size and comprising the principal
body of the text, which I shall refer to as Panel B.
They differ in that Panel B appears to be a continuation
of texts from the pages preceding page XII and following
after page XVIII. Four rows of day glyphs are arranged
horizontally across the middle of each B panel to form a
single 260-day count, whereas the A panels (except for
page XII) each contain a complete 260-day count. However, there is some relationship between Panels A and B,
which will be demonstrated as the analysis proceeds.

The B panels indicate that they are meant to be seen
together. Beginning with page XIV, the pictures are continuous from one page to the next. The serpent is presented in three different postures: the first on pages
XIII and XIV, the second on pages XV and XVI where he is
turned so that he faces to the right of the page, and
the third on page XVII where his head appears at the top
of the panel with his body continuing on to page XVIII.

The day count expressed by the four rows of day glyphs
appearing on the B panels is to be read from left to
right. It starts with 1 Imix at the beginning of the
first row on p. XIII, then follows this row across all
the pages to p. XVIII, to return to the first glyph in
the second row on p. XIII, continuing across it to p.
XVIII, returning again to the first glyph in row 3 on p.
XIII, etc. However, there is a 13-day gap between the
last day glyph of each line on p. XVIII and the first
one on p. XIII (note that the first row ends with day Eb,

the second row begins with day Cimi). As though to inform the reader of this, the two bars and three dots appear in black at the lower right-hand corner of p. XII to fill in the gap. Since there are 52 day glyphs in each row, the count beginning with 1 Imix will at the end of each row finish with day number 13; thereupon, counting the 13 days to fill the gap, one will again end with day number 13, which is shown in red (in the original painted version) bar-and-dot numerals at the lower left-hand corner of p. XII. The count then proceeds to p. XIII starting with day number 1 again. The total number of glyphs in each row totals 52 plus 13, or 65, and 4 x 65 equals 260 days, the full Sacred Round.

The band of day glyphs appears to be superimposed on the serpent pictures; on the last page, the scribe seems either to have run out of space or changed his mind, for the glyphs are blocked out but left incomplete. Again with the exception of p. XII, which seems to function as an introduction or title page, each B panel carries a horizontal band of glyphs along the top border; these contain auguries and information relevant to that page.

After much experimentation with this text, I have found that the serpent positions correlate exactly with the path of the constellation Draco as it would have appeared in the evening sky at regular intervals as the year progressed. What we seem to have here is a combination of a 260-day count, with a count of a solar-sidereal year of 365 days, or count of seasons, represented by the serpent pictures. The information is supplied to the reader on p. XII, the title page (see Fig. 2). The long-nosed rain god Chac is pictured wearing the year sign as a headdress. The serpent is attached to a glyph band containing (from left to right): (1) a variant of the glyph for kin, the term for day or sun; (2) unknown; (3) Kan cross; (4) the akbal (also referred to as akab) glyph, denoting darkness or night. I have proposed elsewhere (Hatch 1971) that earlier in Mesoamerica the Kan cross was used to connote the solar year; I suggest that the information conveyed here is that in the text to be presented the sun (implied in glyph 1 in the panel) is correlated with the sidereal year (represented by glyph 2), and the night (glyph 4) with the solar year (glyph 3). Note that the serpent body attaches to both

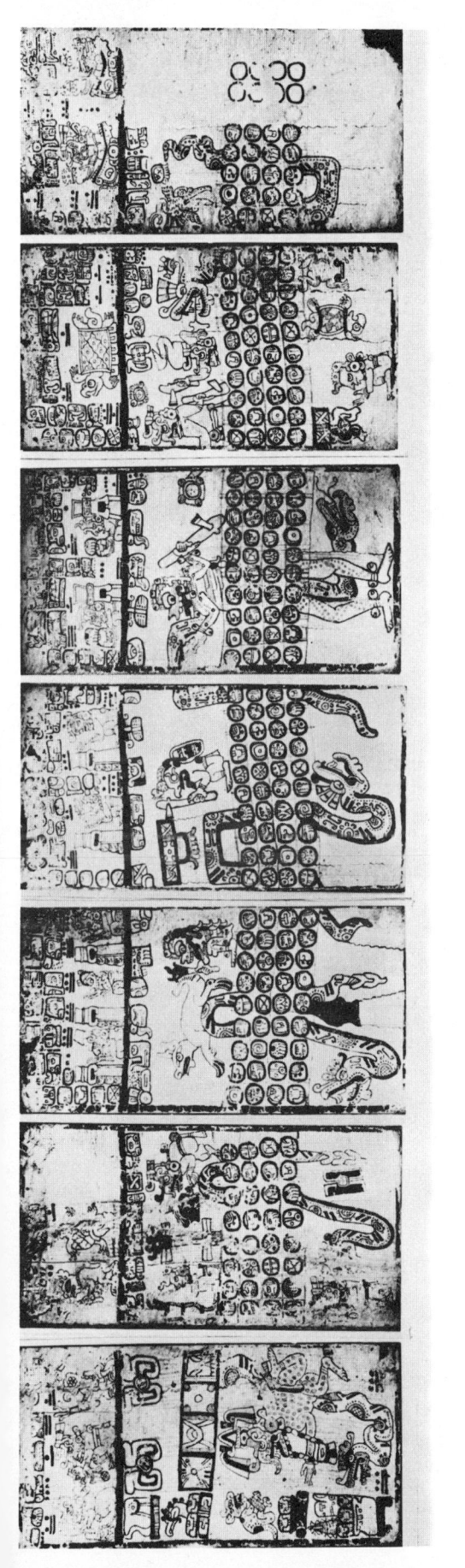

```
Page XII          Page XIII         Page XIV          Page XV           Page XVI          Page XVII         Page XVIII
Title Page        Dec. 21           Feb. 4            Mar. 22           May 6             June 21           Aug. 5
Sept. 21          7:20 p.m.         7:20 p.m.         7:20 p.m.         7:20 p.m.         7:20 p.m.         7:20 p.m.
   19°                                  64°              109°              154°              199°              244°
```

Comments

```
Dec. 21  - Feb. 4   = 45 days      Dec. 21  - Mar. 22  = 91 days
Feb. 4   - Mar. 22  = 46 days      Mar. 22  - June 21  = 91 days
Mar. 22  - May 6    = 45 days      May 6    - Aug. 5   = 91 days
May 6    - June 21  = 46 days      June 21  - Aug. 5   = 91 days
June 21  - Aug. 5   = 45 days
```

1 Pages XII to XVIII of the Madrid Codex (Gates 1911). Subtitles denote the date and time represented by that page; the number expressed in degrees gives the right ascension of the celestial hour circle in the vertical plane of the local terrestrial meridian for that date (correct for A.D. 1400).

2 *Page XII of the Madrid Codex.*

the akbal glyph, referring to the night, and the Kan cross, here defined as the solar year; this comprises the very subject relating to the serpent in the next pages. Each pair of these day-vs.-night glyphs in the panel is placed beneath a kin, or sun, symbol between "wings". Painted vertical lines (only faintly visible in the photograph) join the figures, perhaps to show influence or relationship.

The count of this text apparently begins after the title page; the patterning is then such that every other page refers to a solar event (pages XIII, XV, and XVII), the intervening ones to an evening sidereal observation (pages XIV, XVI, and XVIII). The distribution of kin signs between wings seems to indicate that the compound here pictorially represents the path of the sun; kin between black and white wings (p. XII) convey equal length of day and night or equinox; with black wings (p. XIII), longer nights or winter solstice; with white wings (p. XVII), longer days or the summer solstice.

Seeing that this text correlated so perfectly with the path of Draco, it seemed improbable, even perhaps unbelievable, that this constellation, composed of faint and inconspicious stars, could have had any significance for the Maya. However, a check with Neugebauer's Star Tables (P. V. Neugebauer 1912) supplied the answer: the star Eta Draconis had an absolutely unique characteristic in that for 2400 years it alone among all the stars of the sky had a virtually unchanging right ascension, a trait due to its particular relationship to the pole of the ecliptic. During the millennia between 1800 B.C. and A.D. 500, a time certainly encompassing the development of the Maya calendar, the annual date of its meridian transit varied throughout by less than one day (see Table 1). Compare this with 75 days for Alpha Ursae Majoris, a star with about the same declination. Such a trait would have made it an excellent calendar star, for it would have provided an accurate measure of the sidereal year for a long period of time, serving as a yardstick for comparison not only between solar and sidereal year, but also for the precessional lag apparent in other stars. I have argued elsewhere (Hatch 1971) that there is some evidence that early Mesoamericans were observing midnight meridian transits for calendrical purposes. Such a practice would be consistent with the conceptual framework of an obsession with the four cardinal directions possessed by native Mesoamericans of that time.

The diagram in Fig. 3 shows that the star Eta Draconis was a circumpolar star before 1400 B.C. and never set for observers along terrestrial latitude 20° N. Between 1400 B.C. and A.D. 1400, the lower transit of Eta Draconis occurred below the Yucatan horizon. The declination of the diurnal circle of Eta Draconis around 1400 B.C. was +70°; around A.D. 1400 it was +63°, a remarkably slow change for a duration of almost 3 millennia.

If the midnight meridian transit of Eta Draconis were employed to measure the year, two dates would have become important: Its upper meridian midnight transit on 23 May (a date approximately the start of the rainy season) and its lower meridian midnight transit on 22 Nov. The latter transit would have occurred below the horizon and would not have been visible at latitudes lower than 20° North (see Fig. 3), but it could have been calculated by simple counting a half year from its upper transit in May. That such may have been the case is suggested by the following quote from one of the Chronicles (Herrera, Dec. IV, Bk. VIII, chap. 6, cited in Thompson 1931, p. 353):

They (the people of the coast of Honduras)...counted only by nights, and thus they put the night before the day, and counted 20 nights, or 20 dawns. They adjusted the day of the sun, taking note of its height, or when it was on its downward course...Thus they arranged things. They began their year 40 days before ours, for they took 2 months of theirs before ours began (atras).

Subtracting 40 days from 1 Jan. gives the date 22 Nov., the traditional date of lower meridian midnight transit of Eta Draconis. Although, by the time the Chronicles were recorded, the right ascension of Eta Draconis had gradually increased by 3 to 4 days (due to precession of the equinoxes), the original date which had been stable for so long may have continued to function as the official beginning date of the sidereal year count.

It is interesting that the path of Eta Draconis during the year correlates perfectly with the postures of the serpent on these pages of the Madrid Codex. Furthermore, the outcome of interpreting the serpent calendar in such

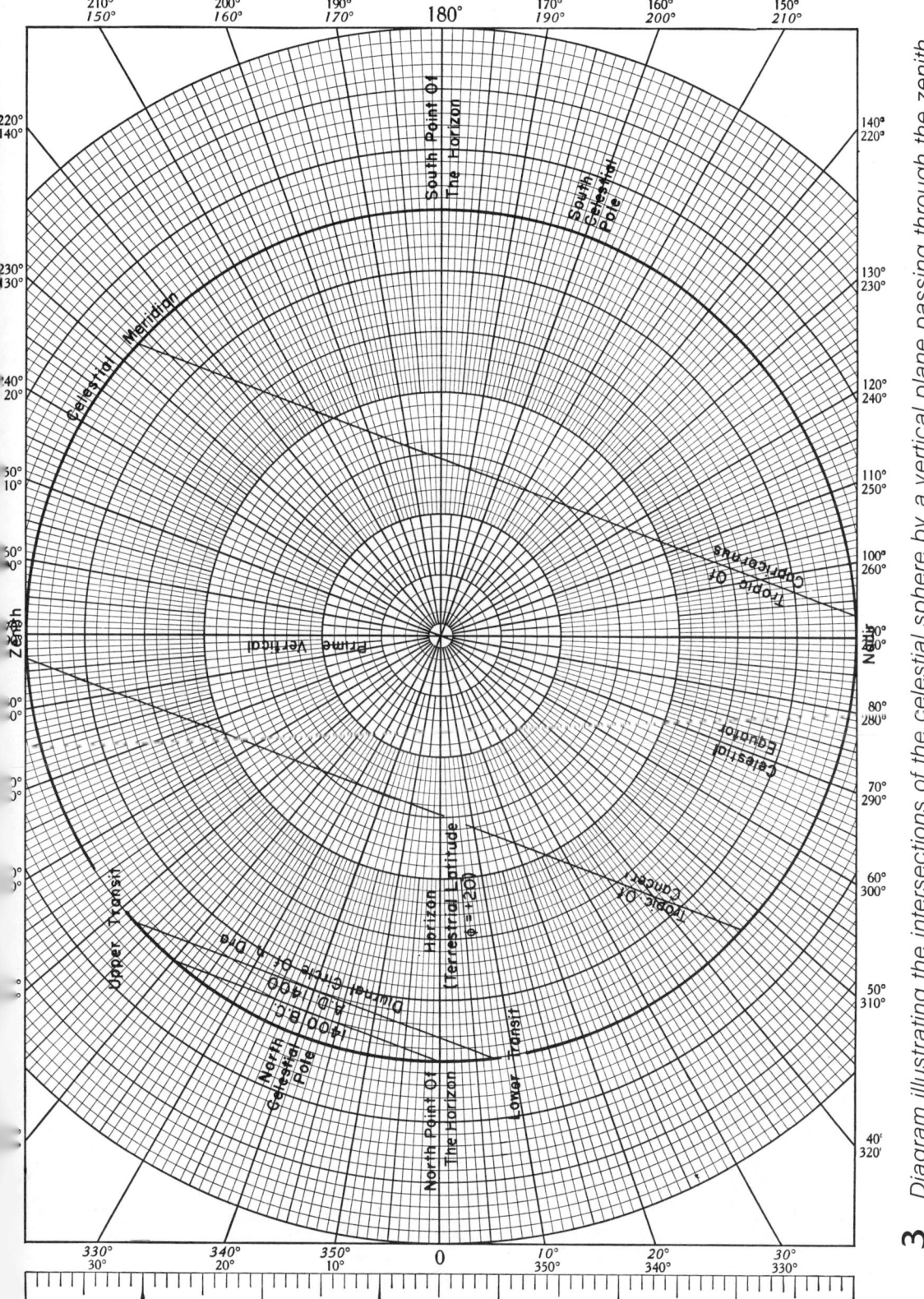

3 Diagram illustrating the intersections of the celestial sphere by a vertical plane passing through the zenith and both celestial poles.

Table 1

Right Ascensions and Equivalent Dates of Midnight Meridian Transits of the Stars Eta Draconis and Gamma Cygni between the Years 2000 B.C. and A.D. 1600

Year	Right Ascension	Date of Midnight meridian lower transit	Right Ascension	Date of midnight meridian upper transit
-2000	241°.94	23 Nov.	270°.05	21 June
-1800	241.43	22 Nov.	271.81	23 June
-1600	241.04	22 Nov.	273.57	25 June
-1400	240.74	22 Nov.	275.33	26 June
-1200	240.54	22 Nov.	277.09	28 June
-1000	240.44	22 Nov.	278.86	30 June
-800	240.41	22 Nov.	280.62	2 July
-600	240.45	22 Nov.	282.40	3 July
-400	240.57	22 Nov.	284.16	5 July
-200	240.75	22 Nov.	285.94	7 July
A.D.0	241.00	22 Nov.	287.71	9 July
200	241.30	22 Nov.	289.49	10 July
400	241.65	22 Nov.	291.27	12 July
600	242.04	23 Nov.	293.05	15 July
800	242.26	23 Nov.	294.83	17 July
1000	242.98	24 Nov.	296.61	19 July
1200	243.51	24 Nov.	298.39	20 July
1400	244.06	25 Nov.	300.18	22 July
1600	244.67	25 Nov.	301.96	24 July

a manner results in the following scheme: from solstice to equinox is a span of 91 days, and between meridian transit of Eta Draconis to its eastern elongation (holding the hour constant) is also 91 days. Since the calendar here is adjusted for evening rather than midnight observations of Eta Draconis, the result is that the count from one page to the next constitutes a span of exactly half of 91, or 45 to 46 days (see Fig. 1). Astronomically, it represents a rotation of the sky by 45 degrees from page to page. From p. XVIII back to XII (September equinox) is 46 days, followed by a jump of 92 days between pp. XII and XIII to complete the 365-day year; then the count begins once again. What may be manifested here is the 4 x 91 equals 364-day count, approximating the solar year. I tried at length to interpret the count from <u>actual</u> celestial observations after sunset which would naturally vary in time according to season; the attempt never succeeded. The only one that fits is the 45-46 day count, which invariably results in a fixed hour of observation: 7:20 p.m. This particular detail was somewhat baffling, until I checked times for the end of twilight at latitude 20° North, the probable latitude of the composers of the Madrid Codex (Thompson 1972, p. 16; Miller 1973). The extremes range from 6:36 p.m. on 16 Nov. to 8:08 p.m. on 4 July (see Table 2). The exact halfway point on this range is 7:22 p.m., which occurs about the second week in March. In other words, 7:20 p.m. is the end of twilight at the time of the vernal equinox, midpoint in the sun's journey; to astronomers accustomed to observing midnight meridian transits, the variation in time of first darkness could be dealt with by establishing the average end of twilight, or vernal equinox, as the reference point, thus keeping the day counts in even intervals by not varying the time of observation.

To speak of "time of observation" does not imply the use of mechanical clocks. There are various ways of keeping track of time; during the day the obvious method is to follow the path of the sun, at night the course of a circumpolar star, or some star which rises in the evening, crosses the meridian at midnight, and sets in the morning. The citation already given from Herrera (p. 10) makes it clear that the Mesoamericans he was talking about timed the day by the sun. That they measured the night

Table 2

End of Astronomical Twilight for Latitude + 20°

(The local mean time of the instant when the true geocentric zenith distance of the central point of the Sun's disk is 108°).

```
Jan.  0 ... 6h50m p.m.
     10 ... 6 56
     20 ... 7 01
     30 ... 7 07
Feb.  9 ... 7 11
     19 ... 7 15
Mar.  1 ... 7 18
     11 ... 7 21
     21 ... 7 24
     31 ... 7 28
Apr. 10 ... 7 32
     20 ... 7 36
     30 ... 7 41
May  10 ... 7 46
     20 ... 7 52
     30 ... 7 41
June  9 ... 8 03
     19 ... 8 05
     29 ... 8 07
July  9 ... 8 06
     19 ... 8 04
     29 ... 7 59
Aug.  8 ... 7 51
     18 ... 7 44
     28 ... 7 34
Sep.  7 ... 7 24
     17 ... 7 14
     27 ... 7 05
Oct.  7 ... 6 56
     17 ... 6 49
     27 ... 6 43
Nov.  6 ... 6 38
     16 ... 6 36
     26 ... 6 36
Dec.  6 ... 6 38
     16 ... 6 42
     26 ... 6 47
Jan.  5 ... 6 52
```

Earliest end of twilight:

Nov. 16 ... 6h36m p.m.
Nov. 21 ... 6 36
Nov. 26 ... 6 36

Latest end of twilight:

July 4 ... 8h08m p.m.

by the course of a star is suggested by the Motul Dictionary definition of the word "Ah ppizakab" as: "Morning star that appears early (in the night) and runs all through it as though it were measuring it," (translation by Thompson 1938, p. 600). Thompson relates the word to "Apizocab" in the Moran Dictionary where it is translated as "measurer of the night." The word "ppiz" is definitely related to "measure" according to the Motul Dictionary. Also significant here is the Perez Dictionary definition of the word "ppizibkin" as "reloj del sol" ("sun clock").

An attempt will now be made to do a page-by-page analysis in detail of the serpent text of the Madrid Codex. The day count represented in it is typed in below the photographs on Fig. 1, giving the date I propose each page illustrates; the number expressed in terms of degrees (shown below each date) gives the right ascension of the celestial hour circle in the vertical plane of the local terrestrial meridian for that date (correct for A.D. 1400), which seems to be the subject of the picture. Should the date of A.D. 1200 be preferred, a glance at Table 1 will show that the right ascension of Eta Draconis differs by only half a degree from that at 1400. For any year in the two centuries following A.D. 1200, the date of the midnight lower meridian transit of Eta Draconis is actually closer to 25 Nov. than the 24th; thus the dates as I have listed them in Fig. 1 would remain valid. The same would also be true should the Codex be found to have been written as late as A.D. 1500, but 1100 or 1600 would lie beyond the limits of the scheme as I have proposed it. A more critical detail with respect to the date of the Codex will be examined in the context of p. XV below.

As an aid to the reader, Fig. 4 shows a circular diagram of the pattern of the serpent calendar as we of European cultural heritage might conceive of it. Each radius lies at a 45° angle (or 45-46 days) from the next to demonstrate the pages of the Codex in sequence, the page number given by Roman numeral.

To assist in the discussion of the context of pages XII to XVIII, I have supplied a homemade astronomical chart to accompany each individual page to show how the sky would have appeared on the associated date at 7:20 p.m. ca. A.D. 1400. The horizon for three charts has been calculated for latitude 20° with the aid of data from P. V. Neugebauer's Astronomische Chronologie (1929, vol. 1, pp. 146-149). The chart shows only those stars important to this discussion having declinations of +20° and higher. The reader must imagine the horizon to extend beyond the chart as a closed curve to include the rest of the visible sky extending to the celestial equator and beyond as far as the southern horizon.

On the astronomical charts I have drawn by a dotted line the constellation of Draco as we conceive of it. It can be seen that the important star Eta Draconis lies near the center of the back of the serpent; Beta and Gamma Draconis form his head. The fact that the serpent posture is so similar to that pictured in the Madrid Codex, with the body positions correlating so well with the apparent locations of these stars in Draco in connection with the day counts, suggests that the Maya may have had a conception of the constellation roughly approximately ours. This is not to imply that all the stars of our Draco correspond to stars the Madrid Codex astronomers and artists saw in the form of a sacred snake. What I hope to demonstrate is that in the scheme of the codex, Eta Draconis seems to occupy a pivotal position, with the head of the serpent in the general vicinity of Beta and Gamma Draconis, the tail lying somewhere between Ursa Major and Ursa Minor.

The procedure in examining the pages of the Codex will be as follows: (1) discussion of the illustration in Panel B, with reference to the accompanying astronomical chart; (2) interpretation of the augural band across the top border of Panel B; and (3) summary and interpretation, as far as possible, of Panel A.

Page XII (Fig. 2)

Panel B

The main illustration on this page has already been discussed and it was indicated that it has the function of a title page, introducing the subject of the serpent calendar to be presented on the following pages. The five glyph compounds arranged in a vertical column on the

Astronomical Calendar

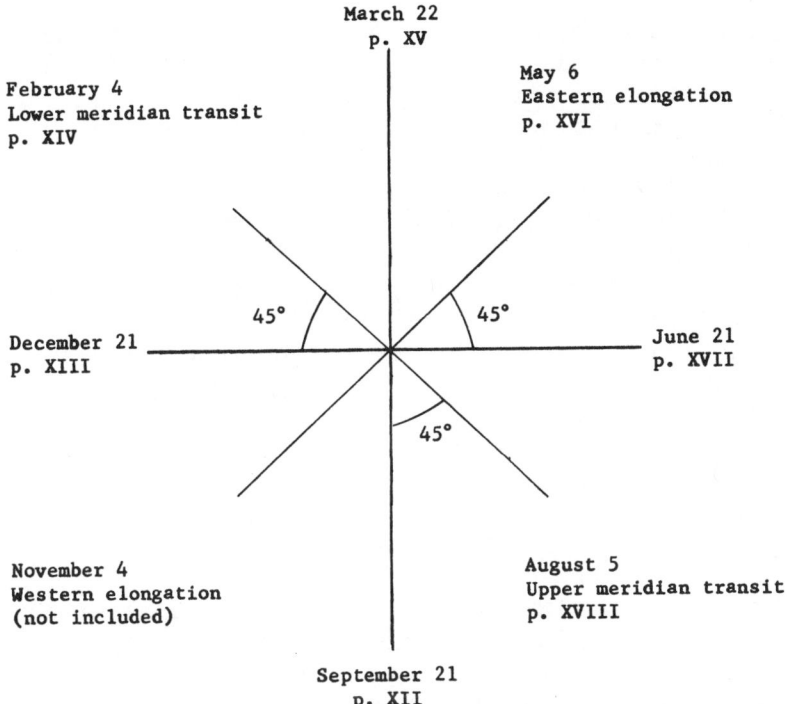

4 *Relationship of solar year to transits of Eta Draconis at 7:20 p.m. (Average end of twilight for 20° North Latitude.) A.D. 1400.*

left-hand side of the panel remain to be examined. I am somewhat reluctant to discuss glyphs, since I am not an expert in Maya epigraphy. However, to ignore them completely would be to offer a merely half-finished analysis; therefore, I will present a quite tentative and exploratory examination of the glyphs, drawing upon the work of other authorities and trying to fit meanings to the context. Since there is a great deal of disagreement among Maya epigraphers, I will cover only what seems to be the most important details, and I caution the reader that all of what is here offered is subject to more study and comparison with other investigations. Nevertheless what is vital at the moment is the examination of the glyphs on these pages in the light of their association with the patterning of the illustrations which so strongly imply their part in the astronomical-calendrical day counts.

In analyzing the vertical column of glyph compounds at the left-hand edge of Panel B, the order of the discussion will begin with the glyph at the bottom of the column and proceed upwards to the top of the panel. Glyph numbers cited with the prefix "T" refer to the Thompson (1962) catalog series.

Glyph Compound 1: A vase with "u" (T-1) prefix below the lip, resting on a stand. It contains three kan (or maize) glyphs (Thomson 1971, p. 75), surmounted on the left by what may be an ear of corn (Thompson 1962, p. 256), and, on the right, by the glyph associated with God B (Schellhas 1904, p. 16), also defined by Thompson as Chac, the long-nosed rain god (Thompson 1972, p. 32). The authorities seem to agree that the compound represents offerings of maize to God B.

Glyph Compound 2: Some type of container (?) holding two kan glyphs. Again the context suggests offerings of maize.

Glyph Compound 3: Picture of spotted animal with legs tethered by a rope. It is hard to be certain as to whether a spotted dog or a jaguar is intended; however, the rounded ears and curved fang would seem to identify it as a jaguar. I have no explanation for its appearance here; many implications are possible.

Glyph Compound 4: What is portrayed are two kan glyphs and a turkey head resting on a glyph that may have the connotation of "offering" or "reverence" (T-103, see Thompson 1972, p. 151); it is identical to the affix usually shown postfixed to the glyph of God B. Thompson (1972, p. 105) interprets the turkey head emerging from the maize sign as "turkey tamales" or "turkey and maize broth (kol); presumably the meaning would be "offerings of turkey tamales."

Glyph Compound 5: This may be a very important compound for the meaning of the introductory page, but there appears to be no satisfactory explanation for it in the literature. The key element is a small circle, outlined by dots, shown in what may be a tripod bowl or stand which rests on a rectangular platform (or glyphic element). An undulating line forms the left edge of the compound above the platform. The compound is enigmatic, but logic suggests that the only information lacking thus far on this page, which is critical to understanding the text to follow, is that the day count is calculated for evening observations at end of twilight when stars first appear. There must have been a word for such a concept in Maya; numerous word compounds denote the time of day, including sunset, after sunset, dawn, etc., according to the position of the sun, utilizing modifiers with the word kin. Likewise one finds phrases such as "to become dark, twilight, very dark, etc." using the word akab with modifiers. Neither the glyph for kin nor akab is present in this compound; however, the word ek, meaning "black" or "star" is also frequently used to convey ideas such as "to become dark, darkness, blackness, etc." Two such examples are ek nacat, defined as "nightfall, twilight" (Motul Dictionary), and ekzamental, meaning "to become twilight or getting dark" (Perez Dictionary). However, the most interesting possibility is the word ekbizental, "to become dark," or ekbizenil, "the darkness of night" (both from the Perez Dictionary). The word is listed in the Motul Dictionary as ekbicen, translated as "something dark." In both dictionaries the word-stem biz relates to the meaning of "worm-hole ridden, or something full of tiny holes." Conceivably, then, the word bizen when combined with ek could supply the basis for the image of a circle enclosed by black dots, to convey the idea of "dotted with stars" (i.e., the night sky). Note that Chac on this page is also spotted, perhaps to relate him to the starry sky.

  Extending the above interpretation a bit further, the tripod bowl or stand containing the dotted circle in this compound could very well express the word oc, "stand for a bowl which serves as legs to support it in an upright manner" (Motul Dictionary). The prefix oc, however, can also signify "enter," as in ocnakin, ocbalkin, and okin, okini, oknakini, all translated as "nighttime, after sunset, entrance of the night" (Perez Dictionary). Compare ochquin, "sunset" (literally "enter sun") and u yochib quin, "west" (literally "the sun's entrance") from the Moran Dictionary (Thompson 1938, p. 600).

  Thus it is possible that the potstand is used to give the phonetic rendering of "oc," with the whole glyph compound representing the idea of "enter (or beginning of) star-dotted (sky)" or "starry night". In other words, it tells the reader that the time in question is the end of astronomical twilight. Other occurrences of the circle within dots in the text do not seem inconsistent with the interpretation, as will be seen as the analysis proceeds. (Also see p.338 for more discussion of the glyph.)

Panel A

I will not include Panel A in the examination of page XII as there is no certainty that it is related to Panel B. It does not involve a complete 260-day count as the other A panels do; instead, it totals a count of about 30 days (exact number is unknown since the bar-and-dot numeral at the upper right-hand corner is effaced). Such a count seems to bear little relevance to the introductory character of Panel B, so it seems best to omit Panel A as part of the whole text that is being dealt with here, and consider Panel B as the real starting point of the text involved in the ensuing pages.

Page XIII (Fig. 5a)

Panel B

The date illustrated is 21 Dec. at 7:20 p.m. In astronomical terms, the hour circle separated by 19° from the autumnal equinoctial colure is in the vertical plane of the observer's terrestrial meridian (see Fig. 6). The kin sign between black wings is to give the information that it is the winter solstice, the nights longer than the days. The serpent is shown facing to the left; he wears rattles on his tail. In Colonial and modern Yucatec Maya the word tzab means both "rattlesnake rattles" and the Pleiades constellation (Redfield and Villa Rojas 1967, p. 206; Motul and Pio Perez Dictionaries). Fig. 6 shows that Draco is to the west (therefore the serpent looks to the left); the Pleiades, indicated by the star Eta Tauri, are approaching the meridian and are very conspicuous in the sky. It can be seen that the Pleiades could easily have been considered to hang at the end of the serpent's tail by imagining the

5 *(a,b) Pages XIII (a, left) and XIV (b, right) of the Madrid Codex.*

dotted line of the serpent to continue in a straight line until it connects with Eta Tauri.

Chac, the rain god, is shown at the upper right-hand corner in what seems to be the act of holding an inverted pot (?). In the lower left-hand corner he is shown again, as though walking off the page. He does not reappear in the series until page XVII, the summer solstice and beginning once more of the rainy season.

It is not clear just what the double-ended cross at the upper center of Panel B may represent, and suggestions at this point would be no more than speculation. The little figure at the upper left-hand corner appears to be more human than god-like (a calendar priest?). The same figure, whoever he is, appears once more in the text on page XVII, the summer solstice, analogous but diametrically opposed to the date of winter solstice. On p. XVII he is shown in a pose in which he may be explaining or demonstrating the glyph compound to which he is pointing (see p.324 for explanation). Notice that the circle enclosed in dots appears at the back of his head in both cases. It is not inconceivable that the sign may associate him with the starry night, in order to designate him as a calendar priest or astronomer.

Augural Band (proceeding from left to right): ight):

Glyph 1: Partially effaced, but T-130, Thompson's bil affix is recognizable. According to Thompson (1972, p. 150), the affix carries the meaning of vegetative growth, functioning as a participle suffix.

Glyph 2: Compound with kan glyph signifying offerings of maize (see p. 17), with the possessive u (T-1) prefix, translated by Thompson as "his," "its," etc. (1972, p. 151).

Glyph 3: T-122 and 564 with u prefix. The glyph T-564 is defined as kak, "fire" (Thompson 1962, p. 186), and T-122 as "flame" (1972, p. 151) or "brilliance" (1971, Fig. 43, no. 70). The idea of the heat of the brilliant sun would fit the context, since it is the time of general subsidence of the rains and the onset of the dry season, perhaps heralded by the arrival of the winter solstice.

Glyph 4: Name glyph of God B, the rain god or Chac, illustrated twice in the picture below.

Glyph 5: Crossed bands are shown on a black background. I have argued elsewhere (1971) that earlier in Mesoamerica the crossed bands were employed to represent the constellation Cygnus, or more specifically, the meridian midnight transit of Gamma Cygni which in ancient times signaled the summer solstice. There are four occurrences of crossed bands in these pages of the Madrix Codex, and in every instance they continue to support the interpretation. Here the black background can be interpreted as meaning west, since black was customarily associated with this cardinal point. Cygnus is in the west at the date and time represented by this page (Fig. 6). Other meanings suggested by authorities on Maya epigraphy do not seem to contribute any other significant meaning for the appearance of crossed bands at this particular place in the text, although Thompson's interpretation of it could be argued to fit in a general way. He defines it as kat (T-552, 1972, p. 150) meaning "across, astronomical or carnal conjunction". In this light it could be suggested that the glyph shows a relationship between God B and God C (to either side of it in the augural band), but the illustration below conveys no information about it.

Glyph 6: Name glyph of God C.

Panel A

The text in Panel A is too effaced to enable one to read the count; however, the patterning is such that it appears to be the same as the succeeding pages in having a vertical column of day glyphs to the left-hand side of the page, and a count totaling 52 days in bar-and-dot notation in a horizontal arrangement across the panel. The count will be explained below. Clearly identifiable at the center of the panel is an upended turtle, its head just above the connecting border with Panel B. In Colonial and modern Yucatec Maya, the word ac refers to both turtle and the constellation Gemini (Redfield and Villa Rojas 1967, p. 206; Motul Dictionary). Fig. 6 shows that Gemini is to the east, and it might be possible to interpret the picture here as something to the effect that "Gemini is entering the picture" (i.e., rising in the east and becoming visible). The subject will be dealt with again in discussing p. XVII of the Codex where at the time of the summer solstice, Gemini leaves the picture. To the left of the turtle is a dog; there is no further information in this text to support what meaning it may convey. Its position next to the turtle may indicate that it also represents a constellation; other meanings are just as likely.

Plate XIV (Fig. 5b)

Panel B

The date is 4 Feb. at 7:20 p.m.; the hour circle marked "Right Ascension 64°" is in the vertical plane of the observer's terrestrial meridian. Eta Draconis is at lower meridian transit, below the northern horizon (see Fig. 7). Since the date of p. XIII, a total of 45 days have elapsed. The serpent is shown in the illustration still wearing rattles on his tail; the Pleiades continue to be conspicuous in the sky, having crossed the meridian and now lying just west of it. It is the last appearance of rattles on the serpent, since the tail is lacking them at the lower right edge of the page which adjoins, p. XV. By the next date the Pleiades will be well beyond the meridian.

Above the serpent is depicted a deer, and at its side a death figure bears a torch. It is not clear to me just what the deer is meant to convey. A number of meanings could fit, but there is insufficient evidence to give adequate support to any of them. One possibility could be that in Maya the phrase "death of deer" is a metaphor for the word "drought" (Thompson 1971, p. 270). The subject may be the dry season but such an idea can only be tentative. The meaning of the inverted vase (?), shown just at the inside edge of the serpent tail, also remains enigmatic. Interesting to observe in the course of these pages of the Codex is that the decorative motifs above the eye of the serpent vary with each picture, as well as the eye itself.

Augural Band

Glyph 1: This compound is based on two main elements: first, the sign for zac (T-58), meaning white, and the akbal glyph (T-504), also defined as akab, meaning night or darkness. Here the akab is encircled in dots and rests on a rectangular motif. Literally, the compound reads "zac akab" or "white dark". Thompson (1972, p. 68) accepts Barthel's interpretation of it as "Lord of the Dawn". I suggest that the glyph zac may stand for north, as white is the color associated with it, and that the compound in some way conveys the idea that Eta Draconis (perhaps as measurer of the night) lies exactly due north. Such an interpretation is supported by the meaning of the glyphs themselves (the dotted outline on akab may relate to it to the starry sky), and fits the context as well. The name "Lord of the Dawn" need not be eliminated as a secondary association, but it is extremely important that Eta Draconis is beyond the North Pole at the moment, below the horizon, at lower meridian transit. Such is the subject of the page and the entire day-count scheme rests on this timely event. It would be logical for such a statement to be present at the beginning of the augural band on this page.

Glyph 2: Name glyph of God K? (Schellhas 1904, p. 32). Zimmermann (1956, p. 74) identifies it as such and calls it a bird head.

Glyph 3: Death (augury).

Glyph 4: Glyph for red (chac) prefixed to jaguar head. Thompson (1965-70, no. 4) identifies the compound as chac bolay, "red" or "great fierce beast". I have suggested in a previous publication (Hatch 1971) that the jaguar was associated earlier in Mesoamerica with the constellation Ursa Major, more specifically, with Alpha, Beta, Gamma, and Delta Ursae Majoris, the "bowl" of the "Big Dipper". It is possible that in the compound here the glyph for red represents "east" (being the color linked with that cardinal point), and that the translation is "jaguar (or Ursa Major) is east," which is true for this date and time (see Fig. 7). Should this be correct, it becomes the third instance, in my analysis, that a color associates a cardinal point with a constellation in the augural bands.

Glyph 5: The glyphic element shown appears to be the prefix to the first glyph in the augural band of page XV, and will be discussed in that context.

Panel A (See Table 3 for Schematic Diagram)

Table 3 demonstrates how the complete round of 260 days

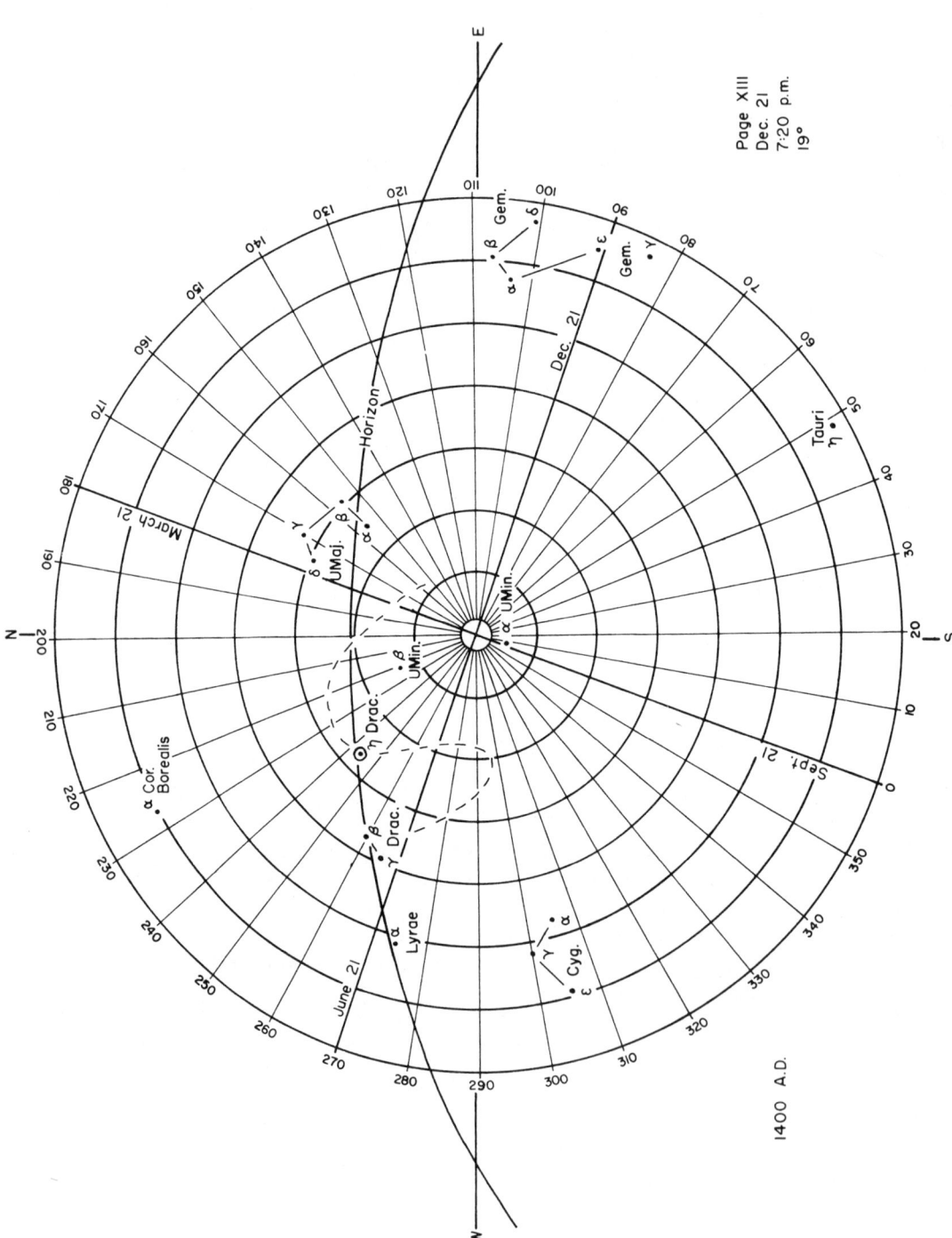

6 *December 21 at 7:20 p.m., the date represented by page XIII of the Madrid Codex; the hour circle with*

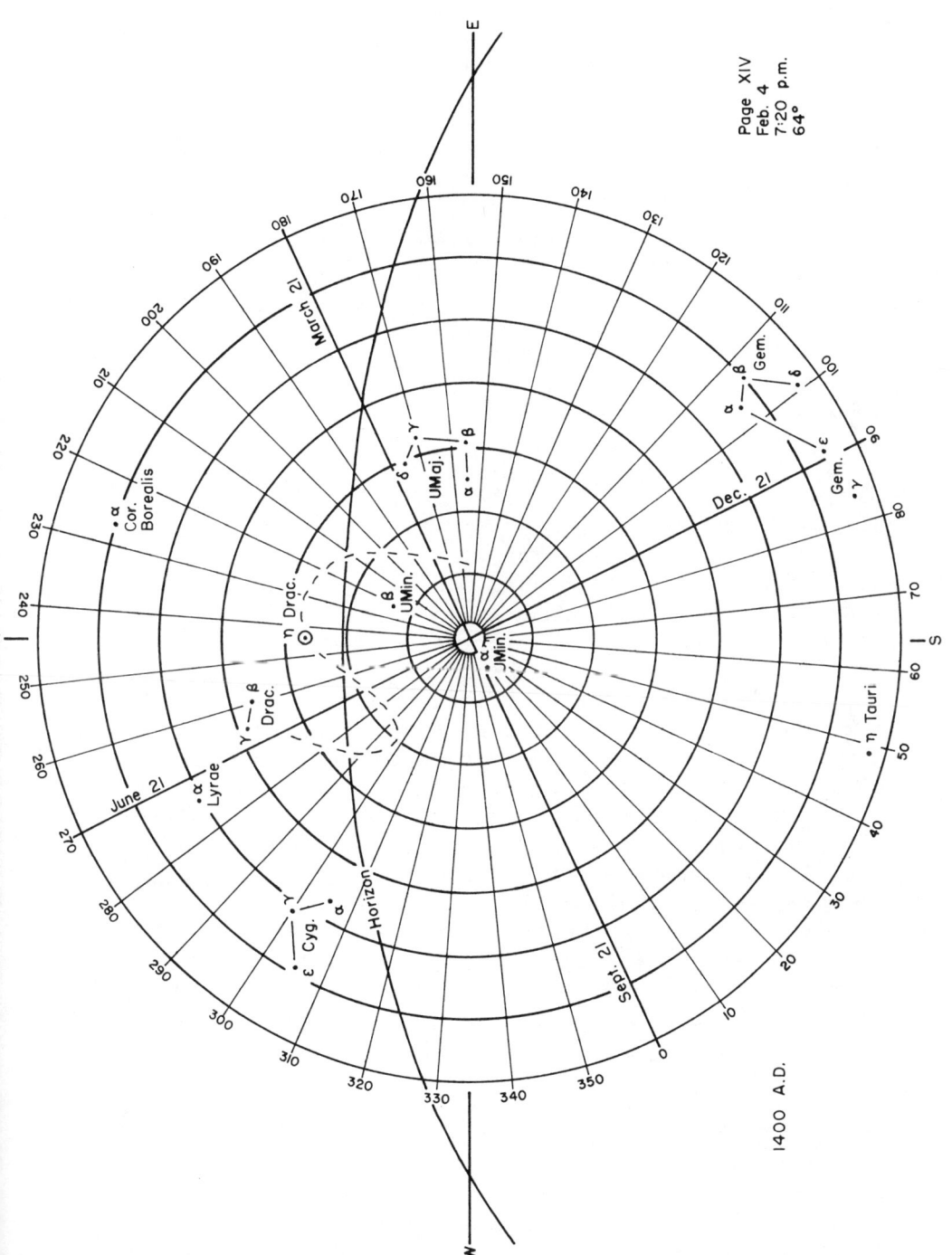

7 *February 4 at 7:20 p.m., the date represented by page XIV of the Madrid Codex; the hour circle with right ascension = 64° is over the terrestrial meridian of the Yucatan observer.*

is to be counted. One starts with the first day sign at the top of the column at the left-hand side of the page with the number appearing above the column (thus beginning with 4 Ahau). Then one counts forward by the amount of the next bar-and-dot numeral in black (here the moon sign stands for 20, plus 2 gives 22) to reach a day and number in red (giving 13 Ik, day name is suppressed); counting forward by the next numeral (13) one reaches 13 Men; counting forward again by the next numeral 17 (mistakenly written in the Codex as 18), one reaches the day 4 Eb. Since this is the second day glyph in the left-hand column, one again proceeds in the same manner to count off the days to reach 4 Kan, the third glyph in the column, etc. Upon reaching the day Lamat at the bottom of the column, the next count of the days will reach 4 Ahau once again. The day count across the panel totals 52 days, and multiplying this by 5 (the number of days in the column) gives the complete 260-day count of the Sacred Round.

The three seated figures illustrated can be identified (from left to right) as God D, the Death God, and the Maize God. Each holds an upraised hand before a design which may symbolize the sky. The notion that it is meant to be the sky is based on the fact that to the left side of the design are three cauac glyphs, placed one above the other, as though to represent rain (the connotation of the cauac glyph according to Thompson 1971, p. 87). Since they appear to the side of, rather than within the opening of the crescent which outlines the sky, the implication would seem to be that the rains are not forthcoming. It is the onset of the dry season at this time of year.

Glyph 1: U prefix with glyph T-96 defined as yol (Thompson 1972, p. 151), meaning "in the heart of". Knorosov (1967, p. 87) gives it the value "ich," though also interpreted as "in". The head below this glyphic element is partially effaced, and identification is uncertain.

Glyph 2: Thompson (1972, p. 84) defines this glyph as kak (T-563), meaning "fire," and the two kak glyphs plus yol in Glyph 1 he reads as yol kak kak, "in the heart of fire". What additional information is supplied by the head in Glyph 1 is unknown, but the phrase could refer to the heat of the dry season (or burning milpas?).

Table 3
Page XIV, Panel A (Madrid Codex)

	Glyph 1	Glyph 2	Glyph 5	Glyph 6	Glyph 9	Glyph 10
4						
Ahau	Glyph 3	Glyph 4	Glyph 7	Glyph 8	Glyph 11	Glyph 12
Eb	(22)	Day 13	(13)	Day 13	(18) (Should be 17)	Day 4
Kan						
Cib	Picture		Picture		Picture	
Lamat						

Glyph 3: Name glyph of God C.

Glyph 4: <u>Tun</u> with <u>il</u> (T-24) postfix and vertical bar (for the numeral 5) to the left. The significance of "5 tunil" is not known to me. One would expect here some reference to God D who is seated directly below the compound.

Glyph 5: Same as Glyph 1 except for the head which, though badly effaced, appears to be different.

Glyph 6: Same as Glyph 2.

Glyph 7: Name glyph of God C.

Glyph 8: Name glyph of Death God, seated below.

Glyph 9: Same as Glyph 1, with head signifying death.

Glyph 10: Same as Glyph 2.

Glyph 11: Name glyph of God C.

Glyph 12: Name glyph of the Maize God (Schellhas 1904 p. 24).

Page XV (Fig. 8a)

Panel B

The date is 22 Mar. at 7:20 p.m.; the hour circle whose right ascension is 109° is over the terrestrial meridian of the Yucatan observer. The serpent has turned so that he now faces to the right of the page; Beta and Gamma Draconis (the head of the serpent) have already made their lower meridian transit, and Eta Draconis is rising in the east (see Fig. 9). The posture contrasts to previous pages where Draco was west of the meridian and the drawings showed him facing to the left. Because we conceptualize in terms of circles and spheres, we think of circumpolar stars as moving in an unchanging counter-clockwise motion; here the drawing demonstrates a slightly different construct in terms of a flat earth and straight lines. In this framework a constellation is seen to be traveling in one direction as it crosses the sky above the north celestial pole, but then it must reverse its route to complete the return trip below the pole back to its original starting position. Thus in the months preceding 22 March at 7:20 p.m., the constellation was visualized as being in a

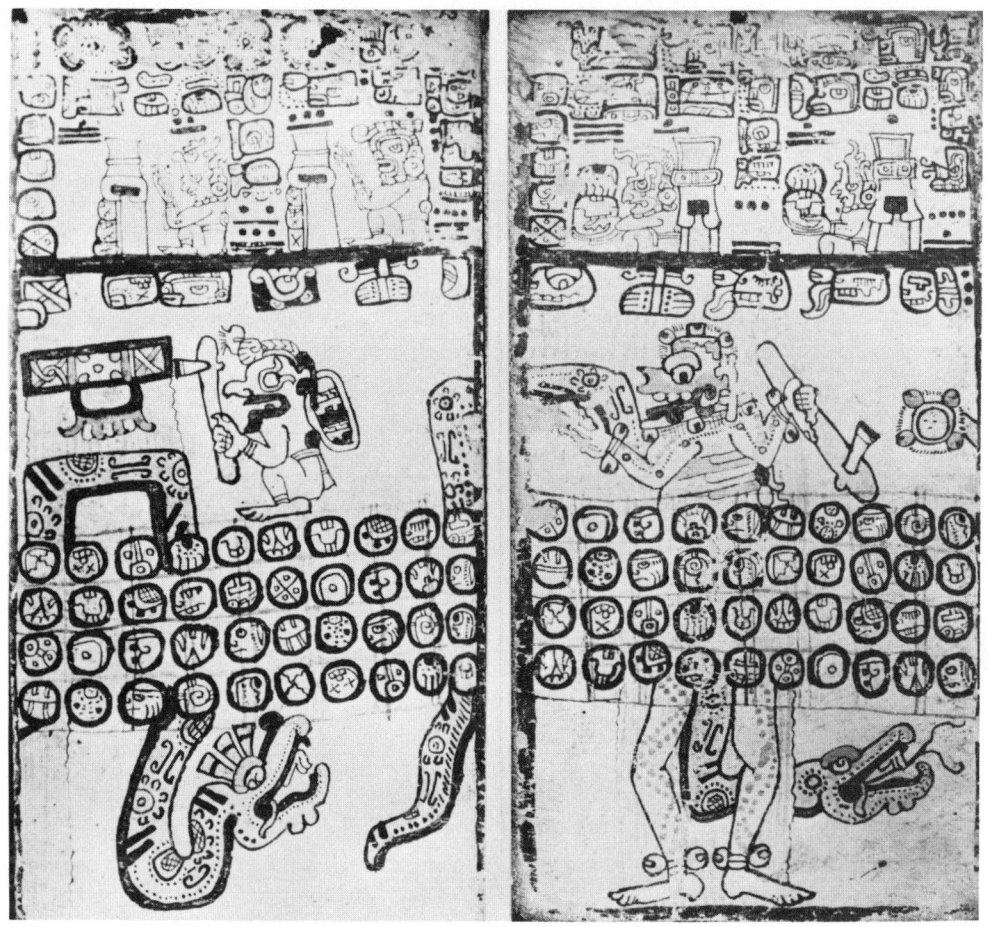

8 (a,b) Pages XV (a, left) and XVI (b, right) of the Madrid Codex.

position in its journey in which it was heading from west to east as it made its lower meridian transit. However, as Draco now rises above the horizon, its position shifts to a direction heading from east to west in its circular path going toward upper meridian transit. Thus the head and tail are now shown reversed from the earlier view in order to illustrate the difference.

    A figure wielding an axe (designated as God M by Schellhas 1904, p. 35) crouches before a panel composed of three glyphs. The patterning in this panel seems to fit a scheme of a division of the 365-day year into three equal parts of 122 days. The Kan cross at the center can be defined as representing the solar year (as on p. XII), and the crossed bands as the constellation Cygnus. The first glyph in the panel is unidentified but it resembles very closely the second glyph of the panel on p. XII which was proposed as standing for the context of the

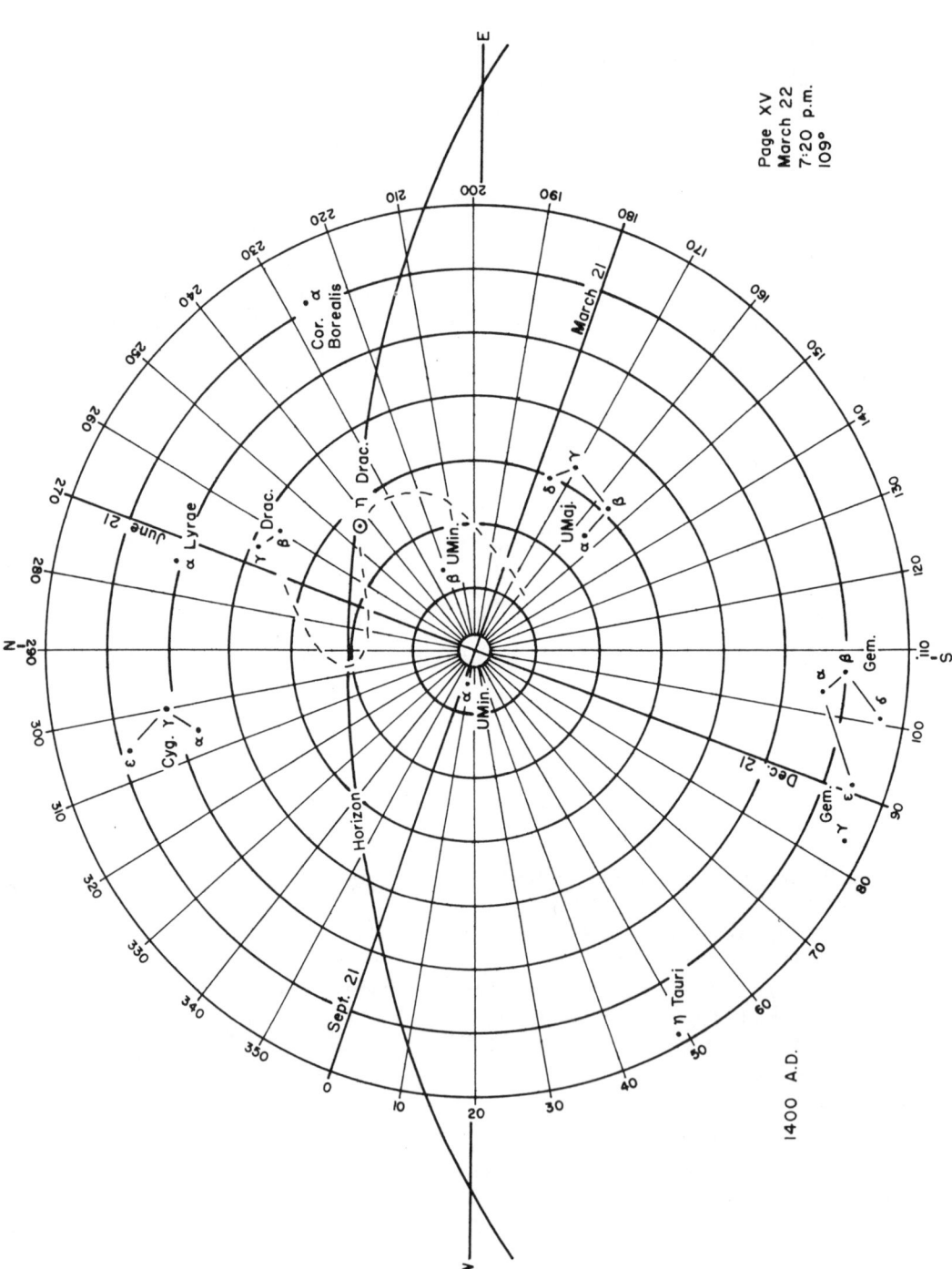

9 March 22 at 7:20 p.m., the date represented by page XV of the Madrid Codex; the hour circle with right ascension 109° is over the terrestrial meridian of the Yucatan observer.

sidereal year, and the interpretation also fits in this case. The scheme is based on starting the count of the solar year at 22 Mar., the vernal equinox (a logical time because the sun sets due west and rises due east for one day only during a time of clear skies). The date 22 March would then stand exactly midway between the start of the count of the sidereal year, counted from 21 Nov., and the midnight meridian transit of Gamma Cygni, thus:

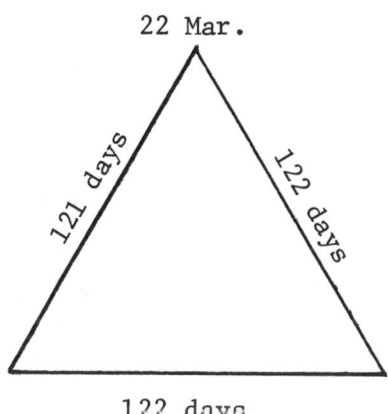

21 Nov.   122 days   22 July

The midnight upper meridian transit of Gamma Cygni ca. A.D. 1400 occurred exactly on the night of 22 July (see Table 1) and it is this detail which lends strongest support to the late date I have proposed for the Madrid Codex. The right ascension of Gamma Cygni at A.D. 1300 was $299°.28$ (21 July) and for 1500 it was $301°.07$ (23 July). Thus, if my analysis has any basis in fact, the date of the Madrid Codex should lie between A.D. 1350 and 1450. However, it is possible that if the Codex were composed at A.D. 1200, the authors may simply have been projecting ahead, or making a statement that was more or less correct. Nevertheless, based on my own research it seems to me that when the Maya made statements about counts, they sought to make them as precise as possible, and that when they inform the reader that an event is occurring, the data will show that they were absolutely correct about it.

Returning once more to the subject of the preceding diagram, every four years the leap day could have been noted appropriately at the date of vernal equinox (although intercalary days were not actually inserted into the Maya calendar), so that the count between 21 Nov. and 22 Mar. would total 122 days instead of the 121 as shown, and the three dates would thus exemplify uniform lengths of time. Since the intervals are equal to each other only according to the dates as listed, the scheme must use a date of 21 Nov. instead of 22 Nov., the traditional start of the sidereal year count based on the midnight meridian transit of Eta Draconis. However, the discrepancy may be more apparent than real, for if the sidereal year were counted in terms of elapsed time from midnight to midnight, the calendar on three pages would have to be properly adjusted for counts based on observations at the end of twilight. Thus 21 Nov. would be correct (22 Nov. entering at midnight). The evening of 22 July would be the day following the observation at midnight of the meridian transit of Gamma Cygni (it would be the evening previous to the transit if the date of the Codex is as late as A.D. 1500).

At this point I would like to digress for a moment and return to the glyph panel on p. XII, in light of the above discussion. It will be recalled that the glyphs were analyzed in the following manner: Glyph 1 as a variant of day or sun; Glyph 2 as a count of the sidereal year; Glyph 3 as a count of the solar year, and Glyph 4 as night (connecting to the serpent body). The interpretation can now be carried further in terms of day counts not very different from those in the glyph panel on p. XV. It was pointed out that each pair of day-night glyphs in the panel on page XII was placed below a *kin* sign between black and white wings (for September equinox, date of the introductory page). The day counts could be structured as follows (the arrangement here duplicating the order of the glyph panel below the *kin* signs):

September equinox (Kin sign)		March equinox (Kin sign)	
Glyph 1	Glyph 2	Glyph 3	Glyph 4
(Sept. 21)	(Nov. 21)	(Mar. 22)	(May 22)

Between September 21 and November 21 there is a total of 61 days (half of 122); from March 22 to May 22 there is

Astronomical Calendar

also a total of 61. Such a scheme would relate the solar count to the sidereal one in a way that is symmetrical (basically a division of the 365-day year into even thirds). In both cases the glyph panels involve dates which are pertinent to the traditional times of midnight meridian transits of Eta Draconis, but adjusted for evening counts (e.g., November 21 instead of 22). For dates of observation of transits at evening (7:20 p.m. ca. A.D. 1400) the counts become part of a division of 365 days into fourths (91 days). So although the body of this text is in terms of the 91-day count adapted for evening meridian transits and the solstice and equinox, the glyph panels on both pp. XII and XV refer the reader to the 122-day structure based on midnight transits and the equinox.

Augural Band

Glyph 1: The crossed bands is shown with prefix appearing at the end of the augural band on p. XIV. The prefix is very similar to Zimmerman's (1956) Glyph 73 (T-62) except that the lower loop below the circle (identical to the loop shown above it) is missing. Barthel (1954, p. 46) points out that this prefix accompanies pictures in the Dresden Codex which represent an activity in the process of being performed, rather than the completed act. If the crossed bands here stand for the constellation Cygnus (as it seems to in the glyph panel of the picture below it), the compound may convey the idea that Cygnus is heading toward the meridian (Gamma Cygni is within $10°$ of it, see Fig. 9), not yet having completed its lower culmination.

Glyph 2: The day sign Cauac with head of God C. I would like to interpret this compound as "rains from the north". It was demonstrated that in Panel A of p. XIV (and here as well in Panel A) the Cauac sign seems to be used in the context of rain from the sky. Thus it is consistent that it may in the augural band also refer to rain, and the head of God C, used as the main part of the glyph for North, associates it with that direction. During the dry season in Yucatan, it is possible to have rains from the north, known as "northers," and such information would be useful to a calendar priest.

Glyph 3: Glyph associated with God M (Schellhas 1904, p. 35), shown in the picture below with an axe, and bearing a burden on his back.

Glyph 4: Glyph T-563, identified by Thompson as "kak" meaning "fire," with affix T-122 which he interprets as "flame" (1972, p. 100) or as "brilliance" (1971, Fig. 49, nos. 69, 70). The meaning of brilliant sun/dry season heat would fit here.

Glyph 5: Prefix T-24 (il) belongs to the first glyph of the augural band on p. XVI, and will be discussed in that context.

Panel A (See Table 4 for Schematic Diagram)

The 260 days are counted in the same manner as on p. XV, starting with day 4 Ahau; the pictures are also similar. On the left the Maize God sits in front of what I have suggested to be the rainless sky; on the right the scene is duplicated with the figure of God C as the actor.

Glyph 1: The central element in the compound is the circle surrounded by dots (already examined on p.299 and interpreted as meaning night, end of twilight, or more literally, star-dotted sky); it is flanked by four glyphs (T-25), originally designated by Thompson (1971, p. 38) as the xoc or "count" bracket, and more recently (1972, p. 37) as having the sound value "ca". The interpretation of "count" is applicable here because the vernal equinox might be considered the completion of four counts of 91 days, or fourth station in the path of the sun, correlated with the course of the night-sky or sidereal year. At this time the Maya astronomer could take note of the leap day and thus keep track of the accumulation of discrepancies between the traditional calendar date and that which they should have reached, a situation similar to the Julian calendar in the Middle Ages. The reference here to counts is consistent with the text of Panel B which was preoccupied with the same subject, but which presented the year divided into three equal parts (instead of four) based on the vernal equinox.

Glyph 2: Effaced, but probably the same as Glyphs 6, 10, and 14, as the same pair seems to introduce each

Table 4
Page XV, Panel A (Madrid Codex)

4	Glyph 1	Glyph 2	Glyph 5	Glyph 9	Glyph 10	Glyph 13
Ahau	Glyph 3	Glyph 4	Glyph 6	Glyph 11	Glyph 12	Glyph 14
Eb	(16)	Day 7		(8)	Day 2	
Kan			Glyph 7			Glyph 15
Cib			Glyph 8			Glyph 16
	Picture			Picture		
Lamat			Day 7			Day 4
			(13)			(15)

phrase. If so, the reading would be "sacrifice".

Glyph 3: Name glyph of the Maize God.

Glyph 4: The several elements together are interpreted as "abundance of maize" (Thompson 1972, p. 32). The main elements are Kan (maize) and Imix (day sign used in the context of "new" or "beginning"). The phrase may read something like "End (or start) of the four counts; sacrifice to the Maize God for abundance of maize(in the coming year)."

Glyph 5: Same as Glyph 1.

Glyph 6: Glyph T-568; accepted by Thompson, Beyer, Lizardi, and Barthel (Thompson 1962, p. 194) as symbol of sacrifice.

Glyph 7: Name of glyph of the Death God.

Glyph 8: Death (augury).

Glyph 9: Same as Glyph 1.

Glyph 10: T-568, sacrifice.

Glyph 11: Name glyph of God C.

Glyph 12: T-567 and 130, augury for good tidings (Thompson 1962, p. 191).

Glyph 13: Effaced, probably the same as Glyph 1.

Glyph 14: T-568, sacrifice.

Glyph 15: Numeral 4 with head of jaguar (?), suggested by Thompson (1965-70, no. 4) to mean Can Bolay (fierce beast of prey), and interchangeable with Chac Bolay.

Glyph 16: T-648 and 25, interpreted as kaz (Thompson 1962, p. 255), meaning "misery" or "evil" (augury). He also associates the jaguar with warfare, and such an interpretation may fit in light of the possibility that glyph compounds 15 and 16 in this panel indicate threatening auguries as a time of war. Conceivably, the dry season would have been most propitious for waging warfare, since it was perhaps, a time of hunger and raiding, and the omens forecast evil and misery. One augury, however (Glyph 12), predicts good fortune and may be balanced against the negative influences.

Page XVI (Fig. 8b)

Panel B

The date is 6 May, 7:20 p.m.; the hour circle whose right ascension is 154° is in the vertical plane of the local terrestrial meridian. The picture shows a death figure chopping the serpent in half; at this moment Eta Draconis has reached its eastern elongation, half-way between its lower and upper transit (see Fig. 10). The concept may be that the serpent has reached the halfway point in his journey about the north celestial pole; his head will be visible in the evening sky from now until November, the entire course of the rainy season. The Pleiades have set and are no longer visible; note the statement in Glyph 1 of Panel A involving the symbol for snake rattles. I suggest that the head of the serpent may be associated with the rainy part of the year, his tail with the dry season.

A close look at the neck of the serpent will show that the symbol of the circle enclosed in dots appears for the first and only time in the series just behind his jaw and again where his body disappears behind the glyph band composing the 260-day count. Fig. 10 reveals that Beta and Gamma Draconis have just risen above the horizon and are visible in the evening sky, and the dotted circle may demonstrate that the serpent head is present now among the observable stars in the sky. On 21 Dec. his head was just setting at 7:20 p.m., and since then has not been in sight at this hour.

Another fact which must be important, although there is insufficient evidence at this point to explain it, is that the serpent head is stripped of his eyebrow paraphernalia, and his forked tongue issues from his mouth (not illustrated earlier). Above him is a shield, held by Chac on p. XVII. Fig. 10 shows that the "bowl" of the Big Dipper is at the meridian, a conspicuous celestial event which may have been significant. I offer a very tentative suggestion to explain the relationships. I have elsewhere proposed (Hatch 1971) that the "bowl" of the Big Dipper was associated with the jaguar earlier in Mesoamerica. Thompson considers the jaguar to be a

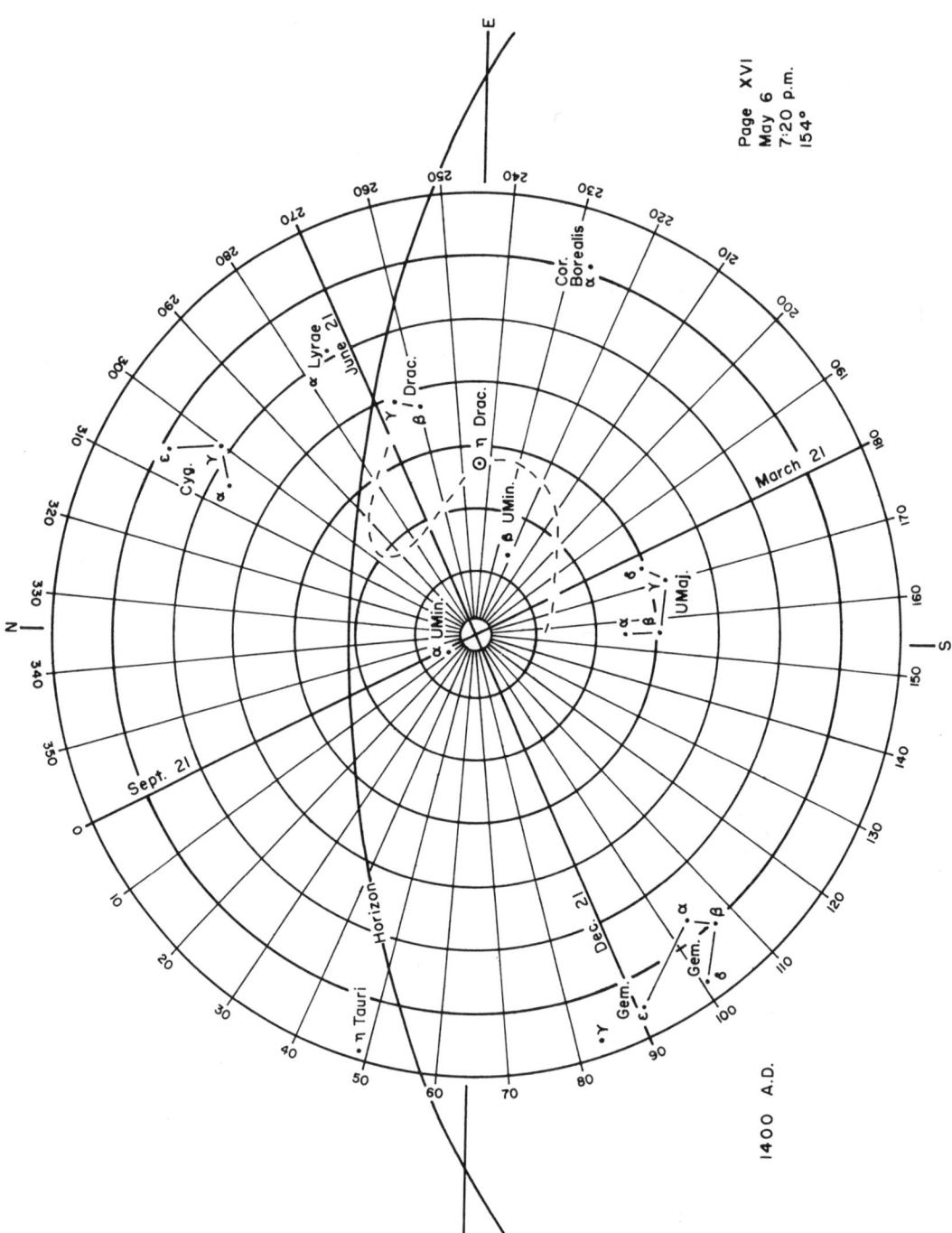

10 May 6 at 7:20 p.m., the date represented by page XVI of the Madrid Codex; the hour circle with right

symbol of war in Maya iconography; thus the shield in
this illustration could be associated with Ursa Major in
its connection with warfare and the jaguar. Though it
is impossible to substantiate the connection at the moment, there may have been some idea of a mythological
contest between the two constellations, or perhaps the
meridian transit of Ursa Major was considered influential in some way. This idea is not far removed from the
theory I proposed suggesting that the event had been
important at an earlier date in Mesoamerica for its
timely calendrical associations. At the present, the
meaning here remains inconclusive.

Augural Band

   Glyph 1: With the il (T-24) prefix on p. XV, the
compound forms the name glyph of the Death God.
   Glyph 2: T-122 and 563, fire and brilliance compound
(see discussion of Glyph 4 in the augural band on p. XV),
again possibly a reference to the heat of the dry season
which is still in session.
   Glyph 3: Death (augury).
   Glyph 4: Death (augury), probably repeated here for
emphasis.
   Glyph 5: Name glyph of God B, or Chac, the long-
nosed rain god. He will be present on the following
page and, presumably, the presence of his glyph here
at the end of the augural band is to signal his appearance and the oncoming rainy season. The rains begin
toward the middle of May; by the next page of the Codex
the rains are in full session. At the end of the augural
band one can see the three black dots which accompany
the augury for good tidings of the first glyph on p. XVII.

Panel A (See Table 5 for Schematic Diagram)

   The 260-day count is carried out in the same manner
as on pages XIV and XV, beginning with 4 Ahau. The picture shows, on the left, the figure of God D seated before a temple holding two glyphs, imix and kan. The

suggested meaning of these two glyphs together is "abundance of maize" (Thompson 1972, p. 33), though perhaps they could be read as "new maize". To the right, the scene is repeated with the Maize God.

Glyph 1: The main sign is effaced but T-207 and the il postfix are clearly visible. The glyph T-207 is defined by both Thompson and Kelley (Barthel 1969, p. 20) as tzab (meaning snake rattles in Yucatec Maya) and by Knorosov as tzub. Tozzer (1907, p. 41) also interprets the sign as snake rattles. It seems almost unmistakable that the glyph must relate to the illustration in Panel B where the tail of the serpent is being chopped off; Fig. 11 demonstrates that the Pleiades have just set.

Glyph 2: Name glyph of God C.

Glyph 3: Name glyph of God D, seated immediately below the glyph compound.

Glyph 4: T-62 and 507 with postfix, il favorable augury (Thompson 1972, p. 35).

Glyph 5: Effaced.

Glyph 6: Name glyph of God C.

Glyph 7: Name glyph of Death God.

Glyph 8: Death (augury).

Glyph 9: Partially effaced but rattlesnake rattles are distinguishable as the prefix, with the u postfix below.

Glyph 10: Name glyph of God C.

Glyph 11: The head of God C is clearly shown here with the il prefix. One would expect the head of the Maize God instead, whose name glyph customarily has the il prefix attached, and who sits immediately below the compound.

Glyph 12: Kan imix compound, signifying "abundance of maize".

Glyph 13: Unknown, and partially effaced.

Glyph 14: Name glyph of God C.

Glyph 15: Good tidings.

Glyph 16: Name glyph of God K.

Astronomical Calendar

Table 5

Page XVI, Panel A (Madrid Codex)

4 Ahau Eb	Glyph 1 Glyph 3 (15)	Glyph 2 Glyph 4 Day 6	Glyph 5 Glyph 6 Glyph 7 Glyph 8	Glyph 9 Glyph 11 (11)	Glyph 10 Glyph 12 Day 1	Glyph 13 Glyph 14 Glyph 15 Glyph 16
Cib Lamat		Picture	Day 3 (10)		Picture	Day 4 (should be 16)

Page XVII (Fig. 11a)

Panel B

    The date is 21 June, 7:20 p.m.; the hour circle numbered 199° is over the terrestrial meridian of the Yucatan observer. The picture shows the serpent head now in the upper part of Panel B, above the 260-day count glyph band instead of below it as on the other pages. Fig. 12 shows that Beta and Gamma Draconis, forming the head of the serpent, have completed their eastern elongation and are moving toward the meridian where they will be high in the sky (thus accounting for the placement of the serpent head at the top of the panel). On the dates of the preceding pages, Beta and Gamma Draconis were near the horizon, or below it. Chac is shown twice in the upper left half of the panel; the rainy season is well under way. The serpent wears a semi-circular motif behind his head at his neck, possibly representing the shell as a water symbol. The turtle is about to depart from the page; there is a lengthy text about this reptile in Panel A. Fig. 12 reveals that the constellation Gemini is in the process of setting into the horizon. The fact that the frog is shown

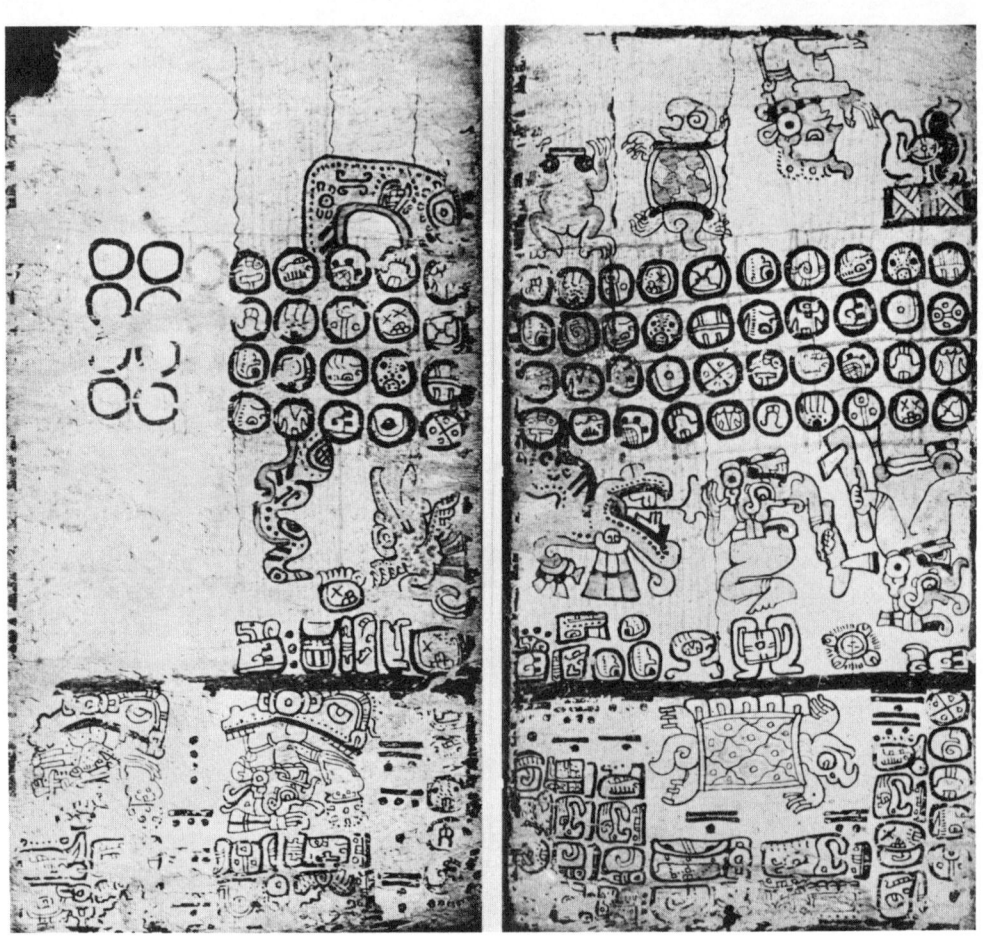

11 *(a,b) Pages XVII (a, left) and XVIII (b, right) of the Madrid Codex.*

exactly parallel to the turtle may indicate that it is a constellation near Gemini, but there is no further information by which the association can be substantiated.

The seated figure at the lower left-hand corner of the panel provides an interesting detail, for he seems to be the same personage that appears in the upper left-hand corner of p. XIII, the time of the winter solstice; it was suggested there that he may depict a calendar priest, in view of the fact that he appears to be a human rather than an anthropomorphic figure. He points to or gestures with his hand to the two crossed bands glyphs set above the kin sign between one black and one white wing. In the hypothesis I earlier presented (Hatch 1971), it was argued that the meridian transit of Gamma Cygni was an important signal of the summer solstice in the first millennium B.C. In the text now under discussion, it has been shown that the crossed bands glyph fits in every case with a reference to the path of the constellation. Here on p. XVII the intention of the scribe would seem to be to inform the reader that the midnight meridian transit of Gamma Cygni (centerpoint of Cygnus) is moving toward the equinox (symbolized by the sun between black and white wings) and away from the summer solstice. Table 1 demonstrates that Gamma Cygni made its midnight upper meridian transit on the night of 22 July ca. A.D. 1400.

It seems to me that the consistency of the fit of the crossed bands glyph with the path of Cygnus goes beyond coincidence. If the relationship has been correctly analyzed, it becomes apparent that the Maya, for whatever reason, were profoundly interested in this particular constellation (or star). Statements involving the crossed bands appear four times on the seven pages. I would like to bring to the reader's attention the possibility of another such reference in an unrelated Maya text. In the Addendum to Teeple's "Maya Astronomy" (1930, p. 115), he illustrates a glyph from Stela 3 at Santa Elena Poco Uinic, Chiapas, Mexico suggesting that it may be a solar eclipse glyph. The glyph (Fig. 13) is virtually identical to the compound on p. XVII of the Codex to which the squatting figure gestures. Both consist of two juxtaposed crossed band glyphs, above a kin sign; in Teeple's illustration, the two "wings" which appear to either side of the kin in the Codex version (to denote equinox) are lacking. Although the stela is badly eroded in parts, the Initial Series date 9.18.0.0.0 11 Ahau 18 Mac is clear; the dates on it have been analyzed by Blom, Palacios, and Thompson (summarized in Thompson 1944, p. 133), and all agree on the readings. The glyph in question immediately follows the date 5 Cib 14 Chen, which is connected to the Initial Series by the Distance Number 4.4, thus recording 9.17.19.13.16 (17 July, 790 according to the 584,283 correlation). Teeple relates the glyph (interpreted as "sun entering its house") to a total solar eclipse which was visible from the spot where the monument was afterward erected, citing the date of the solar eclipse as 16 July, A.D. 790. However, in this detail Teeple made an error. The eclipse to which he refers occurred on Julian Day Number

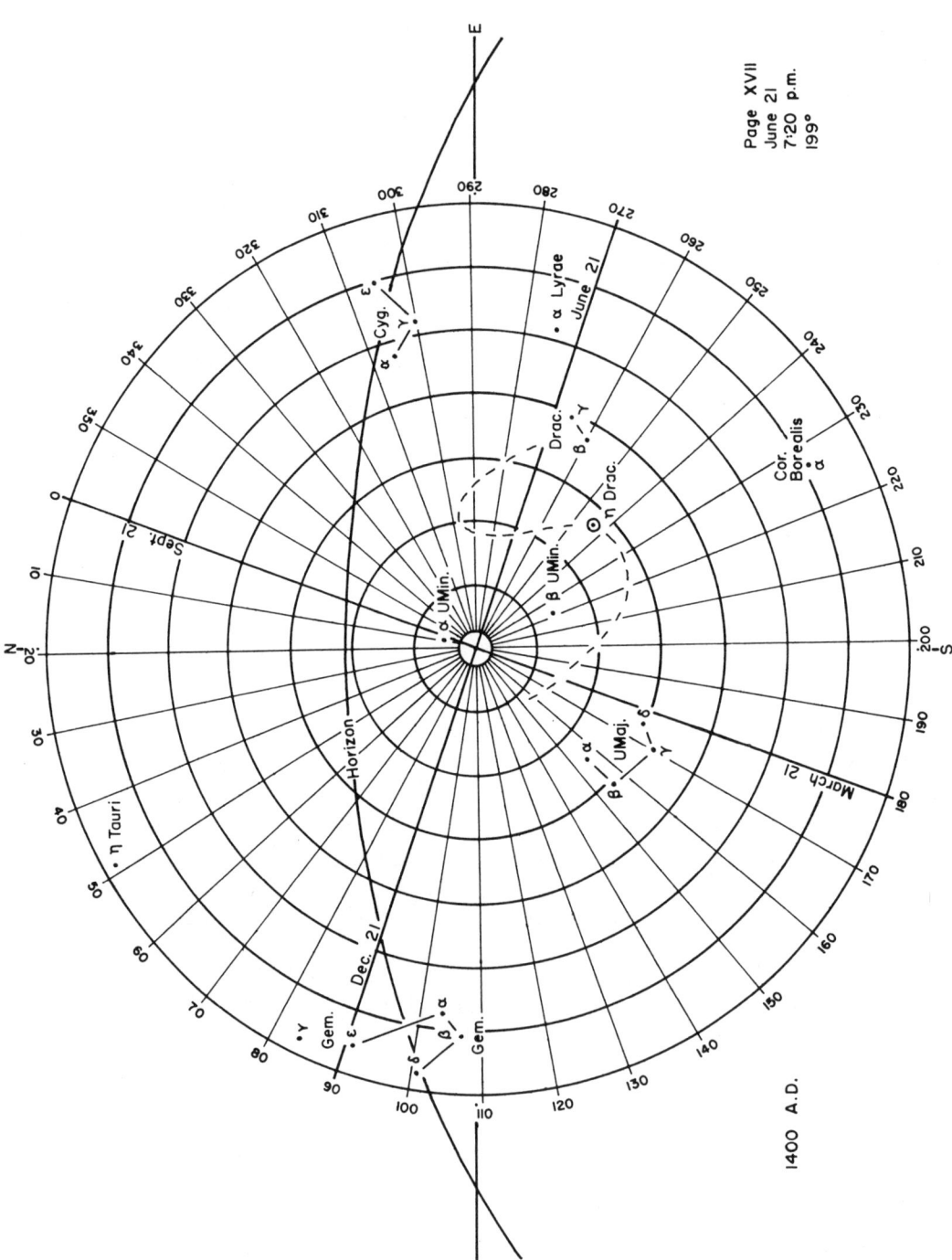

12 July 21 at 7:20 p.m., the date represented by page XVII of the Madrid Codex; the hour circle with right

Astronomical Calendar

13 *Drawing of glyph compound from Stela 3, Santa Elena Poco Uinic, Chiapas, Mexico. (After Teeple 1930, p. 115).*

2,009,802 (Oppolzer 1962, p. 192), equivalent to 16 July, A.D. 790 in the Julian Calendar, but to 20 July, A.D. 790 in the Gregorian. Teeple neglected to allow for the four day Gregorian correction.

The important point, though, is not the matter of the relationship of the glyph to the solar eclipse, but to the constellation Cygnus. The date of the midnight meridian transit of Gamma Cygni for A.D. 790 was 17 July (see Table 1), the date also associated with the glyph on the stela. Now obviously the Gregorian date given for Stela 3 is based on the assumption that the 584,283 correlation is the correct one. Nevertheless, I feel it is worthwhile to bring the matter into the discussion because of the likeness between the two glyphic renderings, and because if my analysis has any truth to it, it means that my theories, resting on astronomical data, and the 584,283 correlation, based on historical evidence, agree to the day.

Augural Band

Glyph 1: Combined with the three dots appearing on p. XVI, the compound conveys good tidings. Note that the good tidings augury appears twice in this band, and once on the following page, as opposed to death symbols which were shown in augural bands on previous pages in connection with the dry season. Similarly, the "brilliant fire" (T-122 and 563) glyph is no longer present.

Glyph 2: The shield is illustrated which was an important symbol on p. XVI. If it is true that it is connected with the constellation Ursa Major, its depiction here would be to indicate that it is still very conspicuous in the sky (within 30° of the meridian). However, it no longer seems to have an ominous relationship to the serpent; the latter once more has his eye decorations and nose scroll.

Glyph 3: <u>Kin</u> between white wings, to indicate that the day is now longer than the night, date of the June solstice.

Glyph 4: Glyph T-588b, meaning uncertain. Thompson finds that on the monuments it appears with the 819-day count (1962, p. 217). Kelley (1962, p. 27) shows the head with a postfix beneath it consisting of three dots flanked by a circle to either side, and gives the meaning (also accepted by Knorosov) as "offer, give, pay, or sacrifice". Thompson is cited as translating it as "it moves in order, it is set".

Glyphs 5, 6, 7: Three <u>caban</u> glyphs, which usually represent "earth".

Glyph 8: Name glyph of God B.

Glyph 9: Name glyph of God K, placed just below the glyph of God B.

Glyph 10: Good tidings.

Panel A (See Table 6 for Schematic Diagram)

The 260-day Sacred Round is counted in the same manner as on the previous pages, except that the count begins with 1 Ahau. The turtle is pictured and each phrase of four glyphs except the final one includes a reference

Table 6
Page XVII, Panel A (Madrid Codex)

	Glyph 1	Glyph 5	Glyph 6	Glyph 9	Glyph 13	Glyph 17
Ahau	Glyph 2	Glyph 7	Glyph 8	Glyph 10	Glyph 14	Glyph 18
Eb	Glyph 3	(12)	Day 11	Glyph 11	Glyph 15	Glyph 19
Kan	Glyph 4			Glyph 12	Glyph 16	Glyph 20
Cib	Day 12	Picture		Day 6	Day 6	Day 1
Lamat	(11)			(8)	(13)	(8)

to it. Also included in every phrase is the spiral (T-19) glyph, identified by Thompson (1958, p. 301) as koch, meaning "divine punishment," "borne on one's shoulders," "fate," or "disease". None of these definitions seems especially applicable to this panel, though perhaps "fate" or "borne on the shoulders" would be appropriate. The glyph seems very important here, yet does not appear elsewhere on the seven pages; one gets the impression that it must have a specific implication for the turtle and his exit from the scene. Knorosov's sound value "mu" for the glyph (1967, p. 89, also accepted by Kelley 1962, p. 41) would fit the Motul and Perez Dictionary definitions of the word-stem muc as "hidden, buried," and such a possibility seems worth consideration. Nevertheless, the matter must remain unsettled for the present.

Glyph 1: Name glyph of Death God.
Glyph 2: Spiral glyph prefixed to cauac glyph, a sign that without the prefix usually stands for rain, year, or season.
Glyph 3: U prefix, turtle head.
Glyph 4: Death (augury).
Glyph 5: Unknown.
Glyph 6: Spiral glyph as main sign, prefix effaced.
Glyph 7: U prefix, turtle head.
Glyph 8: T-62 and 507 with postfix il, favorable augury (Thompson 1972, p. 35).
Glyph 9: Can Bolay (?) (See discussion of Glyph 15 in Panel A of p. XV).
Glyph 10: Spiral glyph as main sign with ti (T-59) prefix.
Glyph 11: U prefix with turtle head.
Glyph 12: Death head. It lacks the prefix usually attached when employed as an augury.
Glyph 13: Partially effaced, name glyph of God C?
Glyph 14: Spiral glyph with ti (T-59) prefix. A bracket (u prefix?) stands before the compound; it lacks the interior elements which would define it as u.
Glyph 15: Same bracket as above, with turtle head.
Glyph 16: Same bracket with T-573 and 130, defined by Thompson as "hel" (1971, p. 161), denoting "change" or "succession".
Glyph 17: Effaced.
Glyph 18: Cauac glyph, meaning rain, year, or season.

Astronomical Calendar

Glyph 19: Partially effaced, drawing not clear.

Glyph 20: Spiral glyph with comb affix (T-25, also known as the count bracket). Without being able specifically to translate the glyphs, one nevertheless gets the general idea from the last three phrases (Glyphs 9-20) that something (spiral) of the turtle has come to an end (death symbols), a change or succession ("hel" glyph) ensues, and another count (comb affix in Glyph 20) begins.

Page XVIII (Fig. 11b)

Panel B

The date is August 5 at 7:20 p.m.; the hour circle numbered 244° is over the terrestrial meridian of the Yucatan observer. Eta Draconis is at upper meridian transit (see Fig. 14), and the illustration therefore shows the serpent's belly as the main focus. The reason for the presence of the bird (identified as a vulture by Tozzer and Allen 1910, Plate 19, no. 13) is not known.

Augural Band

Glyph 1: Glyph of the day sign cauac, usually a reference to rain, year, or season.

Glyph 2: T-227, identified by Thompson as a seated man (1962, p. 60) and used to represent the idea of "seated" (1972, p. 151).

Glyph 3: Name glyph of God B. Glyphs 2 and 3 may therefore read "seating of God B," or "God B, seated".

Glyph 4: Glyph for tun with il (T-24) postfix.

Glyph 5: Augury for good tidings.

Glyph 6: (Placed below the other five glyphs) Cauac glyph with "beard" or "flare" beneath it; significance here not known.

14 *August 5 at 7:20 p.m., the date represented by page XVIII of the Madrid Codex; the hour circle with right*

Panel A

The panel differs suddenly from the others, in that the column of day signs at the left-hand side of the page read, from top to bottom, thus: 4 Oc, Ik, Ix, Cimi, and Eznab. Although it seems that the full 260-day count is intended here (since the count begins and ends with the same numeral, 4), I am unable to follow it because the counts do not seem to arrive at the day numbers which are given. It may be partially due to the damaged condition of the page. The pictures show, on the left, Chac seated in the open jaws of God C, holding some object outlined by dots in an open jar. At the right, the Death God appears in the same context.

There are two glyphic phrases composed of four glyphs, each phrase shown above one of the pictures. The analysis will begin with the phrase on the left of the panel.

Glyph 1 (top left): U prefix with glyph for yax, meaning "new," "fresh," or "green" (Thompson 1972, p. 151). Elements above the yax sign are effaced, but from what can be seen of them, the compound does not appear to be that for "south".

Glyph 2 (top right): Ti prefix with head of God C (?).

Glyph 3 (lower left): U prefix with name glyph of God B.

Glyph 4 (lower right): Ahau glyph with il as prefix and postfix, signifying ahaulil, "his rule" or "ruler" (Thompson 1972, p. 149). Presumably, the reference is to God B as ruler.

Glyph 5 (right side of the panel, glyph at top left): Same as Glyph 1.

Glyph 6 (top right): Same as Glyph 2.

Glyph 7 (lower left): Name glyph of Death God.

Glyph 8 (lower right): Death (augury).

Supplementary Ethnographic Data

Having now completed the page-by-page analysis of pp. XII to XVIII of the Madrid Codex, I would like to make some references to other ethnographic data before bringing the discussion to a close. Fig. 15 shows a photograph

15 *Clay impression of serpent and star symbol incised on deer bone. (Courtesy of Joya Hairs).*

of the design from an incised bone, reproduced on a clay impression from the collection of Joya Hairs, Guatemala City, who was kind enough to supply me with the print and the following information. On the basis of style, Proskouriakoff dates it as Post-Classic, and comments that the design at the back of the serpent resembles the Mixtec symbol for star. It will be noted that the serpent is very similar to that pictured in the Madrid Codex. He wears eye decorations, rattles on his tail, exhibits his forked tongue, and his posture is rather like that in the Codex. It seems too coincidental that a star symbol should be placed at the center of his back, in the same location as Eta Draconis apparently occupies according to the scheme displayed in the Madrid Codex. It is not difficult to conclude that the serpent incised on this bone is meant to represent the constellation Draco, extending to the Pleiades with the rattles on his tail, and exhibiting the star Eta Draconis greatly emphasized.

In view of the fact that this text in the Madrid Codex involves what is usually considered to be the Chicchan serpent, one is led to Wisdom's ethnographic material on the Chicchan serpent among the Chorti (1940, pp. 392-396). This is not to imply that I am generalizing about what the serpent may mean in the ethnography; certainly a great deal more study is needed in this respect. Nevertheless, some comments may be permissible. The manifestations of the Chicchan serpent among the Chorti are very complex and without doubt have degenerated and become confused in the centuries that have elapsed since the Conquest. Yet a few threads may linger of the old ideas. There are four Chicchans in Chorti cosmology, each associated with a cardinal direction. The one at the north is the chief (noh tcix tcan) who gives orders to the other three, his assistants. (The emphasis on the north in regard to Chicchan may be a result of the importance in former times of the meridian transit of the constellation.) The Chicchans are responsible for both beneficial and harmful conditions of earth and sky. They live in streams during the rainy season and in the hills during the dry part of the year, and are believed to inhabit the sea and lakes all year around. Earthquakes are thought to be caused by Chicchan turning over in the ground. The idea of the serpent being underground during the dry season may be related to the old pattern of Draco being below the horizon during the evenings at this time of year.

The calendar year among the Chorti (Wisdom 1940, pp. 12-13) is roughly divided as follows: a near dry season extending roughly from November to February, called "harvesting season"; a dry season which lasts from February to May, called q'in ("sun," "day," or "clear season"), and a rainy season lasting from May to November. The divisions correlate to some degree with the time divisions illustrated by the calendar in the Codex, but are mainly the result of weather patterning. It is interesting, however, how very conveniently the dates associated with the path of Draco worked into the agricultural scheduling. There is little in Wisdom's accounts, other than hints from the relation of the Chicchans with the seasons, weather, sky, earth, and cardinal points, to tie them up definitely with any astronomical events.

One particular detail from Landa seems worth mentioning. A passage concerns the festival of Kukulcan on 16 Xul (Tozzer 1941, p. 157). Thus far there has been no suggestion in any way of a link between Kukulcan ("feathered serpent" in Maya) and the serpent shown in the Madrid Codex calendar. Yet Landa tells of the departure of Kukulcan from Yucatan, stating that some Indians said "that he had gone to heaven with the gods, and on this account they regarded him as a god, and fixed a time for him in which they should celebrate a festival to him as such, and this was celebrated throughout all the land until the destruction of Payapan. After this (city) was destroyed, it was celebrated only in the province of Mani." He then goes on to say that they solemnized the feast "not like the previous ones" on the 16th of Xul. On that date all came quietly to the temple of Kukulcan with offerings of food and drinks. "The lords and those who had fasted remained there without returning to their houses for five days and five nights in prayer...When the five days were ended and past they divided gifts among the lords...They said and considered it as certain that Kukulcan came down from heaven on the day of these (five days), and received their services, their vigils and offerings. They called this festival Chic Kaban."

According to Landa's calendar, the first of Pop fell on 16 July; however, the research which substantiates the 584,283 correlation indicates that Landa had overlooked the one day correction for the leap year, and thus his calendar actually had its start on 15 July. By this calculation, the 16th of Xul fell on 8 Nov.; adding 10 days for the Gregorian correction gives the date 18 Nov. as the date that Landa tells us the festival of Kukulcan began (it actually started that evening). He then informs us that the festival went on for five days and five nights, lasting until the first day of Yaxkin. Counting the 18th as the first night, the fifth night would have been that of 22 Nov., traditional date of the lower meridian transit of Eta Draconis at midnight and start of the sidereal year. So it is this night he refers to in saying "they considered it as certain that Kukulcan came down from heaven on the last day of these (five days), and received their services, their vigils and offerings." However, it must be kept in mind that there is no certainty that this festival was fixed in the year, for Landa makes no mention as to whether it was _always_ on the 16th of Xul (indeed, he may not have known). If it were, the day would have gradually shifted through the year, since the 365-day working calendar of the Maya had no allowance for the leap year correction, in spite of Landa's statement to the contrary (Tozzer 1941, p. 134). Yet it still seems to be enough of a coincidence merely to point out that here we see the commemoration of a mythical event involving a serpent deity, who comes to earth from heaven on the traditional date of the lower midnight transit of Eta Draconis.

Landa informs us that the name of the festival for Kukulcan was _Chic Kaban_, but unfortunately gives no translation of the title. The analysis by Tozzer (1941, p. 157, no. 802) relating the name to the _chic_ or coaticlown is not very convincing. It seems more reasonable to suppose that the name _Chic_ may refer to Chicchan, or the _Chic_ serpent (_chan_ means "serpent" in Maya). There is no clear meaning for the word _chic_, but numerous words prefixed by it give the impression that _chic_ may be associated with something going around, becoming visible after being hidden, and marking or signaling, as, for example, in the following:

chicaantal: to appear, be visible (Perez Dictionary).
chicil: signal (Perez).
chicilbezah: to signal, set landmarks to (Perez).
chicpahal: to find, become visible, perceive from a distance, reappear from being lost (Perez).
chictah: to shake, with a motion up, down, and back and forth several times to stir a liquid or mix it with something else (Perez).
chictahal: to find by searching (Motul). (The implication here is that one is going around looking for it.)
chic-ha: to rinse out with water (Perez).
chic haa tah: to rinse out a vessel with water (Motul). (The action implied is that of swishing the water around inside the container.)

The intention in demonstrating these words is to show that the name Chicchan may involve the idea of the serpent going around in the sky in a rotary motion, becoming hidden, then reappearing, passing landmarks in his journey. The idea may also be present in the name Chic Kaban. Thompson (1972, p. 37) has observed that Landa was very confused between glottal and nonglottal stops and often mixed up the sounds for ca and ka. Therefore, he may have meant caban, meaning "earth," rather than kaban. Thus we have the interesting possibility that the name Chic Kaban means something like "coming down to earth" or "appearance on earth," applicable both to the idea of Kukulcan coming down from heaven, and the lower meridian transit (below the horizon) of Eta Draconis. It begs the question, however, as to the actual relationship between Kukulcan and Chicchan.

Before bringing this analysis to a close, I would like to mention one more item in relation to the data contained in Landa's account. The reader is referred back to the discussion of p. XII regarding the symbol of the circle enclosed in dots, above the potstand. It will be recalled that I suggested that the potstand gave the sound value "oc" (meaning "enter"), and that the dotted circle stood for something like ekbizen, so that together the idea was transmitted as "beginning of star-dotted (night)" or "end of twilight". In Landa's "alphabet" (Tozzer 1941, p. 170) the following is given for the letter "O":

Astronomical Calendar

It is this information which leads Knorosov (Barthel 1969, p. 39) to assign the sound value "O" to the symbol of the circle enclosed by dots. A close look at the sign will show that the drawing is of a circle surrounded by dots below some object (jar? aperture?) which contains a dot. Certainly, what is meant is that one of the dots from the sign below is enclosed by the entity above it, and that what is being illustrated is the act, not the name of the sign. It is completely analogous to the potstand on p. XII of the Codex, for the word-stem oc (from ocol) gives the meaning "to enter in" ("entrar, introducirse, meterse," Perez Dictionary). In Landa's picture one of the dots (from the star-dotted sky?) has entered the enclosure above it. When Landa asked for the sound "O," the scribe undoubtedly pondered a bit, and then with perfectly good logic offered him an example of oc.

Concluding Remarks

To conclude, I would like to state, on behalf of the argument contained herein, that it seems too much to simply assign as coincidence the following:

1) That the correlation is so close between the illustrations in the Codex and the annual path of Draco.
2) That Eta Draconis showed unusual stability for a long period of time, making it an excellent candidate as a calendar star.
3) That the ethnographic names for Gemini and the Pleiades fit the illustrations as well as the context.
4) That the resulting scheme comprises intervals of absolutely regular counts.
5) That the time of observation coincides with the average end of astronomical twilight at the latitude appropriate for the Codex.
6) That all the above together provides a logical framework in terms of day counts typical for the Maya as well as being representative of actual seasonal events.

Furthermore, I would like to point out what seems very obvious:

1) It would have been logical for an astronomer-priest to want to adjust auguries and prognostications of the 260-day count to those of the year of the seasons.

2) The absolute symmetry expressed in the relationships as outlined is a feat worthy of record.

3) The resulting scheme fulfills to no small degree literary standards of rationality, practicality, and beauty.

Acknowledgments

Sincere thanks are extended to the Wenner-Gren Foundation for Anthropological Research for awarding me the funds to make this research possible. Gratitude is also rendered to my professors John A. Graham and Robert F. Heizer, University of California, Berkeley, for their assistance to me as a student and for reading this report. I am greatly indebted to Alexander Pogo, Carnegie Institution of Washington, for his kind and generous help with the astronomical parts of this paper. To my colleagues Thomas R. Hester and Eric Blinman who drafted the astronomical charts, and Christopher R. Corson, who has aided in numerous ways, I also extend my appreciation. Special thanks go to Joya Hairs, Guatemala City, who allowed me to publish her photograph and information regarding the incised bone (Fig. 15). Finally, I wish to acknowledge and express gratitude to Knowles A. Ryerson, who helped inspire and sustain the interest that led to this research. Of course, I take sole responsibility for the ideas expressed in this paper.

17

Olmec Mosaic Pendant

Alexander Marshack

Peabody Museum of Archaeology and Ethnology

Harvard University

Cambridge, Massachusetts

A unique, relatively complete pyrite mosaic pectoral or pendant of the so-called "mirror" type (Fig. 1), excavated from the Olmec site of Las Bocas in west Puebla, Mexico, around 1963-1964 and dated ca. 1000 B.C., is the earliest Mesoamerican mosaic excavated to date. It is, surprisingly, also one of the most complex artifacts to come from a prehistoric, pre-classic context and the most complex mosaic to come from Mesoamerica. It has provided the opportunity for a new type of systematic analysis and measurement based on a sequential reconstruction of its structure and mode of manufacture (Marshack 1972, a, b, and c).

The analytic data secured in this study pose significant questions concerning the origin, development, and sophistication of the Olmec culture, precursor of the later Mesoamerican classic civilizations (Bernal 1969; Coe 1968).

Olmec skill in carving and sculpting stone is well-known, but precision of the kind evidenced in the mosaic has not till now been documented in prehistoric Mesoamerica. In the standard evaluation of the later classic tradition of mosaic manufacture, Kidder, Jennings, and Shook (1946, p. 126) declare:

Pyrite, with a hardness of 6.5 and with no natural cleavage planes to facilitate subdivision of the crystals, could not have been other than most difficult to work...

Copyright © 1975 by Alexander Marshack

1 Main face of nearly complete pyrite mosaic pendant from the Olmec site of Las Bocas, Mexico. The mosaic, ca. 1000 B.C., is the oldest known from Mesoamerica and the first of this type to be excavated. The original, highly polished yellow-silver pyrite. Length, 14 cms.

Olmec Mosaic Pendant

Nothing produced in aboriginal America seems to us to rival these plaques in the matter of skilled and meticulous craftsmanship. One hesitates even to guess at the number of man hours that must have been expended in the making of each one...It is to be supposed that they were turned out by members of guilds like those of the jade, feather and goldworkers which existed among the Aztec.

The earlier Olmec tradition was apparently already highly developed with a comparable implication of specialized, elitist manufacture. Pyrite, however, tends to decompose rapidly in humidity and mosaics of this material in good condition are rare even from the Classic period. Survival of the Las Bocas example is probably due to the accident of a dry, protected burial in Puebla, northwest of the humid Olmec lowlands. The pyrite pieces exhibit the first stages of discoloration but are otherwise undeteriorated.

A Preliminary Cognitive Analysis

The present analysis follows and is theoretically dependent upon the sequential analysis of an equally rare contemporary calendar notation, obtained ethnographically by Gossen (1974), which had been maintained until 1968 by a Maya speaking shaman of the village of Chamula in Chiapas, Mexico. This notation, marked with charcoal on a wooden board (Fig. 2) in a village that is self-consciously separated from Spanish speakers, seems to be the remnant of an ancient shamanistic tradition of daily calendar notation and record keeping for ritual, divinatory, and agricultural purposes, in a form far simpler and perhaps more ubiquitous than the elaborate priestly codices and tables that were kept in the late ceremonial centers. A wood and charcoal notation of this type could not become an archaeological artifact and, in fact, none are known.

Analysis of the Chamula board was undertaken in 1972 as a "blind test" of the methodological and theoretical concepts that had been developed for study of the European prehistoric notations where neither the language nor other supporting semantic contents could be known but in which certain cognitive contents could nevertheless

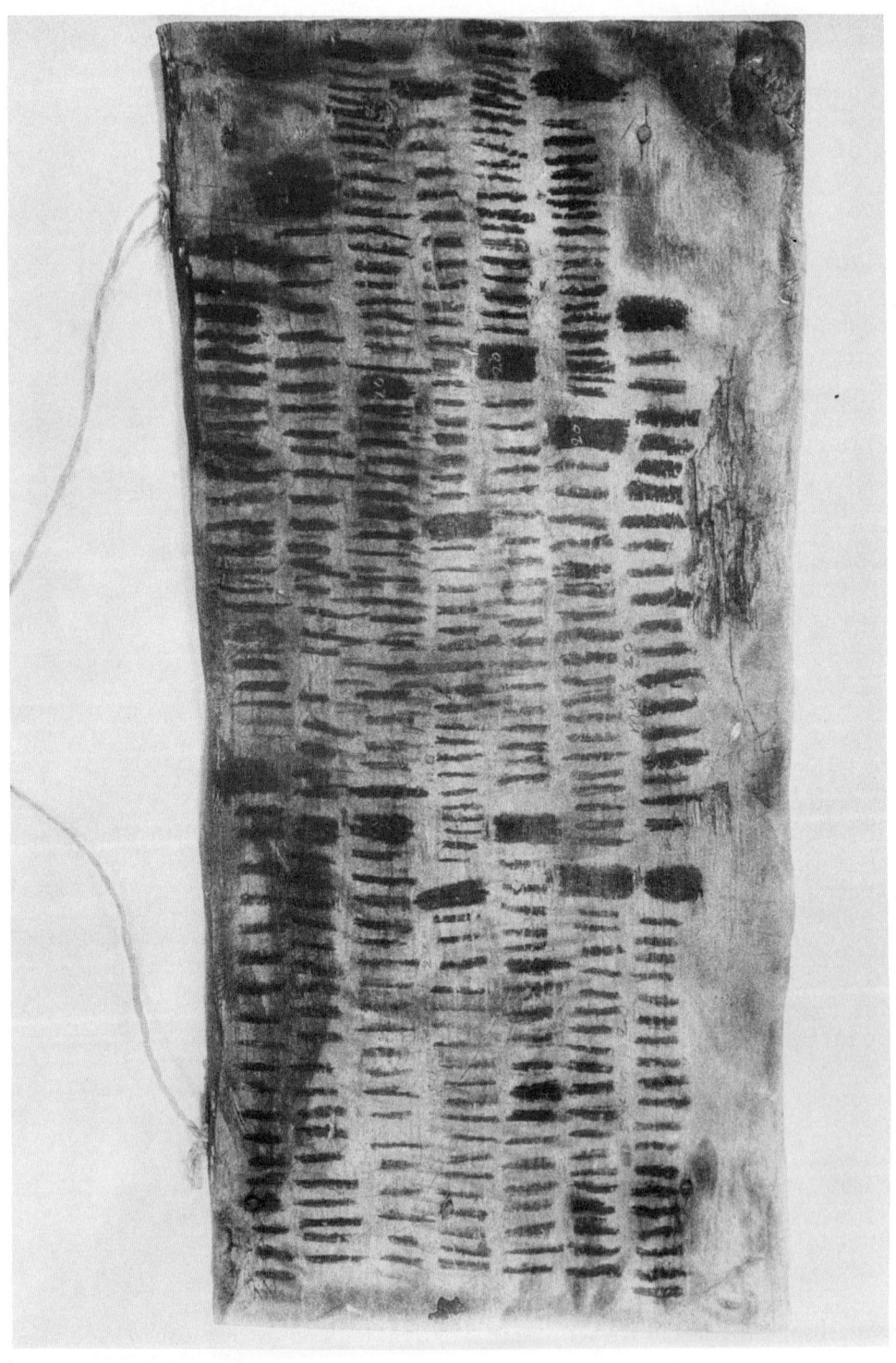

2 Infrared photo of modern calendar notation for one calendar year from the Maya speaking Chamula of Mexico. Analysis showed that the board, kept to old Maya month count, had been used for generations. Internal analysis indicated the possibility of performing sequential and cognitive analysis of artificial arithmetic sequences and led to study of the Las Bocas mosaic.

be determined (Marshack 1972, a, b, and c). The board was analyzed as though it were an archaeological artifact without reference to any ethnographic data.

The Chamula board was marked with 18 Maya months of twenty days each and included one anomalous period of 5 days to give a sum of 365, the number of days in the tropical solar year. The breakdown into sets could be determined by certain internal notational clues, including a heavy terminal mark for each 20 day set, a mark that both closed and summed the period. There were also different rhythms of spacing for each set. The number of such clues evidencing the notational, non-decorative intent were many (Marshack 1974).

The sum of 365 could not be accidental and could have been derived only from an original solar, solstitial or equinoctial observation and count. The odd placement of the anomalous 5 days in mid-sequence, however, which originally could have fallen only at the beginning or end of a notational year count, indicated that the marking represented a devolved or altered calendar and that solar observations were no longer being maintained or marked. One could assume, therefore, that the year structure on the board was being maintained as a tradition, that is by sequential naming and narrative and perhaps by formalized counting.

In the second stage of analysis, infrared photography and microscopic examination made it possible to reconstruct the sequence of marking and the details of usage. Infra-red indicated that the board had been used for a long time, that at the end of each year it had been wiped clean and a new year begun. The polish of wiping indicated that this had occurred for perhaps many generations. The person marking the board was, therefore, working in a tradition and not in an individual, idiosyncratic manner. Infra-red indicated that each month was marked with many different charcoals of different hardness, thickness, and blackness but that these charcoal sub-sets were irregular, indicating that there was no pattern in the number of marks that could be made with one piece of charcoal. Sometimes the same print ran for a number of marks. This suggested that one mark was made each day but that a piece of charcoal could be kept for many days and that sometimes a number of days were marked at the same time. Despite the seemingly random structuring of the sub-sets, the combined data confirmed that this was a notational marking being performed to a plan and a carefully maintained structure.

It was possible also to determine the direction and sequence of marking. The notation began at top left, proceeded left to right on each line and descended with the straight edge of the board as the rough guide and alignment. There was no such alignment at right. There was the tendency to maintain the rhythm of marking within each set, though different charcoals were used and the marking was done over a period. There was a tendency also to begin with a tighter rhythm of spacing when three months of 20 days were planned for a line, indicating pre-planning, but there was also a tendency to squeeze the last set at right as though often the marker found he had miscalculated and was short of space. The largest month was the last, made in the knowledge that no month followed and there was no problem of space.

This type of analytic data is typical of the cognitive solutions one usually finds in writing systems and calendric notations marked within a limited space, whether these are based on non-astronomical arithmetic units such as the Maya or on observational, astronomical, but non-arithmetic sets based on the lunar phases (Marshack 1972, a, b, and c). Since the Chamula board represented a daily record apparently kept in the habitation and was completed to day 365, I assumed that any artifact of this class found archaeologically would almost always be incomplete to the last day of marking. It would therefore be impossible to determine that such an artifact was calendric and based on a 365 day tropical year since the sets of 20 were arithmetic, artificial, and unrelated to an observational month. I assumed that the Chamula board had been intentionally completed to day 365 for the anthropologist since on any other day it would have been incomplete, on day 366 it would have been wiped clean.

The above analyses were confirmed at a later date by the anthropologist who had obtained the board (Gary Gossen, 1972, private communication). I stated in the paper on the Chamula board that one might never find evidence of such charcoal notation archaeologically but that one might find the day-count of the year structure in permanent, symbolic form. After delivery of the

Chamula paper, Furst (1972, private communication) suggested that I try to locate and analyze the Las Bocas mosaic as a possible symbolic calendar composed of day units. Furst suggested that the Olmec mirror "simply must be of calendric significance. The total would be about 350 to 360 (365) polygons as near as can be made out." The Las Bocas mirror was first described by Furst in his unpublished thesis on the shamanistic contents of west Mexican archaeology, 1966. He wrote: "...the Olmec plaque represents a remarkable technological achievement in that its tightly fitting mosaic is composed of no less than 330 individual pieces of minute size, roughly ten times the number of polygons normally found in an Early Classic plaque of comparable size."

The Las Bocas Mosaic

The pyrite mosaic (Fig. 1) has lost some pieces along the edges. This meant that a reconstruction and estimation would have to be made of the original inlay. At first glance the mosaic seemed to be a mere carpeting of the askew rectangular form without any recognizable image or design motif. The pattern of inlay, however, was totally different from the style used in the later Mesoamerican mosaics where irregularly shaped pyrite polygons of diverse size were used to form a random pattern to fill the area of a symbolic circular form (Kidder, Jennings, and Shook 1946, Fig. 53). The form of these later pyrite pieces made any sequential arithmetical structuring impossible.

Engraved and sculpted figures from the Olmec culture show that circular pendants, as well as square pendants with rounded corners, were worn symbolically (Joralemon 1971) and these may on occasion have been inlain with pyrite. There is no Olmec depiction, however, of a pendant or pectoral as a skewed rectangle.

The shape of the Las Bocas mosaic is apparently unique, yet analysis indicates that it was intentional. This intentionality was first suggested by Ekholm's determination for the author that the base was ceramic and not stone. Gordon Ekholm of the American Museum of Natural History determined that the base was ceramic. The presence of aberrational, non-geometric shapes for carved stone pendants and amulets, based on the original random shape of the stone, is well known (See A.V. Kidder, <u>Jades from Guatemala</u>, pp. 1-8, A.V. Kidder, <u>Certain Archaeological Specimens from Guatemala</u>, pp. 9-27.). It was therefore manufactured and further shaped and did not represent the accidental form of a stone plaque. The base had been bevelled toward the rear in an early style of bevelling and this accented the intentionally skewed rectangle (Fig. 3).

A simple analysis of the base (Fig. 4) indicates that the angle made by the vertical left edge with the horizontal bottom forms a perfect right angle, without variation of a second of arc. This could only have been derived intentionally and by use of a right angle frame. The upper horizontal is slightly curved and its angle to the vertical at the corner is 89°, suggesting that the intent was for a right angle alignment in only one corner. Squaring the skewed form and measuring the lower edge indicated that the major arc along the top began almost precisely at the 2/3 mark and that the skewing of the hypothetical rectangle occurred primarily in the last third. This, too, did not seem accidental, considering the availability of the straight edge and right angle. The base had been covered with red cinnabar, and the holes for a cord indicated that it had been worn, apparently as a symbolic object of considerable importance.

The continuing analysis revealed that in the sequence of manufacture the ceramic base had been made first, to a pre-planned size and geometry, and that the inlay was then structured to this form, overrunning it at the very end of the inlay process.

The few pieces missing from the mosaic on the front were primarily from the edges where they could easily have been knocked off in the digging, handling or shipment since excavation. (Two pieces are now missing that were present when the mosaic was examined by Furst in 1965, and half a dozen are now loose and easily removable.) Most of these missing pieces could be accurately reconstructed, particularly along the straight edges.

Analysis of these edges for the purposes of reconstruction revealed aspects of the inlay process. These turned out to be again different than the processes of inlay

3 The rear of the Las Bocas mosaic indicating the skewed shape, the holes by which it hung, the bevelling towards the back in an early style. Back is ceramic and was heavily covered with red cinnabar.

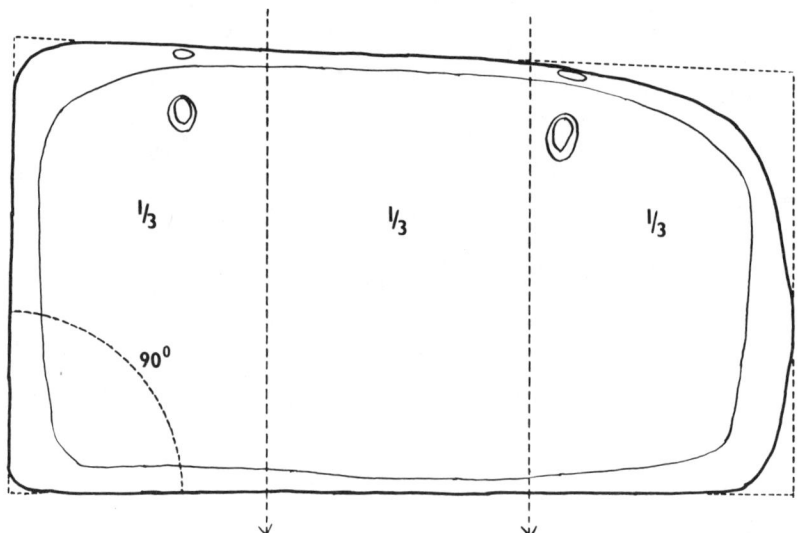

4 *Schematic rendition and analysis of the shape and structure of the mosaic base indicating the one perfect right angle in the form. The arcing at top right begins precisely at the 2/3 mark and skews the shape, apparently to cut the available surface for inlay in this third. Evidence for these two geometric findings appears also in the inlay.*

used in the later circular mosaics which have been published. The technical and cognitive processes and strategies involved in the Las Bocas pectoral suggested that this was a mosaic of a different class or type.

The long straight edge forming the 90° angle in Fig. 4 was composed of pieces whose sides were ground to a right angle with the surface (Fig. 5a). The fit between pieces was as tight here as along the surface. The pyrite sides of these blocks were inset 2 mm from the ceramic edge, creating a lip or ledge and indicating that the ruling or aligning rod had rested here and that it was against this that the blocks had been perfectly aligned. The ledge had then been covered or filled with a lime cement (as determined by Gordon Ekholm) to the ceramic edge and this was overpainted with cinnabar. Where the lime had broken away the pyrite sides showed the presence of cinnabar, suggesting the presence of the

5 a Extreme close-up of three pieces along the exposed upper edge indicating the precise shaping and the rough polishing of the right angle sides. The precision of the fit between the odd-sized pieces is shown by the absolute right angle join between two pieces and the equally close fit between two pieces slightly off a right angle. These pyrite pieces are inset 2 mm from the ceramic edge and were finished so carefully to take an aligning frame. The paint is a combination of lime and cinnabar.

color in the burial where the pendant was found. Loose cinnabar was also detected microscopically in cracks of the mosaic surface.

The lower edge was sided differently, with pieces that were roughly worked so that they created an uneven surface against which an aligning rule could not be placed (Fig. 5b). There was no ledge and the pyrite blocks came to the edge of the backing, extending here and there a fraction of a millimeter beyond this edge. These pieces had no lime covering and were thickly covered with a cinnabar paste.

The short side at left had pieces polished like those in Fig. 5a, but there was no ledge and this suggested that the pieces had been inlain against a permanent straight edge while a freely moving aligning rule had been used for the upper horizontal.

The differences in these three edges suggested that the inlay had begun with an effective right angle in the corner at top left in Fig. 1 and had proceeded downward and to the right.

The arced edge at right, surprisingly, extended beyond the ceramic base a full 2 mm by a build-up of lime cement (Fig. 6a, b) and the final inlay was made into this unsupported structure. The chance retention of one pyrite piece in this overhang made it possible to delimit the extended arc and to draw the mosaic outline to the form and curve of the base. A break in the ceramic base in the rear where this overhang occurs indicates that most of the pieces had been broken loose by a heavy blow from the rear. Had this lime overhang and the single piece not been accidentally retained it would have been impossible to surmise such an awkward and dangerous completion to a precisely and laboriously manufactured artifact.

Using the completed arc as reconstructed around the single extended pyrite piece it was possible to divide the area into the probable number of missing pieces (plus or minus one), based on the manner in which corners and small areas of various size and shape were inlain elsewhere. A study of the mosaic indicated that it would not have been difficult for the maker to end his mosaic evenly at the ceramic edge, using less pieces, but that for some reason he had felt it necessary or advisable to extend it beyond.

5 b Extreme close-up of three pieces along the exposed lower edge indicating their equally tight fit and the formation of a straight line surface but the rough relatively unworked sides. The downward inlay process ends with these pieces. The rough surface was heavily covered with cinnabar.

6 *a Front view of the mosaic inlay within the arc of the last third of the pendant indicating the singly extant piece extended beyond the ceramic base and the underlying lime structure which held the rim of extended inlay.*

6 *b Detail of the arc from the rear, photographed at a slight upward angle to show the extention of the lime and the single piece inlain beyond the protective base. No such technique of inlay is known in Mesoamerica and it was apparently used here in the desire to support a previously projected number of pieces. A blow has broken the ceramic edge and it was this that knocked off most of the lime and its pyrite pieces. (Note: the extended single pyrite piece is missing in Fig. 3 since it is now loose and so was removed when the mosaic was turned.)*

Upon completing the reconstruction of all the missing pieces, a count of the probable number in the original inlay gave a sum of 354 (Fig. 7). This is the number of days in 12 observational lunar months (12 x 29.5 = 354). I had been expecting a possible 365 day symbolic solar year. If this sum were more than chance it was possible that the arc had been extended by adding a lime structure in order to provide the pieces needed to complete a required count. Having completed analysis of the edges it became necessary to turn to an internal analysis of the inlay to see whether the sum, the skewed shape, and the overhang could be intentional and interrelated.

Internal Analysis

Examination of the mosaic with a low power binocular microscope (10 to 60 X) indicated that no two of the 325 extant pieces were the same size or shape, yet they fit precisely (Fig. 8). There is not enough space between pieces for cement or the insertion of a razor's edge.

The mosaic pieces are primarily six-sided blocks with four sides more or less at right angles to the polished quadrilateral or rhomboidal surface. The underside of each piece is intentionally unworked and irregular to grip the lime cement. Since the pieces themselves are of different depth, the lime is relatively thick to encompass the varying sizes. There are only seven exceptions to this pattern of quadrilateral forms, and they will be discussed later.

Analysis of the manner in which these pieces were inlain revealed an extraordinary arithmetical sequence and pattern. Each two pieces meet or join along a horizontal straight line (Fig. 9). These two, forming a unit, are then joined to another two, along a vertical straight line. These four, as a unit, are joined to another similarly formed set of four below, along a horizontal. This set of eight forms the basic unit of the mosaic. It is matched to a similar set of eight at the right along a vertical straight line. The set of sixteen is matched to a similar set of sixteen below, along a longer horizontal straight line. This set of thirty-two, joined vertically, form a unit of sixty-four, separated from a similar set

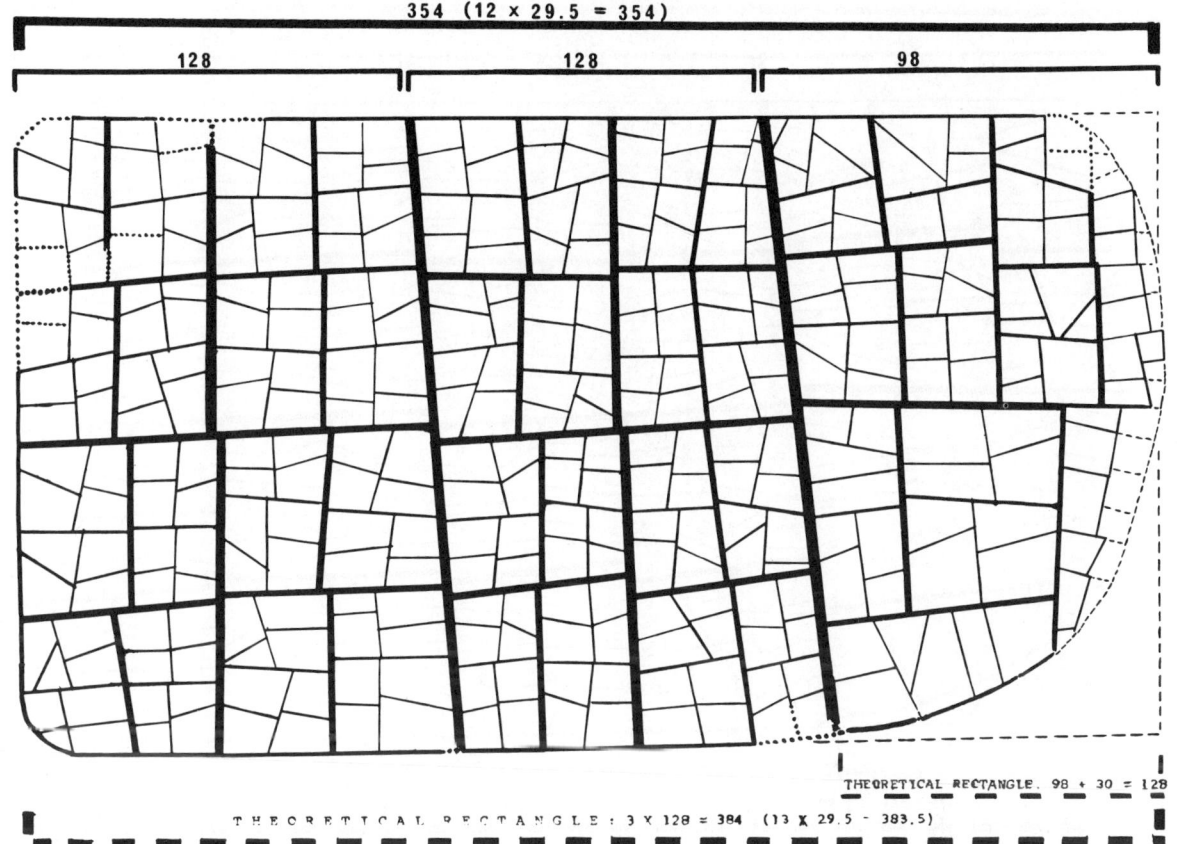

7 Rendition of all the pieces in the Las Bocas mosaic schematically indicating the arithmetical division into three sections totalling 254 pieces, the sum for one lunar year. The two sections of 128 are divided into 16 sets of 8 pieces, subdivided into sets of four, divided into sets of two. The sets of 8 combine to form sets of 16 which combine to form sets of 32 and these combine to form sets of 64. All sets are aligned along a straight line, except at the curve. The theoretical rectangle which is cut by the curve at right angles suggests an intentional subtraction of 30 pieces from the basic 128, a subtraction from a theoretical 13 month lunar year of 384. All the elements, sums and divisions are arithmetical and apparently intentional.

8 Detail of the unbroken portion of the inlay in the upper portion of section II indicating the varying size and shape of the pieces, the tight fit of the pieces, the perfect straight line of all edges and the divisions into basic sets of 8.

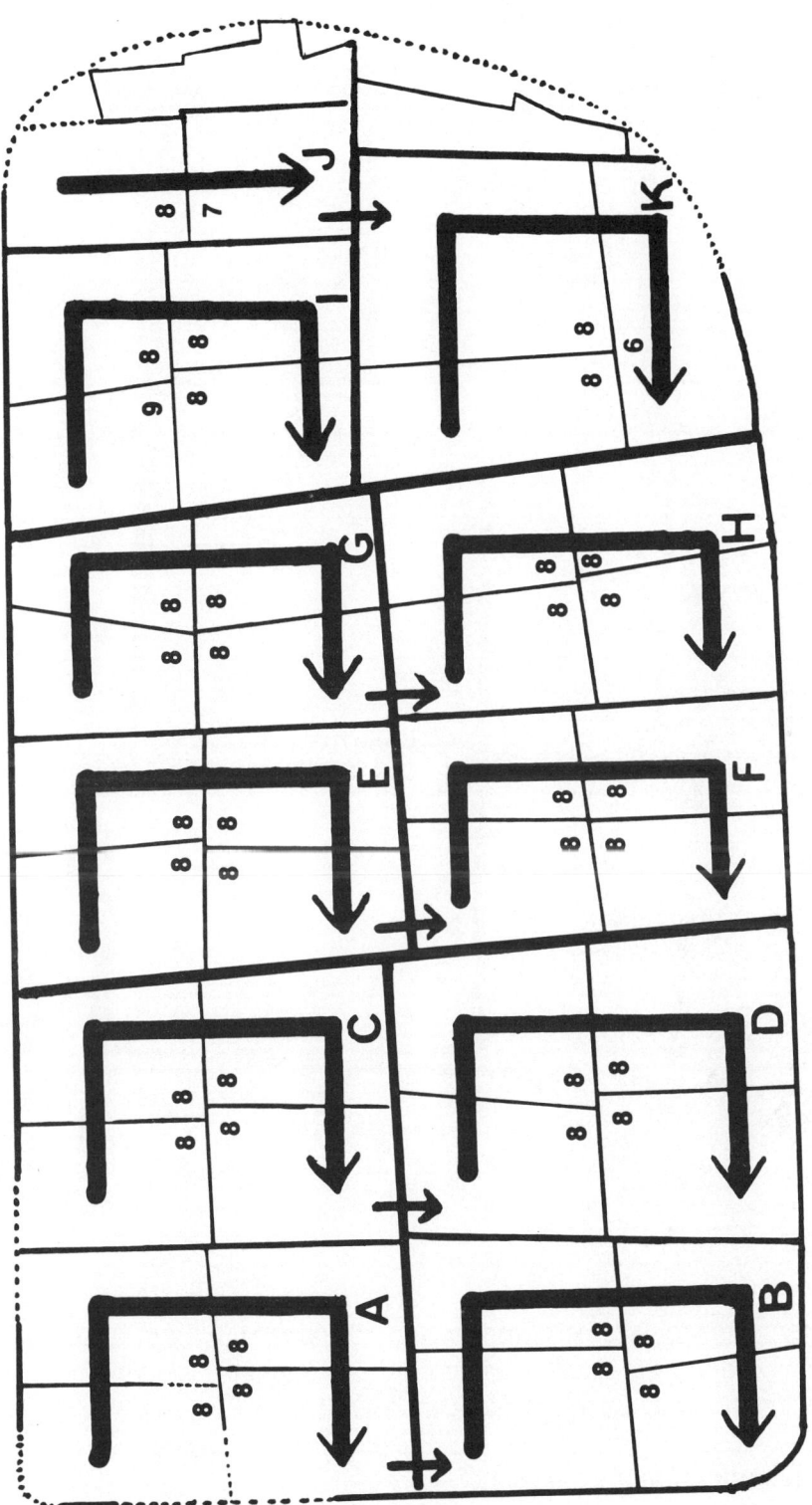

9 Schematic rendition of the probable order and direction of accumulating sets of eight. The sets accumulate in a downward, serpentine manner as the mosaic develops. In the third section the first set opens with an extra piece. The last set in the upper half corrects this with the absence of one piece. This process of inlay, like the process of alternate alignment, allows the maintenance of a continuing count and necessary adjustments.

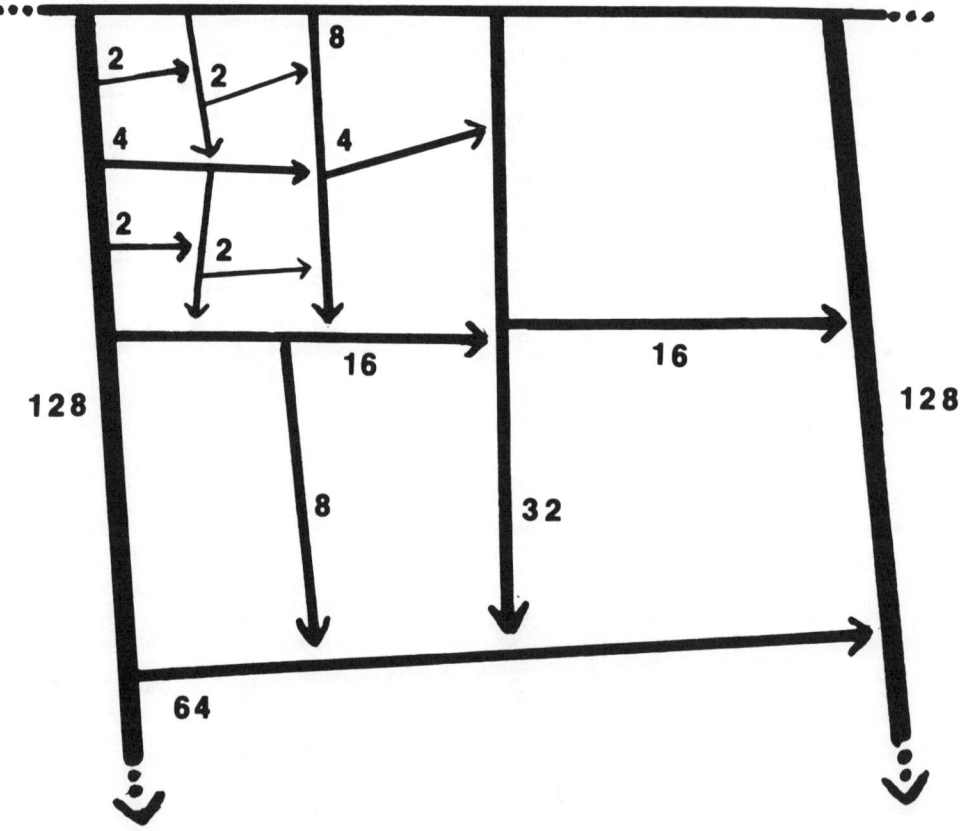

10 *Schematic breakdown of the upper central portion of section II in Fig. 8 indicating the alternation of the horizontal and vertical aligning straight line. Each longer line joins sets of a higher order. The patterning is one more element making possible the maintenance of an arithmetic count.*

Olmec Mosaic Pendant 361

of sixty-four below, along a still longer <u>horizontal</u>. This superordinate unit of 128 is then separated by a long <u>vertical</u>, running from edge to edge, from the next section of 128 pieces at right. A similar vertical straight line separates this second 128 from the final, third section. The total of the two arithmetically similar sections is 256.

If the pectoral were not skewed, this basic pattern could continue in the third section to give another 128. This would have given the mosaic three equally formed sections with a sum of 384 pieces. Instead the patterning in the third section begins to break down as the inlay proceeds. It becomes almost aberrant as one approaches the curved edge. Despite this, as we shall see, there is evidence of continuous arithmetic counting, control and adjustment.

Reconstruction of the missing pieces along the arc as described above gives a sum for this third section of 98. The total, therefore, breaks down into 128 + 128 + 98 = 354. The projected total for the theoretical rectangle which was skewed by the arcing is the precise sum for the number of days in an observational thirteen month "long" lunar year (13 x 29.5 = 383.5). Such a long year of 13 moons encompasses either two solstitial or two equinoctial solar observations at an interval of 365 days. Such a long lunar year could also represent the intercalary year needed to bring solar and lunar years into phase.

The number of hypothesized pieces missing from the theorized 384-piece rectangle is precisely 30, the sum for one observational lunar month. It seems possible, therefore, that the intentional skewing of the form was meant to decrease the area of the final third, a geometricized subtractive process that included both the area and a projected sum. Further analysis confirmed the interrelation of geometry, arithmetic, and technology.

Secondary Analysis

The only way that a sequence of inlay such as that evidenced in the first two sections can be maintained is by use of at least one stationary and one freely moving straight edge against which accumulating pairs and sets can be aligned. The original set of two in the corner at top left must have been placed into and against what was an effective right angle frame. Once pieces were in place they served as edges for aligning other pieces and for holding the aligning rule for measuring subsequent sets.

Since pieces placed against others required a straight line face on their open side, but not a face that was at right angles to the opening right angle, any slight deviation from a right angle would be perpetuated through a set of 8. These deviations could accumulate along the free aligning edge, and one can note that adjoining sets of 8 often tend to adjust each other, but only partially. The two first sections of 128 tend to spread slightly toward the right as they grow downward, an indication not only of the direction of inlay but of the presence of a beginning or planning line along the top edge.

The entire process, apparently, necessitated constant counting and adjustment. It would not have been possible, for instance, to begin with a set of 8 in the upper left corner and end with a set of 8 made to the same pattern and of roughly the same size and height in the lower right of section one or two without the most careful preplanning and constant supervision of the process. It is here significant that the upper aligning edge almost precisely divides the mosaic into three equal sections. The shift from this nearly equal tripartite division occurs as the mosaic descends. There are only a few millimeters of variation from this equal division where the long dividing verticals meet the upper edge. The first occurs a few millimeters after this point and the second adjusts this. At the theoretical 2/3 mark the amount of overrun is only 3 mm. The amount of lime structure that was intentionally added to the third section at the curve was 2 mm, as though this was the area needed to make up for the slight overrun.

An analysis of the full sequential process indicates that the pattern of accumulation may have proceeded via a unit composed of four sets of 8 (Fig. 10). Four seemed to be the conceptual organizing frame. Four sets of 2 to make 8, four sets of 8 to make 32 and four sets of 32 to make 128. The first set of 8 in Fig. 10 was joined to one at the right, the second. The third, apparently, was placed below the second and the fourth

Olmec Mosaic Pendant

was inserted to the left, under the first. The movement of the descending sequence by 8's is therefore somewhat serpentine, continually changing or alternating the direction of inlay, each set of 8 attaching to the one just completed. This process is perhaps comparable to the alternation of the aligning edge.

The evidence seems again to be indicative of preplanning and of continuous control, counting, and adjustment in the inlay process.

Anomalies in the Pattern

A number of anomalies in the mosaic appear to be intentional in much the way that the anomalous skewing and overhang were.

Among the 325 extant pieces there are seven that are triangular (Fig. 11). Since pyrite is exceedingly hard and each such piece must be fitted to a neighbor in a set of two, the shaping of a triangular piece must have been specially noted, particularly when at times it required the shaping of a 5-sided block as its mate. Since the break in the pattern of manufacture was obvious, it may have been intended. Each triangle significantly functions in the set of two as does an ordinary piece, without breaking the alignment. If they are intentional, these triangular pieces may have indicated points in the sequence of inlay or counting that had relevance either for a lunar count or for one of the other artificial arithmetical systems supporting the inlay.

A test of these triangles is quite simple. Assuming the sequence of inlay to be that which is outlined in Figs. 9 and 10, each triangle and its mate, forming a unit of two, can be numbered precisely. The triangles are numbers 60, 124, 257, 260, 267, 301, and 324.

If we assume a usual ending for an observational lunar year at last crescent and a beginning of the next year at invisibility, then the numbers 60, 267, and 324 are observational last crescents or first invisibilities following months 2 ($29.5 \times 2 = 59$), 9 ($29.5 \times 9 = 265.5$), and 11 ($29.5 \times 11 = 324.5$). Day 354, of course, is a last crescent as would be day 384.

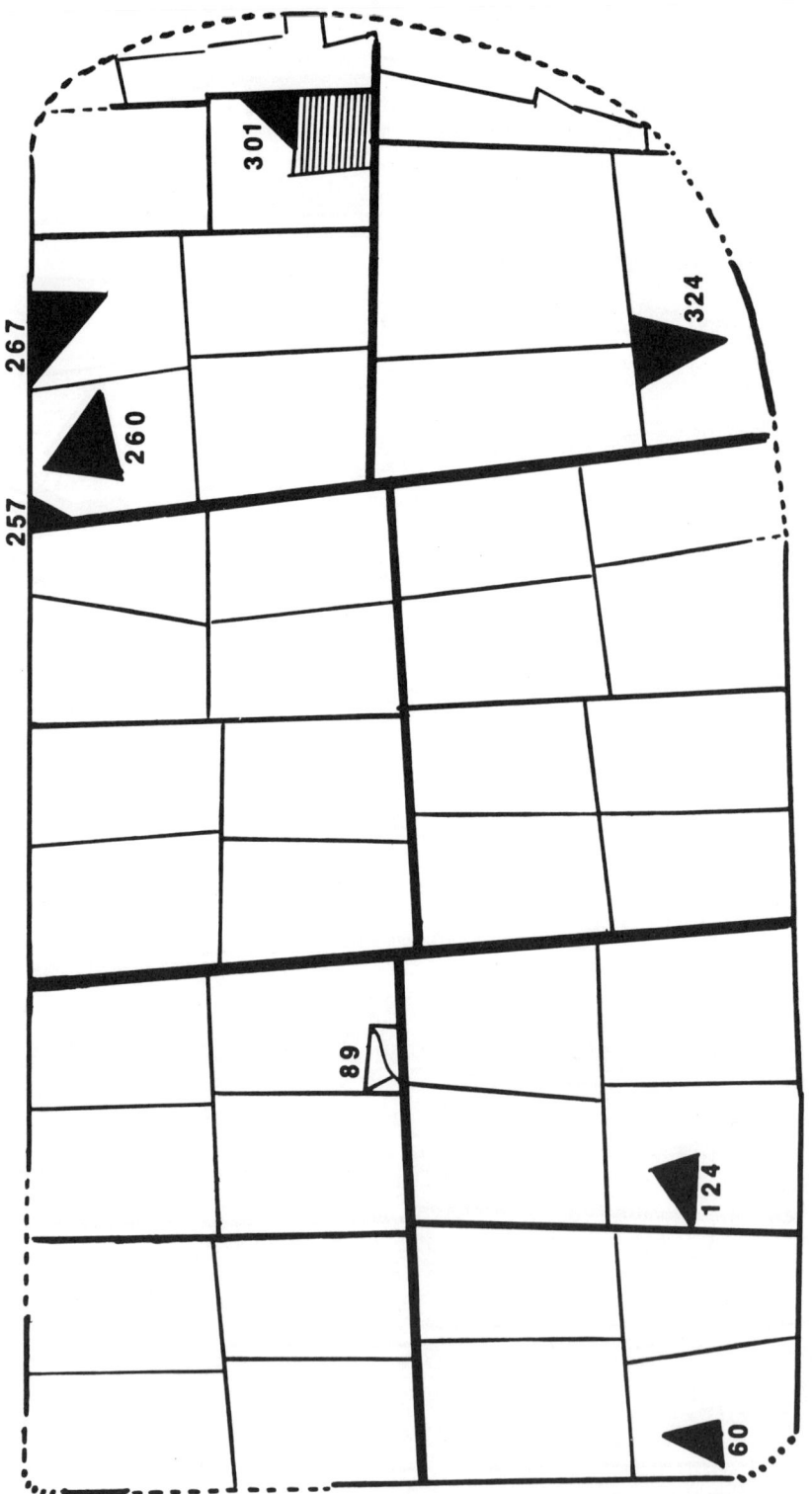

11 Schematic rendition of certain anomalies in the Las Bocas mosaic; those pieces deviating from the pattern. There are seven triangles. One piece was intentionally engraved after the surface had been polished. Piece #257 is an added "cueing" piece for the start of section III, piece 302 seems to correct that addition by being a single piece where there should be two. The other triangles seem to be related to arithmetic or lunar patterns.

Triangle 257 begins the aberrant third section and may be a cueing mark to indicate the beginning of a careful, difficult computation. It is inserted to make a normal set of 8 into a set of 9, the only such set in the mosaic. Since this piece #257 throws off the count by sets of 8 within the upper half of this third section, the last piece in this section, #302, does not form a set of two as it should, but is a single large piece, making a set of 7, again the only such set in the mosaic. The added triangle and the subtractive square may thus be evidence of the continuing cognitive, counting, adjusting process.

Triangles 124, 260, and 301 fall neither at lunar points nor change the counting pattern. Two hundred and sixty is the sum for the later, artificial Mesoamerican ceremonial or ritual year, the <u>tzolkin</u> (Fig. 12), a non-astronomical sequence based on the vigesimal count by 20's, one form of which appears in the Chamula board. If the number 260 in the mosaic is evidence of the later Mesoamerican count, then the preceding day 260 in the sequence would have fallen on the day before the mosaic begins, that is, on the day of last crescent ending the prior lunar year.

If the last crescent of the prior year or the invisibility of the new year also fell on a day of solstitial or equinoctial observation, the triple conjunction would have made it both symbolically significant and the starting point of at least three sequences, a lunar, a solar and an artificial year of 260 days. Day 60 would then be, for instance, the second new moon after the equinox or solstice and the time, perhaps, of ritual.

From day 260 to day 384 of the theoretical 13 month lunar year implied by the mosaic is exactly 124 days (260 + 124 = 384). In like manner, from day 260 of the prior year to day 124 of the mosaic would constitute another theoretical 384.

Triangle 301 is the piece following the ending of the period 15 x 20 = 300. It is non-lunar and is possibly significant as the second month after the ending of the artificial year at 260.

The basic inlay in this third section, except for the pieces along the curve at the far right, ends at piece 326. This would be the astronomically correct day of new moon, that is, the second observational day of invisibility, beginning the last, 12th month. The 28

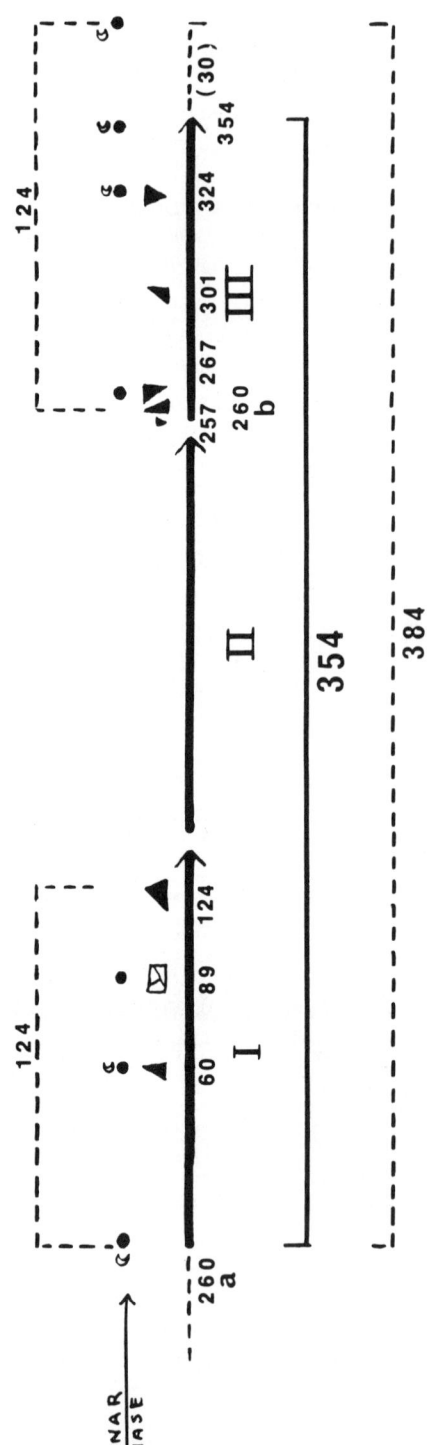

12 *The anomalies of Fig. 11 structured sequentially against a model of the 354 piece mosaic and the theoretical rectangle of 384 pieces. Lunar phase points are shown for triangles that fall at these points. The possible meanings for these pieces are: (a) 60 = 2 months (29.5 x 2 = 59); (b) 89 = 3 months (29.5 x 3 = 88.5); (c) 124 (260 + 124 = 384, theoretical form); (d) 257 (beginning of section III); (e) 260 (artificial calendric number, 13 x 20 = 260); (f) 266 /267 = 9 months (29.5 x 9 = 265.5); (g) 300 /301 (artificial calendric number, 15 x 20 = 300); (h) 324 = 11 months (11 x 29.5 = 324.5).*

Olmec Mosaic Pendant

irregularly set pieces forming the arc would therefore begin at first crescent on day 327 and end at last crescent or first day of observational invisibility on day 354. It is this full last month that was apparently required and which was supplied by adding the lime overhang and its pieces.

There is a last anomaly which was discovered by microscope. Apparently after the mosaic had been completed and the surface had been polished, one piece had been engraved with a straight line, running from the edge of an intrusion into the pyrite crystal to the edge of the block (Fig. 13). It is the only piece so engraved and is apparently intentional. This is piece #89, the day of ending for the third month (29.5 x 3 = 89.5).

These counts, sums, and interpretations are accurate to the degree that the sequence of inlay and the reconstructions are accurate.

The final "cognitive solution" or strategy occurs in the lower half of the third section. Here the two sets of 8 are the longest and largest in the mosiac, clearly an effort to fill a terminal large area with an intended number of pieces utilizing the basic set. In like manner the terminating set of 6 along the bottom edge contains oversized pieces. The final edging, after these large sets, returns to what might be considered normal-sized pieces. It was in the inlay of these pieces that the maker suddenly realized that he needed a bit more space, a 2 mm extension of the base.

The direction and sequence of inlay indicates that the pectoral hung "upside down" for the wearer and for any observer. The pattern, of course, could not be read or be distinguished by an outsider. Yet the wearer, lifting it to his view, would have the mosaic in the position in which it was inlain.

Cultural Bases of the Mosaic

The above analysis, like that of the Chamula calendar board, was undertaken as a blind test of the cognitive-sequential methodology developed for study of prehistoric compositions. In the Chamula analysis, the ethnographic field notes confirmed the findings concerned with the

13 Extreme close-up of mosaic piece #89 showing the engraved straight line running from the intrusion in the pyrite crystal to the edge. There seems to have been an outlining also of the intrusion. This is the only pyrite piece so marked and is apparently a late addition.

notational technique and the deduced cultural contents. The anlysis was able to provide the anthropologist with internal data on the notational tradition that could not have been obtained in the field.

Analysis of the Las Bocas mosaic was conducted without knowledge of the complexities of later, classic Mesoamerican calendrics or glyph writing as it is documented in the codices and carved monuments. Since the mosaic was a prehistoric artifact, no ethnographic data could be obtained. When the analytic results were brought to M. Coe, a specialist in the Olmec, early in 1973, with a presentation of two alternative possibilities for the sequence of inlaying sets of 8, one of which was found more probable and is illustrated in Fig. 10, he indicated that the two alternatives occurred in later Mesoamerican glyph writing. One had been used by the Maya of Yucatan and the other by the Mixtec of western Oaxaca (Fig. 14 a, b).

The Maya system of glyph writing contains the same quadripartite base as the mosaic, that is, four blocks set two-by-two into a "square" and read sequentially as a unit. The sets begin at top left and proceed from left to right for each two blocks on a line. The reading then proceeds by twos and fours downward. The basic blocks of four can be arranged in vertical columns and when completed the second vertical column begins to the right (Thompson 1950 and 1960). The glyphs themselves are squared images set separately as though they were tiles for inlay. The blocks of four represent neither an arithmetic nor a calendric multiple but are a pattern of the writing system.

Thompson, an authority on Maya writing, describes the writing process in terms of the cognitive adjustments continually being made. The mode of solving these problems is the same as in the mosaic.

To fit a long text into limited space, glyph blocks could be halved...or the glyph block could be quartered, allowing two or four glyphs or glyph compounds to occupy the normal space for one. Halving and quartering is often confined to the second half of a text, as though the artist realized he was running out of space. (Thompson 1972, p. 27).

The process is notational and is not found in design patterning or in simple mosaic inlay.

The Mixtec used the same unit of two as the glyph base but proceeded downward in the double column in a boustrophedon or zig-zag manner, alternating the direction of writing and reading for each set of two (Caso 1966, Figs. on pp. 19-20, 119-120). This is the style of inlay suggested for the blocks or sets of 8 in the mosiac (Fig. 9).

The Las Bocas inlay apparently relates to each of these later Mesoamerican systems of writing, and the later systems could each have derived from the former. The recognition by Coe that the Las Bocas mosaic had been inlain by a pattern which could have been an early Mesoamerican template for the sequence of "writing" and "reading" posed a number of possibilities and questions.

If the mosaic sequence could now be used as a grid with glyphs inserted into the pyrite forms it could be "read" in the manner of the later writing. The primary adjustment was that the base unit of two had been formed vertically instead of horizontally. With that direction known by the maker the sequence proceeded in the glyph manner.

This suggests that the Las Bocas mosaic may, in fact, have been based on an early Olmec system of glyphic or pictographic writing or notation, though no document of this type exists. The possibility is supported by a variety of data: (a) while the mosaic tradition continues and evolves in Mesoamerica the inlay process and patterns are in no way related to that of the Las Bocas pectoral, but the later writing systems are; (b) Olmec iconography as engraved and carved on stone indicates the presence of a number of god and symbol images which appear later in Mesoamerican writing (Joralemon 1971); (c) two stone artifacts from the late Olmec period, from Tres Zapotes and Los Tuxtlas in Vera Cruz, each approximately a thousand years later than the mosaic, contain calendric, glyphic and arithmetic data suggesting a long, indigenous development, and these are the same forms that would be used later by the classic Maya. Stela C from Tres Zapotes (the Late Formative Olmec period) is to the present the oldest dated monument in the New World with a carved date of 31 B.C. and an implied starting date of 3113 B.C. (Stirling 1940). The Las Bocas

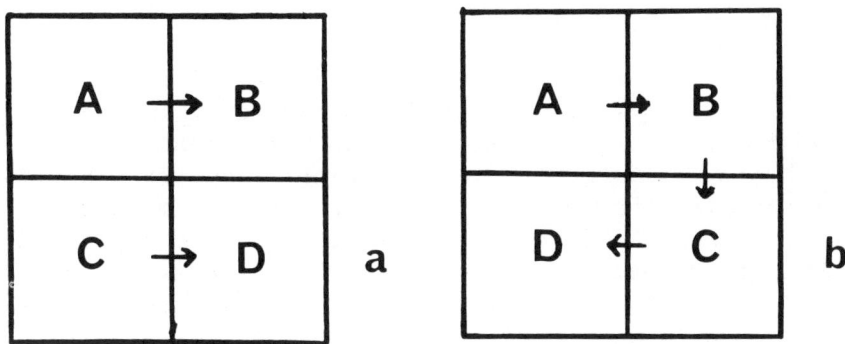

14 *(a, b) The Maya and the Mixtec systems of reading glyphs by double columns. The Maya (a) reads from left to right, the Mixtec (b) reads in a serpentine manner, each glyph attaching to the next in the sequence. The Las Bocas mosaic sequence is apparently related to both these later writing systems.*

mosaic, a thousand years earlier, suggests the presence of already evolved and advanced calendrics, arithmetic and notation, making the starting date perhaps a bit more reasonable as an historic rather than a computational point.

The Las Bocas mosaic, then, stands conceptually between these late Olmec documents and the library of known Olmec iconography since it suggests the presence of a number of corollary supporting symbol systems, calendric, arithmetic and ritualistic.

It should be noted that the Chamula calendar board (Fig. 2) represents an arithmetized sequence of unit notation divided into sets. The board could nevertheless be "read" non-numerically at a number of levels. Any mark and set in the sequence could be named and its place in the ritual, agricultural or seasonal sequence could be described (Gossen 1974). This represents a far more complex, integrated system of pre-writing notation than a mnemonic device for remembering sequences of simple events. It was cohering and was the abstract of a cultural complex. It is therefore possible that the Olmec marker of the Las Bocas mosaic could read the inlay and interpret the anomalies in a manner comparable to that derived from the analysis.

Implications

If the arithmetical, calendrical, geometrical and notational data in the Las Bocas mosaic are present as described, it documents a constellation of cultural skills and strategies that could be used in the maintenance of astronomical and calendric alignments and in the planning of symbolic architectural structures. The presence of right angle planning, center line divisions, and the perplexing intentional skewing of geometrical forms, which appear later in Mesoamerica, make their early appearance among the Olmec (Heizer, Graham and Napton 1968, Fig. 1 and Map; Aveni and Linsley 1972, pp. 528-531). The apparent intentional skewing of a balanced geometric form in the Las Bocas mosaic and the explanation presented in this paper offer an indication of the practical, cognitive, cultural bases of such usage.

The development of a mosaic technology as advanced as in the pectoral poses a problem. Mosaics are known from the Shang period in China, which just precedes the Olmec date of ca. 1000 B.C., but these are not of the Olmec quality or complexity. Did the Olmec precision skills in working stone develop indigenously?

Development of a formalized calendar requires record keeping over a considerable period. Was there a prehistoric form of calendric notation on perishable materials comparable, perhaps, to the notation on the Chamula board. If so, was this a specialized shamanistic lore brought early into the Americas, later to be amended with transoceanic accretions? The mosaic, like the Tres Zapotes stela, offers significant data and as many questions as answers.

This research was initiated in 1972 under an NSF grant for study of the European materials in order to further test and develop the theoretical and methodological bases of the author's cognitive archaeological studies.

Summary

A sequential, cognitive analysis of the earliest known Mesoamerican mosaic, from Las Bocas, Puebla, apparently documents the presence of prehistoric arithmetical, geometrical and technical skills of a surprisingly high order. There is a probability that the mosaic represents the symbolic lunar year, perhaps a particular lunar year in a solstitial or equinoctial and artificial year conjunction. The structure and patterning of the mosaic are similar to those found in later Mesoamerican glyph writing and it may therefore indicate such writing in the Olmec period.

The origins of the technology and lore supporting this highly complex artifact, ca. 1000 B.C., remain unknown.

Appendix

A Mesoamerican specialist has written to question the reality or existence of "a 'thirteen month "long" lunar year' of 383.5 days. I know of no such 'year' being in use in Mesoamerica or in any other part of the world. How can (one) claim that such a 'year' could represent the intercalary year needed to bring solar and lunar years into phase? There have been various solutions to the luni-solar problem, but none, to my knowledge, of this kind..."

The query concerns both the history of the calendar and the nature of observational astronomical periods, matters not usually studied by field archaeologists. The problem is generic, basic for an understanding of the development of any calendar system, including possible early developments in Mesoamerican calendrics.

The use of an intercalary thirteenth month to solve the "luni-solar" problem was, in fact, the most common and widely dispersed of the historical solutions attempted:

a) In early Greece the eight year cycle, the oktaetris, sought to adjust the lunar calendar to solar and seasonal observations. A month of thirty days was interjected three times within the eight year period, forming intercalary years of 383-385 days. This added month carried the name of the normal month preceding it.

The system was later improved by adoption of the Metonic cycle in which seven intercalary months were inserted into a nineteen year period, creating seven years that had counts of 384 days. Use of this cycle persisted into modern times.

b) The early Roman calendar attempted a similar solution, introducing an intercalary thirteenth month, Mercedonius, into the normal month of February every second year. The attempt was inept, since the month was only 22-23 days long, and the process was misused politically with the result that it led to basic reform of the Roman calendar under Julius Caesar.

c) The ancient Chinese used a number of simultaneous calendar systems. The civil lunar calendar was based on twelve regular lunar months of alternating 29 and 30 day lengths with an intercalary month added to adjust solar and lunar observations. Seven such months were intercalated within a nineteen year cycle, a thousand years before a comparable system was adopted by the Greeks. This thirteenth month created a lunar year of 383-385 days. It was an "unused" month within the naming sequence, and served merely as an intercalary adjustment. This system was later adopted by the Japanese.

d) At one point in the Chaldean-Babylonian period the intercalary month was interjected as above to adjust the lunar and solar periods.

e) In North America, diverse Indian groups used the thirteenth month intercalary system, sometimes inserting it as a "moon" to adjust lunar and seasonal sequences and sometimes inserting it arithmetically as a 30-day supernumerary month once every 30 months or two-and-a-half years. As in the Chinese system, this intercalated month was considered a "lost month," and was not included in the naming sequence.

The dynastic calendar of Egypt, adopted in the 3rd mill. B.C., provides indications of an original lunar calendar and mythology. Because the official Egyptian solution was "stellar" rather than solar, the attempt may have a certain comparative relevance to Mesoamerican developments. I present the analysis of Richard A. Parker (1972):

Early farmers in the Nile valley counted the periods of the river's annual flood and withdrawal and of the planting, cultivating, watering and harvesting by lunar months: four for the season of flood, four for the season of planting, four for the season of harvest and low water.

How long this process of calendar approximation went on...we have no idea, but we do know that the first calendar-maker in Egypt took the...concepts...above and by adding one very important observation formulated a calendar, which we can detect in the written records, at least by protodynastic times and probably earlier. The new element was derived from observation...of the brightest star in the sky, the dog-star Sirius...Sirius would apparently change its place in the sky at night as the earth moved around the sun and so shifted the point of observation. Eventually earth, sun, and star would be so nearly in a line that...the star would be swallowed up in the sun's brightness and the star be invisible for a period later recognized by the Egyptians as of seventy days. At the end of this period there would be a night when, just before dawn, the star would again become visible, if but momentarily, in the eastern horizon, an astronomical event known as its heliacal rising. The Egyptians saw Sirius as a goddess...and the heliacal rising of the goddess would usually occur just about the time when the...low water was coming to an end ...The rising of Sothis came to be regarded as the herald of the inundation, and it provided a...peg on which to hang a true calendar...

...The calendar year consisted of three seasons, each of four lunar months. The months had names, taken from important festivals occurring in them, and the seasons were called akhet, 'flood' or 'inundation,' peret, 'emergence,' and shomu, 'low water' or 'harvest'. The great feast of the rising of Sothis, called wep renpet, 'opener of the year'...gave its name to the fourth month of the third season, i.e. the last month of the year. Unlike the other month festivals...in each case assigned to various days of the lunar month such as the first day, first quarter-day, full-moon day, etc., the rising of Sothis was a stellar event with no relation whatever to the moon, and it was necessary therefore to arrange a calendar which would keep this event properly within the month which it named. Experience led to a very simple rule. Since twelve lunar months total on the average 354 days -- some eleven days shorter than the natural year -- whenever the rising of Sothis took place in the last eleven days of the month wep renpet, the following month was not taken as the first of the year, but as an intercalary or extra month, and the next year became a 'great' year of thirteen months, some 384 days, which would keep the rising of Sothis safely in its month.

Such...was Egypt's first calendar on record -- a normal year of three seasons, each of four lunar months, with an extra month every three, or more rarely two, years, and kept in place in the natural year by being tied to the heliacal rising of Sirius...For the predynastic and protodynastic Egyptian, such a year would have been completely adequate...

In time, however, Egypt became a well-organized kingdom, and such a fluctuating calendar, with now twelve months and now thirteen...must have become an administrative handicap....What they did was to create a schematic or averaged lunar year...of 365 days...

This was done either by counting the lunar months over a number of years and averaging them or by counting the days between the heliacal risings of Sirius (Parker 1972, pp. 14-18).

The problem, then, is generic. In my analysis of the Olmec mosaic I indicated that a thirteenth month is suggested but is not present. It seems to represent a theoretical month used in the arithmetical structuring of a symbolic lunar year in this early, formative period.

A different evaluation comes from Gordon Ekholm. He suggests that the inlay was performed in reverse, with the face down and not as it is now seen; that the corners at left, including those of the ceramic and the mosaic, were rounded off after completion of the inlay and that pendants are meant to be seen from in front and not upside down by the wearer. Ekholm also suggests that the pyrite pieces might have expanded through absorption of moisture and this may have pushed the pieces at far right over the ceramic edge.

Actually, it was a totally dry burial that preserved the inlay and lime in a rare, almost perfect state. An accidental chipping of one corner of a single pyrite piece in mail shipment has shown that the pieces are internally unchanged. The slight discoloration occurs only as an exceedingly thin surface film. Ekholm also suggests that the arithmetical structuring may be due to the technique of inlay, though this technique does not occur in later Mesoamerican decorative or symbolic mosaics. The inlay in reverse was based on a design and a shaping to a pattern made with the "face" upward. I do not, therefore, think that Ekholm's technical suggestions affect the basic analysis or the problems posed.

18

Mesoamerican Archaeoastronomy So Far

Elizabeth Chesley Baity

African Studies Program

Department of Political Science

University of North Carolina at Chapel Hill

Chapel Hill, North Carolina 27514

We are triply privileged to have taken part in the first inter-American seminar in archaeoastronomy, and to have as associates the greater number of American workers in this exciting new interdisciplinary study.

Rather than repeating the arguments you have heard, in this summarizing paper I will attempt to (1) stress the implications of these archaeoastronomical findings for archaeologists and other historians, and to (2) note lacunae in this initial inter-American seminar which suggest promising areas for future inter-disciplinary research.

The Implications of Time-Oriented American Architecture

This meeting has strongly indicated that archaeoastronomy offers the possibility of precise dating of structures oriented in accordance with such astronomical events as winter and summer solstices, spring and autumnal equinoxes, and heliacal or other significant rise/set astronomical phenomena. As is also evident from some of the studies presented, the recording of eclipses in aboriginal Mesoamerican records should afford a precise peg for dating the cultural events with which they are associated. In short, archaeoastronomy promises to be a tool of signal importance in the solution of problems of American chronology, ethnic diffusion, and cultural-ecological adaptations.

With all the information accumulated by a generation of skilled archaeologists and textural scholars, however, there is as yet no agreement as to the time and place of origin of American astronomical and calendric skills. Here, to my thinking, lies the most urgent of the unsolved problems facing the workers in the new field, and one touched here only by implication. Hawkins has shown us that Western European and Egyptian astronomers of the third millennium B.C. possessed not only sophisticated astronomical knowledge expressed in both calendrics and architecture but also the drive and attainments of proto-scientists. The other papers presented in this seminar indicate that this was no less true of pre-Columbian American astronomers. An early origin for Mesoamerican astronomical systems is implied by both Marshack and Hatch, whose arguments suggest that as early as the second millennium ancestors of the Olmecs were already keen observers of astronomical events. Marshack dates to around 1200 B.C. an Olmec mosaic pendant or "mirror" from Las Bocas which he interprets as a symbolic 354-day calendar indicating not only an early appreciation of the moon's phases but also a sophisticated attempt to reconcile lunar and solar calendars.

Marshack's hypothesis of early Olmec astronomical knowledge is supported by Hatch's earlier (1971) study of the orientation around 1000 B.C. of the ceremonial complex at La Venta, 8° west of north. She attributes this to a tradition which, in her hypothesis, was already a millennium old, equating the orientation to the setting (around midnight) at the summer solstice of the center point in the bowl of the Big Dipper. She equates Ursa Major with the American jaguar tradition. In the tradition appropriate to 2000 B.C., she notes this center point made a lower transit of the meridian, the constellation then being circumpolar. At the same time, she states, the Pleiades rose in the east, Gamma Cygni transited the meridian going west, and Scorpio set in the west. Thus, these constellations marked in turn by their midnight transits, the solstices and equinoxes.

In this seminar Hatch argues on evidence from the Madrid Codex that Mesoamerican astronomers also appear to have preserved the tradition of a midnight meridian transit of Eta Draconis, Draco being equated in her hypothesis with the American dragon-serpent motif. She correlates the iconography accompanying a 260-day count in the Madrid Codex (pp. 12-18) with constellations as they could have been observed after sunset as the year progressed, arguing that the serpent pictures represent the solar-sidereal year of 365 days while the 260-day sacred calendar count is shown alongside by glyphs, the two bands thereby constituting a celestial counterpoint system which is useful in adjusting the auguries and prognostications of the 260-day count to those of the tropical year. She suggests that the obvious importance of the dragon-serpent was due to the close relationship of Eta Draconis to the pole of the ecliptic, this star having varied its meridian transit by only one day during the time between 2000 B.C. and A.D. 500.

Knowledge and skills such as those implied by these hypotheses could have been acquired only by a long period of development elsewhere or, for those who accept the assumption that Olmec astronomy was indigenous, here in the Americas. I am intrigued, however, by Hatch's attribution to the early Olmecs of a pole-centered astronomy. Such a system is similar to that of the Chinese of the same period (Needham 1959, Vol. 3, section 20). The system is, notably, at variance with the Western Asian system of ecliptic-centered astronomy, which resulted in the creation of the solar zodiac and lunar house constructs. According to the hypothesis of Kelley (1971), these zodiacal constructs were introduced into Mesoamerica from a North Indian source not much earlier than the second century before Christ. Thus, a pole-centered system could well have been the earlier tradition.

With reference to the later Mesoamerican cultures, the evidence presented in this seminar strongly emphasizes the high importance of astronomical observations in the orientation and design of sacred structures and complexes. On the basis of both archaeological site maps and ethnographic studies, conference participants offer various examples of differing types of astronomical orientations. Hartung, while stressing that astronomical skills appear to have played a highly significant role in the orientation of Mesoamerican sacred structures and centers, suggests the need for rigorous criteria to distinguish between lines relating to visual factors and those having to do with astronomical orientations.

Aveni reports on measurements made by means of a transit and corrected site maps and analyzed by computer methods to determine the presence or absence of correlations with the azimuths of heavenly bodies, calculated for a combination of epochs, latitudes, and horizon elevations. This type of field work, which in its meticulous exactness equals that of Professor Alexander Thom in Great Britain, is matched by that of Reyman in the Southwest. It affords a methodology for the formulation and testing of archaeoastronomical hypotheses by objective precision methods: Reyman, like Coe, stresses the importance of a close study of ethnographic sources during the formulation of such hypotheses. Aveni indicates that preliminary measurements of the Caracol at Chichen Itza suggest the need for further work; his data suggest that two window alignments match extreme setting positions of Venus. He finds nearly identical orientations for the Pyramid of the Sun at Teotihuancan and the pyramids of Tenayuca and Tepozteco, all of which align approximately $17°$ north of west. He also suggests a match with a Pleiades setting position. Drawing attention to a possible $17°$ east of north "family of orientations," he stresses the need for further study of axial alignments.

Reviewing the archaeoastronomical arguments presented here, one is struck by a built-in contradiction in that while various of the hypotheses advanced are based upon acceptance of certain calendric correlation numbers, these numbers differ by two or more centuries with regard to the Zero Day implied. An error of two centuries would not invalidate solar rise/set positions, but it could raise serious questions where stellar references are implied, due to the changes caused by precession. Specifically, Owen and Hatch offer hypotheses which are dependent to a greater or lesser degree on the calendric correlation used. Using data from the Dresden Codex (pp. 51-59), Owen applies Makemson's correlation as a fit for four dates on which two solar eclipses occurred 30 days apart with a lunar eclipse between the two, in a series with a significant clustering of 24 lunar eclipses all occurring at "ten pictured intervals". In a post-seminar communication, however, she offers an improved correlation number to fit this time-specific clustering of spectacular events.

Both Coe and Reyman, as well as Cook de Leonard, emphasize the extreme significance of aboriginal documents in the formulation of archaeoastronomical hypotheses. Cook de Leonard gives evidence from ethnography which suggests that a Venus legend is depicted on the Tajin ball-court panels. She suggests the deliberate falsification by Mexican informants of the astronomical information and data correlations given the Spanish.

Another focus of the seminar has been on astronomical icons in Southwestern rock art. While there is agreement by Williamson et al., Brandt et al., Britt, Mayer and Ellis that Southwestern rock art indicates astronomical observations, Ellis does not agree that a likeness of the Crab Nebula explosion of A.D. 1054 July 4 is depicted. She offers a counter-hypothesis that this symplegma of star and crescent was a sun-moon, morning-star combination pertaining to solstice observations. She sees the winter solstice as the hub of the Southwestern ritual year and, like Reyman, advances evidence indicating that astronomical knowledge, orientations, and rituals were introduced into the Southwest by Mexican traders.

Cowan's hypothesis that effigy mounds may have been models or reproductions of constellations overhead at ritually significant times offers a variant principle with regard to sighting lines. Archaeoastronomers in general have been working in terms of single-point alignment systems in which a fixed terrestrial point coincides with a single heavenly body or constellation on the horizon. Cowan's suggestion that effigy mounds may on the contrary have been template matches or models of constellations passing overhead at times of calendric significance is of interest in connection with what appears to have been a similar system used by ancient astronomer-priests in Southeast Asia and China, where the then north pole star and circumpolar quarter points served as a cosmic model for city plans and governmental structure on the cult dictum principle "as above, so below". Smiley (private communication 1973) has advanced a similar suggestion with regard to one of the longest of the Nasca lines, which in his hypothesis may have been a template match to the Milky Way. The idea appears worthy of further research in the Americas and elsewhere.

Regrettably, Kursh and Kreps, though attending the seminar, did not present a report on their most original research indicating the use of linear constellations in Pacific navigation. This "star-path" construct differs radically from the "mandala" system of the Eurasians (to use their own highly appropriate designation) which is based on the cross within the circle. This iconographic construction depicts the heavens as a wheel centering on the North Star and adjacent constellations. Such a polar reference point is, of course, useless for tropical navigation. As they note, the "star-path" linear constellation system, though suitable only for the tropics, enigmatically appears to have been the basis of the three-path Sumerian system of Ea, Anu, and En-lil.

Further Areas for Interdisciplinary Research

The studies given here indicate the necessity of considering the Americas as a whole, with greater attention to areas on either side of Mesoamerica. No reference has been made to the circular solstice indicators of the Cahokia culture or those of Kansas, the Big Horn Medicine Wheel (Eddy 1974), nor to Mesoamerican astronomical influences in the southern mound-building areas (Baity 1973). Above all, the highly significant evidence from Peru suggesting parallels with Western European megalithic astronomy must be compared with that of Mesoamerica. Muller (1972, Fig. 18) compares a large stone ring at Tiahuanaco with strikingly similar European constructions. Even more strikingly, the Tiahuanaco people. appear to him not to have used the astronomical equinox to divide the year into two halves (a common but incorrect solution), but instead used a sun declination of $+0.5°$. It is this value that Alexander Thom finds to have been used by megalithic astronomers in structures apparently used for calendar correction. Zuidema (private communication 1973) stresses winter solstice megalithic-type indicators. He further indicates that from the Temple of the Sun in Cuzco, astronomically oriented lines radiated out into the empire.

This seminar has contribued at least two significant clues to possible Old World/New World contacts: Fuson (private communication 1973) suggests that the compass, or the idea for it, may have come with the jade complex from China or Southeast Asia sometime before A.D. 300. Kelley's lunar house parallels suggest contact with Northwestern India at about this same time.

Aside from these suggestions, the seminar has not demonstrated the new subdiscipline's potential value with regard to the problem of trans-oceanic borrowings, a topic on which highly significant interdisciplinary studies, which surely have a bearing on American astronomical systems, have been edited by Riley et al. (1971) and Ashe et al. (1971). While data gathering must continue, problem-centered research is indicated. For archaeoastronomers, the following topics for further research are suggested:

1. Further precision measurements with transit, and the codification of structural types plausibly used as orientation aides, such as "Group E-Uaxactun-type" and Caracol-type structures, openings used for diagonal sight lines, and for vertical sight lines, the use of near and distant markers, etc.

2. An analysis of the units of measurement used in the Americas, and comparisons with the megalithic yard, the Egyptian remen, the Spanish vara, and other Old World metrical units.

3. Analysis of the geometry involved and comparisons of it with that of the Eurasian megalithic astronomers.

4. Analysis of the astronomical objects and events most closely observed, including studies in-depth of lunar and Venus observations.

5. Analysis and codification of iconographic materials associated with the structures and texts.

For ethnoastronomers working with both textual and archaeological materials, special emphasis is due to:

1. The correlation of the various calendars and the coordination of the calendars with the indicated astrological, astronomical, and ritual events.

2. Studies in-depth of textual references to specific astronomical objects (sun, moon, Venus, Mars, circumpolar and ecliptical constellations) and events (solstices, equinoxes, heliacal and other rise/set phenomena).

3. Further searches for information on methods and tools.

4. Further comparative studies of New World and Old World subjective constructs (solar zodiacs and lunar houses, circumpolar and ecliptic constellation icons, etc.).

5. Comparative studies on linguistic data relative to astronomy.

These are, of course, only a few of the areas for further exploration which have been suggested by the research so interestingly presented at this seminar sponsored by the AAAS and CONACYT. We are very indebted to these organizations for this encounter.

In conclusion, we may justly claim that Mesoamerican findings in archaeoastronomy are significant for other areas as well, not only because of the almost unparalleled abundance of the American textual, archaeological, and ethnological remains, but for the precision measurements made possible by our well-preserved sacred sites. I note in particular the American astronomers' emphasis on vertical sight lines and thus on zenith and meridian passages of sun and stars, an observational system insufficiently studied in Old World archaeoastronomy. The suggested correlation of the Tzolkin with the agricultural season, as well as Reyman's indication of the functioning of astronomical data in the adaptation of high-yield cultigens in the agriculturally marginal Southwestern area, suggest parallel studies of the function of astronomy with regard to the adaptation of Old World high-yield cultigens during the Neolithic and Early Bronze Age expansions. The analyses of subjective constructs such as the solar zodiacs and lunar houses indicate the need for further comparative studies. Aside from its usefulness to prehistorians and other students of early cultural-ecological adaptations, American archaeoastronomy clearly has an important contribution to make to the history of science, whether or not it can be made to contribute decisive evidence on such controversial points as polar reversals, crustal shifts, and axial changes.

Notes on Contributors

Dr. Michael D. Coe is Professor of Anthropology at Yale University and advisor to the Center for Pre-Columbian Art, Dumbarton Oaks, Washington, D.C. Professor Coe has participated in archaeological digs in Mexico, Belize, Guatemala, Costa Rica, and in the U.S.: Massachusetts, Connecticut and Tennessee. In 1966 with the support of the National Science Foundation he began a long-term field project among Olmec sites in Veracruz, Mexico which he has recently concluded. His published books include: <u>The Maya</u> (1966) and <u>Mexico</u> (1962) both published by Thames and Hudson, London, <u>The Jaguar's Children: Pre-Classic Central Mexico</u>, (Mus. of Prim. Art, New York, 1966), <u>Americas First Civilization</u> (Am. Heritage, New York, 1968), and <u>The Maya Scribe and His World</u> (Grolier Club, 1973).

Dr. Ray A. Williamson earned his B.A. in Physics from Johns Hopkins University in 1961 and his Ph.D. in Astronomy from the University of Maryland in 1968. He was Assistant Astronomer at the University of Hawaii's Institute for Astronomy from 1967 to 1969 and is currently Assistant Dean and Tutor of St. John's College, Annapolis, Maryland. His major research interests are the kinematics of diffuse nebulae, the history of the study of interstellar matter and more recently prehistoric Pueblo Indian astronomy.

Dr. John C. Brandt obtained his B.A. in Mathematics from Washington University in 1956 and his Ph.D. in Astronomy and Astrophysics from the University of Chicago in 1960. He began his career as Assistant Professor of Astronomy at the University of California, Berkeley. He is now the Chief of the Laboratory for Solar Physics and Astrophysics of the Goddard Space Flight Center, Greenbelt, Maryland. He has published many papers on the solar wind, comets, and galaxies. For a long time both he and his wife have maintained a peripheral interest in astronomical petroglyphs.

Dr. Florence Hawley Ellis is Professor of Anthropology Emeritus at the University of New Mexico. Two of her chief concerns in a lifetime of research pertaining to the Pueblo Indians have been (1) the deviation of certain living groups from specific prehistoric stems and branches, and (2) the proposition that many rather than only a few of the categories characterizing pueblo culture had their roots in Mexico. "Within a total of 46 years of field work in the area, I found that the several years largely concentrated on representing 7 pueblos in their land claims and some others in their water claims led not only to a considerable body of new information but also to contacts of confidence in which I could speak to the elders on subjects not ordinarily discussed with outsiders. I was not prying into native religion; I was attempting to visualize as far as possible the lines of development from past to present through specific evidence rather than fireside hunches."

Claude Britt, Jr. became interested in Indians and archaeology at the age of five, after his grandfather gave him some artifacts found on his Ohio farm. Growing up in an archaeologically-rich area, his interest was whetted to the extent that by the time he completed high school he had read all the geological and archaeological literature available in local libraries.

After serving in the U.S. Armed Forces in Germany, he received the B.S. degree in geology in 1962 at Bowling Green State University. In 1967 he completed his M.A. degree at Bowling Green with a thesis dealing with the geology of archaeological sites near his hometown. When he came to Arizona to work for the park service in 1966 he first became interested in the planetarium sties of Canyon de Chelly. He completed his residence requirements for the Ph.D. in geochronology and anthropology at the University of Arizona in 1970, specializing in Early Man studies. Since that time he has worked closely with the Navajo people on the reservation both as a park ranger and as a teacher. He has published more than 90 articles on the archaeology of the Southwest and the Ohio Valley.

Notes on Contributors

Dorothy Mayer is a former graduate student in Philosophy at U.C. Berkeley. Her special interest is in the philosophy and history of physical science and mathematics. When Berkeley skies permit, she is an enthusiastic amateur astronomer, but her first love is mountaineering: she came across the Owens Valley petroglyphs while waiting for a storm to subside on the Palisades in August of 1972, and has been interested in petroglyphs ever since.

Dr. Gerald S. Hawkins was born and educated in Great Britain, having received the B.S. degree from the University of London and the Ph.D. from the University of Manchester, both in Physics. He was assistant professor of astronomy at Boston University when, in 1962, he initiated his pioneering work on Stonehenge which led to the publication of the popular book Stonehenge Decoded (Delta 1965). Professor Hawkins has also published in the field of atmospheric physics and planetary astronomy, specializing in the visual, photographic and radar study of meteors. He has served as a consultant to the Smithsonian Astrophysical Observatory. His most recent publication in the field of astro-archaeology is Beyond Stonehenge (Harper and Row, 1974).

Dr. Anthony F. Aveni was educated at Boston University and the University of Arizona, where he received a Ph.D. in Astronomy in 1965. He has been Visiting Associate Professor of Astronomy at the University of South Florida and is currently Associate Professor of Astronomy at Colgate University where from 1971 to 1973 he served as Chairman of the Department of Physics and Astronomy.

His attraction to Americanist studies, which developed as a result of his interest in the History of Astronomy, began in 1970 with the first of 9 field expeditions to Mexico, funded by the National Science Foundation, for the purpose of studying building orientations and their possible relation to astronomical events occurring along the local horizon. He has since published five papers on the subject, including a set of astronomical tables specifically designed for use by investigators working on problems relating to astronomical orientations.

In 1973, assisted by Arquitecto Hartung, with whom he has worked on the problem of orientations since 1971, he co-arranged and co-chaired the first international meeting on Pre-Columbian Archaeoastronomy which resulted in this publication.

Dr. Horst Hartung, a practicing architect, has been, since 1951, Professor for City Planning and Pre-Columbian Architecture at the University of Guadalajara, Mexico. His articles on Teotihuacan and Palenque (1964 and 1966) pointed out the importance of astronomical considerations in Pre-Columbian city planning. In 1966 and 1968 he participated in the International Congress of Americanists, presenting papers concerning the influence of the astronomical knowledge of the Maya in the planning of their ceremonial centers. These studies stimulated the publication of his book <u>Die Zeremonialzentren der Maya. Ein Beitrag zur Untersuchung der Planungsprinzipien</u> (The Ceremonial Centers of the Maya. A Contribution to the Investigation of the Principles of Planning), (AKad. Druck-u. Verlag., Graz, 1971).

Since 1971 he has been doing joint research in the field of Mesoamerican astronomical orientations with Professor Aveni, with whom he co-arranged and co-chaired this symposium.

Dr. Jonathan E. Reyman acquired his A.B. in 1965 at Indiana University and his Ph.D. in 1971 at Southern Illinois University and now holds the position of Assistant Professor of Anthropology at Illinois State University.

He has done field work utilizing archaeoastronomical techniques in New Mexico, Colorado, Arizona, northern Mexico. This fieldwork has centered on two major themes: (1) the use of astronomy for calculating seasonal changes and adjusting the subsistence base (agriculture and gathering activities) in terms of these changes, for both the modern (ethnographic) Southwestern Pueblos and for the prehistoric Anasazi; (2) the diffusion of agriculture and the necessary "tool kit" to make it work (including astronomy) from Mexico to the American Southwest in terms of the long-term and systematic exploitation of the latter area by Mexican trading (<u>pochteca</u>) groups.

As an anthropologist, he is concerned with devising general "laws" to explain human behavior. From his theoretical perspective, these "laws" must be based on and couched in terms of biocultural, ecological, and/or material explanations. His recent grants from NSF and Illinois State University have dealt with these theoretical concerns. These are reflected in his paper and were strongly emphasized in his dissertation, <u>Mexican influence on Southwestern ceremonialism</u>.

Dr. Thaddeus M. Cowan is Associate Professor of Psychology at Kansas State University. He received his undergraduate training at Centre College of Kentucky and took his doctorate in Experimental Psychology at the University of Connecticut. Prior to his present position he taught at Albion College and Oklahoma State University.

In addition to his professional interests in Mathematical Psychology, Perception, and Cognition, he has developed an active interest in Archaeoastronomy in recent years. His publications in this area have been restricted to the megalithic structures in Western Europe, and the contribution in this book represents his first investigation of Archaeoastronomy in the New World.

Nancy Kelly Owen, holds a B.S. degree from Michigan State University and a M.Ed. degree from the University of Texas. In 1955 she became interested in Mayan Studies while accompanying her husband on a sabbatical leave in Mexico. During the next ten years, availability of the Yale libraries, and the encouragement of friends in the departments of anthropology and archaeology developed this interest into a serious study, particularly of the astronomical and calendrical parts of the Dresden Codex. She has presented or published a half-dozen papers on the subject since 1970. In 1974 at McMaster University (Hamilton, Ont.) she presented a paper on "The Dresden Codex and the Velikovsky Catastrophy Dates".

Dr. Charles H. Smiley received his Ph.D. in mathematics from the University of California, Berkeley in 1927. For 39 years until his retirement from teaching in 1969, he was a professor at Brown University. He was Chairman of the Department of Astronomy and Director of the Ladd Observatory for 31 years. He has led sixteen solar eclipse expeditions and has written numerous papers on solar eclipses.

The first of his 19 papers on Maya Astronomy was published in 1960 and contained his correlation of the Maya and Chirstian calendars, based on astronomical evidence in the Dresden Codex.

Dr. David Kelley is an archaeologist with a strong interest in ancient Maya astronomy. He received his B.A. from Harvard in 1949 and his Ph.D. also from Harvard in 1957. While at Harvard he studied Maya writing under Tozzer where he picked up some astronomical knowledge from his calendrical studies. In 1953-54, he dug early agricultural sites in Tamaulipas with R. S. MacNeish and in 1957 he received a Fulbright Research Fellowship to do an archaeological survey of the north coast of Peru. Dr. Kelley has also worked on comparative studies of Eurasian and Mesoamerican calendar systems as well as Mesoamerican writing systems where he focused upon Maya writing and the astronomical problems presented in the Maya inscriptions and codices. His numerous publications on calendrics include a book (with Hugh Moran): The Alphabet and Ancient Calendar Signs (Daily Press, Palo Alto, 1969).

Carmen Cook de Leonard was educated in Mexico and has devoted a large portion of her vigorous and active life to archaeological pursuits. (She is also a teacher, newspaper reporter and newspaper and book editor.) Mrs. Cook excavated at Acalpixcan and Chalcatzingo and has published interpretations of the Olmec rock carvings at the latter site. She has served as director of the Sociedad Alemana Mexicanista and is president of the Centro de Investigaciones Antropologicas de Mexico. Her many plaudits as a publisher include the recent Cien Anos de Arqueologia Mexicana (SAM, Mexico, 1969), the eleventh volume in the Mexico Antiguo series.

Dr. Marion Popenoe Hatch received her Ph.D. in the Anthropology Department of the University of California, Berkeley, doing research on the subject of Maya epigraphy. She first became interested in early Mesoamerican astronomy while investigating possible explanations for the orientation of the Olmec site of La Venta in Mexico, which resulted in the paper "An Hypothesis on Olmec Astronomy," Contributions of the University of California Archaeological Research Facility, No. 13, pp. 1-64, 1971.

Notes on Contributors

Alexander Marshack is a proto-historian, presently serving as Research Associate in the Peabody Museum of Archaeology and Ethnology at Harvard University. His principal area of interest is Upper Paleolithic materials, in which field, since 1966, he has been funded by NSF, the Wenner-Gren Foundation and the Bollingen Foundation to do research. He is the author of numerous publications including a book: <u>The Roots of Civilization: The Cognitive Beginning of Man's First Art, Symbol, and Notation</u> (McGraw Hill, New York, 1972).

Dr. Elizabeth Chesley Baity was educated at the Texas University for Women in Denton (B.A., B.S.), and the University of North Carolina at Chapel Hill (M.A., anthropology, 1962; Ph.D., anthropology, 1968), with special studies in protohistory at the School of Oriental and African Studies at the University of London and the Universities of Geneva and Teheran. With Dr. H. G. Baity and their sons, William and Philip, she has lived in South America and Europe. Her own work in literacy has taken her to Asia, Africa, and Oceania. Among her publications are two books for younger people: <u>Americans Before Columbus</u> (New York: Viking Press, 1951) and <u>America Before Man</u> (revised edition, New York: Viking Press, 1964).

Coming from a family with strong astronomical interests, she researched in Eurasia for a thesis and a dissertation on the Iberian fire bull, and solstice festivals and their protohistoric parallels as depicted on rock art along a continuum extending from Spain through Africa to India. In 1973 (<u>Current Anthropology</u>, 14, No. 4, pp. 389-449), she published her synthesizing review of studies in the subdiscipline of Archaeoastronomy.

References

Alexander, J.K., J.C. Brandt, S.P. Maran, and T.P. Stecher 1971. The Gum Nebula: further evidence from spacecraft and ground-based instruments. Astrophysical Journal, 167, pp. 487-490.

Allen, Richard Hinckley 1963. Star names, their lore and meaning. Dover, New York.

Anales de Cuauhtitlan, see Velazquez, Primo F.

Anderson, Arthur O. and Charles E. Dibble 1951. Florentine Codex, Book 2. Monographs of the School of American Research, no. 14, Pt. 3. Santa Fe.

Andrews IV, E.W. 1968. Torre cilindrica de las ruinas P. Rico, Campeche. Bol. Inst. Nac. Antr. e Hist., no. 31, pp. 7-13. Mexico.

_____ and G. Stuart 1968. The ruins of Ikil, Yucatan, Mexico. Middle American Research Institute, Publication no. 31, pp. 69-80.

Ashe, Geoffrey, ed. 1971. The quest for America. Praegee Press, New York.

Atkinson, R.J.C. 1960. Stonehenge. Penguin Books, Baltimore.

Attneave, F. 1954. Some informational aspects of visual perception. Psychological Review, 61, pp. 183-193.

Aveni, A.F. 1972. Astronomical tables intended for use in astro-archaeological studies. American Antiquity 37, pp. 531-540.

_____, S.L. Gibbs, and H. Hartung 1974. The astronomical significance of the Caracol of Chichen Itza, Yucatan, Mexico. Paper presented at XLI Int'l. Cong. Americanists, 2-7 Sept., Mexico City.

_____ and R.M. Linsley 1972. Mound J, Monte Alban: possible astronomical orientation. American Antiquity, 37, pp. 528-531.

Baade, W. 1942. The Crab Nebula. Astrophys. J., 96, pp. 188-198.

Baer, Philip, and Mary Baer n.d. Materials on Lacandon culture of the Petha region. Microfilm Collection of Manuscripts on Middle American Cultural Antropology, 28. Chicago.

Baity, E. 1969. Some implications of astroarchaeology for Americanists. Proceedings of the 38th International Cong. Americanists at Stuttgart, 1, pp. 85-94.

_____ 1973. Archaeoastronomy and ethnoastronomy so far. Current Anthropology, 14, no. 4, pp. 389-449.

Bannister, B. 1965. Tree-ring dating of the archaeological sites in the Chaco canyon region, New Mexico. Southwestern Monuments Association, Technical Series, 6, pt. 2, pp. 119-214.

Barguet, P. 1962. Le temple d'Amon-Re a Karnak. Impr. de l'Institut Francais d'Archaeologie Orientale, Cairo.

Barthel, Thomas S. 1954. Maya epigraphy: some remarks on the affix 'al'. Proceedings of the 30th International Congress of Americanists, 1952, pp. 45-49. Cambridge, London.

_____ 1969. Intentos de las lectura de los afijos de los jeroglificos en los codices mayas. Seminario de Estudios de la Escritura Maya, cuaderno 2. Mexico.

References

Benedict, Ruth 1935. Zuni mythology. Columbia University Contributions to Anthropology, 21.

Bernal, Ignacio 1969. The Olmec world. Trans. by Doris Heyden and Fernando Horcasitas. University of California Press, Berkeley and Los Angeles.

Bleibtreu-Ehrenberg, Gisela 1970. Homosexualitaet und transvestition. Antropos., 65, St. Agustin.

Bourke, John 1884. The snake-dance of the Moqui of Arizona. Charles Scribner's Sons, New York.

Bowditch, Charles P. 1910. The numeration, calendar systems and astronomical knowledge of the Mayas. Harvard University Press, Cambridge, Massachusetts.

Brandt, J.C., S.P. Maran, and T.P. Stecher 1971a. Astronomers ask archaeologists aid. Archaeology, 21, no. 4, p. 360.

_____, _____, _____, and D.L. Crawford 1971b. The Gum Nebula: fossil Stromgren sphere of the Vela X Supernova. Astrophys. J. Letters, 163, p. L99.

_____, _____, R.S. Harrington, and M.M. Kennedy 1972. A northern California pictograph that may be another record of the Crab Nebula Supernova explosion. Bull. Am. Astrom. Soc., 3, p. 319.

_____, _____, _____, _____, R. Williamson, C. Cochran, W.J. Kennedy, and V.D. Chamberlain 1973. Possible records of the Crab Nebula supernova in the western United States. Bull. Am. Astron. Soc., 5, p. 20.

Britt, Jr., Claude 1971. Canyon de Chelly and the Anasazi. The Redskin, 6, no. 3, pp. 96-99.

_____ 1973. Early Navajo astronomical pictographs in Canyon de Chelly, Northeastern Arizona, U.S.A. These proceedings.

Brown, Peter Lancaster 1971. <u>What star is that?</u> Thames and Hudson. London.

Bruce, Roberto D., Carlos Robles U., and Enriqueta Ramos Chao 1971. Los Lacandones 2, cosmovision maya. <u>Instituto Nacional de Antropologia e Historia, Departamento de Investigaciones Antropologicas, Publicaciones 26.</u> Mexico.

Bunzel, Ruth 1932. Introduction to Zuni ceremonialism. <u>Bureau of American Ethnology</u>, <u>Annual Report</u>, 47, pp. 467-544.

Burland, Cottie A. 1950. <u>The four directions of time: an account of page one of Codex Fejervary-Mayer.</u> Museum of Navajo Ceremonial Art, Santa Fe.

───── 1952. The Toltec-style calendar of Mexico. <u>Thirtieth International Congress of Americanists</u>, 23-26. Cambridge, England.

Burgh, Robert F. 1934. Ruins of the Far View House group. Ruins survey, W.P.A. Project 496, U.S. Department of Interior, National Park Service. Mesa Verde National Park.

Caso. A. 1938. Exploraciones en Oaxaca. <u>Inst. Pan. Am. Geog. e Hist.</u>, Publ. no. 34.

───── 1954. <u>El pueblo del sol</u>. Fondo de Cultura Economica. Mexico.

───── 1958. <u>The Aztecs: people of the sun</u>. Univ. of Oklahoma Press, Norman.

───── 1966. <u>Interpretacion del Codice Colombina</u>, with Mary E. Smith, <u>Los Glosas del Codice Colombina</u>. Sociedad Mexicana de Antropologia, Mexico City.

───── 1971. Calendrical systems of central Mexico. <u>Handbook of Middle American Indians</u>, 10, pp. 333-48. University of Texas Press, Austin.

References 399

Codex Borgia 1904. Seler edition. See Seler, Eduard.

Codex Dresden 1892. Forstemann edition. Die Maya-Handschrift der Koniglichen offentlichen Bibliothek zu Dresden. R. Bertling, Dresden.

_____ 1930. Villacorta edition. Codex dresdenis. Reproducido y desorrollada por J. Antonio Villacorta C., y Carlos A. Villacorta, Guatemala, C.A.

_____ 1932. Gates edition. The Dresden Codex. Reproduced from tracings of the original colorings finished by hand. The Maya Society at Johns Hopkins University, Baltimore.

_____ 1972. Thompson edition. A commentary on the Dresden Codex: a Maya hieroglyphic book. Am. Phil. Soc., Philadelphia.

Codice Chimalpopoca. See Velazquez, Primo F.

Coe, Michael D. 1968. *America's first civilization*. American Heritage and the Smithsonian Institution, New York and Washington.

_____ 1973. *The Maya scribe and his world*. The Grolier Club, New York.

Colton, Harold S. 1959. *Hopi Kachina dolls*. University of New Mexico Press, Albuquerque.

Cook de Leonard, Carmen and Don Leonard 1949. Costumbres mortuarias de los Indios Huaves. *El Mexico Antiguo*. Sociedad Alemana Mexicanista, Mexico.

_____ 1967. Sculptures and rock carvings at Chalcatzingo, Morelos. In *Studies in Olmec Archaeology*. Contributions of the University of California Archaeological Facility, no. 3. U. of California, Berkeley.

Cowan, T.N. 1971. Megalithic lunar observatories: a review. *Journal for the History of Astronomy*, 2, pp. 202-203.

Cristophe, L.A. 1965. Abou Simbel. Mercky, Brussels.

Curtis, E.S. 1926a. The North American Indian, 2. Johnson Reprint Co., New York.

———— 1926b. The North American Indian, 15. Johnson Reprint Co., New York.

Cushing, Frank H. 1941. My adventures in Zuni. Peripatetic Press, Santa Fe.

———— 1967. Reprint: My adventures in Zuni (Filter Press, Palmer Lake, CO) from original article in The Century Magazine, 25 and 26, 1882-1883.

Davies, R.D. and F.G. Smith (Eds.) 1971. The Crab Nebula. I.A.U. Symposium, no. 46. Reidel, Holland.

DeHarport, D.L. 1951. An archaeological survey of Canon de Chelly: preliminary report of the field session, 1948, 1949, and 1950. El Palacio, 58, pp. 35-48.

———— 1959. An archaeological survey of Canyon de Chelly, northeastern Arizona: a Puebloan community through time. Doctoral Dissertation. 1657 pp. Harvard University, Cambridge.

Dobrizhoffer, M. 1822. An account of the Abipones, an equestrian people of Paraguay. John Murray, London.

Dow, J.W. 1967. Astronomical orientations at Teotihuacan, a case study in astroarchaeology. American Antiquity, 32, pp. 326-334.

Duncan, J.C. 1921. Changes observed in the Crab Nebula in Taurus. Proc. Nat'l. Acad. Sci. U.S., 7, pp. 179-180.

———— 1939. Second report on the expansion of the Crab Nebula. Astrophys. J., 89, pp. 482-485.

Duran, Fray Diego 1971. Book of the gods and rites and the ancient calendar. University of Oklahoma Press, Norman.

Dutton, B.P. 1963. Sun Father's way: the Kiva murals of Kuava. University of New Mexico Press, Alburquerque.

Duyvendak, J.J.L. 1942. Further data bearing on the identification of the Crab Nebula with the supernova of 1054 A.D. Pub. Astron. Soc. Pacific, 54, pp. 91-94.

Eddy, John A. 1974. Astronomical alignment of the big horn medicine wheel. Science, 184, pp. 1035-1043.

Ellis, Florence Hawley and Laurens Hammack 1968. The inner sanctum of Feather Cave, a Mogollon sun and earth shrine linking Mexico and the Southwest. American Antiquity, 33, no. 1, pp. 25-44.

_____ 1973. A thousand years of the Pueblo Sun-Moon-Star Calendar. These proceedings.

Escalona Ramos, Alberto 1940. Chronologia y astronomia Maya-Mexica. Mexico, D.F.

Fenichel, Otto 1953. The psychology of transvestitism. In collected papers of Otto Fenichel, First Series. W.W. Norton and Company, New York.

Ferguson, T.S. 1962. One fold and one shepherd. Olympus Publishing Co., Salt Lake City.

Fewkes, J. Walter 1897. Tusayan katcinas. Bureau of American Ethnology Annual Report, 15, pp. 267-312. Washington.

_____ 1917. A prehistoric Mesa Verde pueblo and its people. Smithsonian Institution Annual Report, 1916, pp. 461-488.

_____ 1917a. Far View House--a pure type of Pueblo ruin. Art and Archaeology, 6, no. 6, pp. 166-141.

Flannery, K. 1972. The cultural evolution of civilizations. Ann. Rev. Ecol. System, 3, p. 399.

Forde, C. Daryll 1931. Hopi agriculture and land ownership. *Jr. Royal Institute*, 61, pp. 357-405.

Forstemann, Ernst 1886. *Erlauterung zur Mayahandschrift der Koniglichen offentlichen bibliothek zu Dresden.* Warnatz & Lehmann, Dresden.

———— 1903. *Commentar zur Pariser Handschrift.* (Codex Peresianus) Danzig.

———— 1906. Commentary on the Maya manuscript in the Royal Public Library of Dresden. *Papers of the Peabody Museum, Harvard University*, 6, no. 2. Cambridge.

Fowke, G. 1902. *Archaeological history of Ohio.* Ohio State Archaeological and Historical Society, Columbus.

Frink, Maurice 1968. *Fort Defiance and the Navajo.* Pruett Press. Boulder, Colorado.

Fuson, Robert H. 1969. The orientation of Mayan ceremonial centers. *Annals of the Association of American Geographers*, 59, pp. 494-511.

Garcia Payon, Jose 1959. Ensayo de interpretacion de los bajorrelieves de los tableros del juego de pelota sur del Tajin, Veracruz. *El Mexico Antiguo.* Sociedad Alemanna Mexicanista, Mexico.

Garibay, Angel Ma. 1965. *Teogonia e historia de los Mexicanos.* Editorial Porrua, Mexico.

Gates, William 1911. *Madrid codex* (Tro-cortesianus). Photographed for William E. Gates. Point Loma, California.

———— 1932. The Dresden Codex. *Maya Society Publication*, no. 2, Johns Hopkins University, Baltimore.

Giese, M. and V.E.v. Gebsattel 1962. *Psychopathologie der sexualitaet.* F. Enke, Stuttgart.

Girard, Rafael 1966. *Los Mayas.* Libro Mex., Mexico.

Glifford, E.W. 1936. Northeastern and western Yavapai. University of California Publications in American Archaeology and Ethnology, 34, pp. 247-354.

Goldstine, Herman H. 1973. New and full moon: 1001 B.C. to A.D. 1651. Memoirs of the Am. Phil. Soc., 94, Philadelphia.

Gossen, Gary 1974. A Chamula solar calendar board from Chiapas, Mexico. In Proceedings of the Symposium on Mesoamerican Archaeology, edited by Norman Hammond. Duckworth and University of Texas, London and Austin.

Grant, Campbell 1966. The rock paintings of the Chumash. University of California Press, Berkeley.

_____ 1967. Rock art of the American Indian. Thomas Y. Crowell, Co., New York.

Hagar, Stansbury 1912. Zodiacal symbolism of the Mexican and Maya months and day signs. 17th International Congress of Americanists, pp. 140-59. Mexico.

Haile, Berard O.F.M. 1947. Starlore among the Navajo. Museum of Navajo Ceremonial Art, Santa Fe.

Hartung, H. 1968. Consideraciones urbanisticas sobre los trazos de los centros ceremoniales de Tikal, Copan, Uxmal y Chichen Itza. Proc. 37th Int. Cong. Amer., 1, pp. 121-125.

_____ 1971. Die zeremonialzentren der Maya. Akademische Druck-u. Verlagsanstalt, Graz.

_____ 1972. Consideraciones sobre los trazos de centros ceremoniales maya: influencia de los conocimientos astronomicos en el acomodo de las construcciones. Proc. 38th Int. Cong. Amer., 4, pp. 17-26.

Hatch, Marion Popenoe 1971. An hypothesis on Olmec astronomy, with special reference to the La Venta site. *Contributions of the University of California Archaeological Research Facility*, no. 13, pp. 1-64, Berkeley.

Heizer, Robert F. and M.A. Baumhoff 1962. *Prehistoric rock art of Nevada and eastern California*. University of California Press, Verkeley and Los Angeles.

_____, J.A. Graham, and L.K. Napton 1968. The 1968 investigations at La Venta. *Contributions of the University of California Archaeological Research Facility*, no. 5. University of California, Berkeley.

Hawkins, G.S. 1963. Stonehenge decoded. *Nature*, 200, pp. 306-308.

_____ 1964. Stonehenge: a Neolithic computer. *Nature*, 202, pp. 1258-1261.

_____ 1965. *Stonehenge decoded*. Doubleday and Co., New York.

_____ 1966. Astro-archaeology. *Smithsonian Research in Space Science* Report no. 226.

_____ and S. Rosenthal 1967. 5,000- and 10,000-year star catalogs. *Smithsonian Contr. Astrophys.*, 10, pp. 141-179.

_____ 1968. Astro-archaeology. *Vistas in Astronomy*, 10, ed. by Arthur Beer, pp. 45-88. Pergamon Press, New York.

_____ 1973. Astro-archaeology. The Unwritten Evidence. These proceedings.

_____ 1973. Astronomical alignments in Britain, Egypt and Peru. *Philosophical Transactions of the Royal Society*, 276, pp. 157-167.

Ho, Peng-Yoke, F.W. Paar, and P.W. Parsons 1972. The Chinese Guest Star of A.D. 1054 and the Crab Nebula. *Vistas in Astronomy*, 13, pp. 1-13.

Hochleitner, Franz J. 1970. An attempt at a chronological-astronomical interpretation of the numbers and day signs of pages 2-24 of the Dresden codex. Translated by Monroe Edmonson. Tulane University Press, Tulane, La.

Hoyle, F. 1966. Stonehenge--an eclipse predictor. Nature, 211, pp. 454-456.

_____ 1966. Speculations on Stonehenge. Antiquity, 40, pp. 262-276.

_____ 1972. From Stonehenge to modern cosmology. Freeman Press, San Francisco.

Hubble, E. 1928. Novae or temporary stars. Astron. Soc. Pacific Leaflet, no. 14, pp. 55-58.

Hyde, G.D. 1962. Indians of the Woodlands: from prehistoric times to 1727. University of Oklahoma Press, Norman.

Joralemon, Peter 1971. A study of Olmec iconography. Studies in Pre-Columbian Art and Archaeology, no. 7, Dumbarton Oaks, Washington.

Judd, N.M. 1964. The architecture of Pueblo Bonito. Smithsonian Miscellaneous Collections, 147.

Kelley, David 1962. A history of the decipherment of Maya script. Anthropological Linguistics, 4, no. 8, pp. 1-48. Bloomington, Indiana.

_____ 1971. Cosmology, social organization, and settlement patterns. Paper presented at the 36th Annual meeting of the Society of American Archaeology, Norman, Oklahoma.

_____ 1972. The nine lords of the night. Contributions of the University of California Archaeological Research Facility, no. 16, pp. 53-68. University of California Press, Berkeley, California.

Kidder, Alfred, J.D. Jennings, and E.M. Shook 1946. Excavations at Kaminaljuyu. Carnegie Institution of Washington, Publication, no. 561. Washington.

_____ 1949a. Certain archaeological specimens from Guatemala. Carnegie Institution of Washington, Notes on Middle American Archaeology and Ethnology, no. 92. Washington.

_____ 1949b. Jades from Guatemala. Carnegie Institution of Washington, Notes on Middle American Archaeology and Ethnology, no. 91. Washington.

Kinsey, Alfred C., et.al. 1948. Sexual behavior in the human male. W. B. Saunders Co., Philadelphia.

_____ 1953. Sexual behavior in the human female. W. B. Saunders Co., Philadelphia.

Knorosov, Y.V. 1967. Selected chapters from The writing of the Maya Indians. Translated by S. Coe. Peabody Museum, Harvard University, Russian Translation Series, 4. Cambridge.

Kosok, P. and Maria Reiche 1949. Ancient drawings on the desert of Peru. Archaeology, 2, pp. 206-215.

Kursh, C.O. and T. Kreps n.d. Starpaths: linear constellations in tropical navigation. Unpublished Manuscript. 99 pp. with 21 pp. of illustrations.

Lange, Charles H. 1959. Cochiti: A New Mexico pueblo past and present. University of Texas Press, Austin.

_____ and Carroll L. Riley 1970. The Southwestern journals of Adolph F. Bandelier: 1883-1884. University of New Mexico Press, Albuquerque.

LaPaz, L. 1948. Meteoritical pictographs. Popular Astronomy, 56, pp. 324-330.

Lehmann, Walter 1938. Die geschicte der konigreiche von Colhuacan und Mexiko. (Codex Chimalpopocatl) Trans. and notated by W.L. Stuttgart. Wikohlhammel, Stuttgart and Berlin.

Leland, C. 1898. *The Algonquin legends of New England.* Houghton Mifflin, New York.

Leon, Juan de 1954. Diccionario Quiche-Espanol. Guatemala Editorial Landivar, Guatemala.

Leon-Portilla, Miquel 1963. *Aztec thought and culture.* University of Oklahoma Press, Norman.

Levi-Strauss, Claude 1964. *Le cru et le cuit.* Librairie Plon, Paris.

_____ 1969. *The raw and the cooked.* Harper and Row, New York.

Link, Martin 1968. *Navajo--a century of progress--1868 to 1968.* Navajo Tribal Press, Window Rock, Arizona.

Lockyer, J.N. 1884. *Dawn of astronomy.* Cassell, London. Reprinted by MIT Press, 1964, Cambridge.

_____ 1906. *Stonehenge and other British stone monuments.* Macmillan Co., London.

Loeb, E.M. 1926. Pana folkways. *University of California Publications in American Archaeology and Ethnology,* 19, pp. 149-404.

Lumholtz, Carl 1900. Symbolism of the Huichol Indians. *American Museum of Natural History,* Memoirs 3. New York.

Macgowan, K. 1945. The orientation of middle American sites. *American Antiquity,* 11, p. 118.

Makemson, Maude W. 1946. The Maya correlation problem. Publications of Vassar College Observatory, no. 5, Poughkeepsie, New York.

Mallory, G. 1893. Picture-writing of the American Indians. *Tenth Annual Report of the Bureau of Ethnology to the Secretary of the Smithsonian Institution,* 1888-1889, by J.W. Powell, director. Government Printing Office, Washington. (Reprinted in 2 vols. by Dover Publications, New York, 1972).

Maran, S.P., J.C. Brandt, and T.P. Stretcher (eds.) 1973. The Gum Nebula and related problems. National Technical Information Service NASA SP-332, Springfield, Virginia.

Marcus, J. 1973. Territorial organization of the lowland classic Maya. Science, 180, pp. 911-916.

Marquina, I., and Ruiz, L. 1934. La orientation de las piramides. 25th Cong. Int. Americanistas. Univ. Nac. de La Plata, Buenos Aires, 2, pp. 101-106.

_____ 1951. Arquitectura prehispanica. Memorias del Instituto Nat'l. de Anthropologia e Historia, 1, Mexico.

_____ 1964. Arquitectura prehispanica (revised edition). Instituto Nacional de Antropologia y Historia, Memorias 1, Mexico City.

Marricott, Alice and Carol Rachlin 1968. American indian mythology. Crowell, New York.

Marshack, Alexander 1972a. Cognitive aspects of Upper Paleolithic engraving. Current Anthropology 13, pp. 445-477.

_____ 1972b. The roots of civilization. McGraw-Hill and Weidenfeld and Nicolson, New York and London.

_____ 1972c. Upper Paeolithic notation and symbol. Science 178, pp. 817-828.

_____ 1974. The Chamula calendar board: an internal and comparative analysis. In Proceedings of the Symposium on Mesoamerican Archaeology, edited by Norman Hammond. Duckworth and University of Texas, London and Austin.

Maudslay, A.P. 1889-1902. Archaeology, biologia Centrali Americana. R.H. Porter and Dalau and Company, London.

_____ 1912. A note on the position and extent of the great temple enclosure of Tenochtitlan. 18th Cong. Int. Americanistas, London, pp. 173-175.

Mayall, N.U. and J.H. Oort 1942. Further data bearing on the identification of the Crab Nebula with the supernova of 1054 A.D. Pub. Astron. Soc. Pacific, 54, pp. 95-104.

_____ 1962. The story of the Crab Nebula. Science, 137, pp. 91-102.

McAllester, David P. 1956. The myth and prayers of the great star chant and the myth of the Coyote chant. Navajo Religion Ser., 4, Santa Fe.

McKusick, M. 1964. Men of Ancient Iowa. Iowa State University Press, Ames.

Meadows, A.J. 1972. Science and controversy. MIT Press, Cambridge, Massachusetts.

Meeus, Jean, C.C. Grosjean, and W. Vanderleen 1965. Canon of solar eclipses. Pergamon Press, New York.

Meinshausen, M. 1913. Uber sonnen-und mondfinsternisse in der Dresdener Mayahandschrift. Zeitschrift fur Ethnolgie, 45, pp. 221-227.

Miller, Arthur G. 1973. Archaeological investigations of the Quintana Roo Project: a preliminary report of the 1973 season. Contributions of the University of California Archaeological Research Facility, no. 18, pp. 137-148. Berkeley.

Miller, Williams C. 1955. Two possible astronomical pictographs found in northern Arizona. Plateau, 27, no. 4, pp. 6-12. Museum of Northern Arizona.

_____ 1955a. Two prehistoric drawings of possible astronomical significance. Astron. Soc. Pacific Leaflet, no. 314, pp. 1-8.

Millon, R. 1968. Urbanization at Teotihuacan: the Teotihuacan mapping project. Proc. 37th Int. Cong. Amer., 1, pp. 105-120.

Mindeleff, Cosmos 1897. The cliff ruins of Canyon de Chelly, Arizona. <u>Sixteenth Annual Report of the Bureau of Ethnology</u>, 1894-95. By J.W. Powell, Director, Government Printing Office, Washington, D.C., pp. 79-198.

Minkowski, R. 1971. Comments on supernova remnants and ancient nova. In <u>The Crab Nebula</u>, R.D. Davies and F.G. Smith (eds.). Reidel, Holland, pp. 242-247.

Molloy, J.P., R. White, T.P. Culbert, and D.W. Kayers 1973. The Casa Grande archaeological zone precolumbian astronomical observation. Preprint.

Moore, Patrick 1965. <u>Naked-eye astronomy</u>. W.W. Norton and Co., New York.

Moran, Hugh A., and David H. Kelley 1969. <u>The alphabet and the ancient calendar signs</u>. <u>Daily Press</u>, Palo Alto.

Motul Dictionary 1929. <u>Diccionario de Motul, maya-espanol, atribuido a Fray Antonio de Ciudad Real y Arte de lengua maya por Fray Juan Coronel</u>. J. Martinez Hernandez, ed. Merida.

Muller, Rolf 1972. Sonne, mond und sterne urer dem reich der Inka. Springer Press, Berlin, Heidelberg, and New York.

Museum Notes 1938. Sacred places and shrines of the Navajo. <u>Notes of the Museum of N. Arizona</u>, 11, no. 3, pp. 31-33. Flagstaff.

Needham, J. 1959. Mathematics and the science of the heavens and the earth. <u>Science and Civilization in China</u>, 3. University Press, Cambridge.

_____, Wang Ling, and D.J. de Solla Price 1960. <u>Heavenly clockwork</u>. University Press, Cambridge.

Nequatewa, Edmund 1931. The place of corn and feathers in Hopi ceremonies, Museum Notes, 3, no. 9. Reprint in Hopi customs, folklore, and ceremonies, Museum of Northern Arizona Reprint Series, no. 4, 1954.

Neugebauer, Paul Victor 1912. Tafeln zur astronomische chronologie, 1. J.C. Hinrichs, Leipzig.

———— 1929. Astronomishe chronologie, 1 and 2. W. de Gruyter & Co., Berlin and Leipzig.

Newcomb, Franc J. 1967. Navaho folk tales. Mus. of Navaho Ceremonial Art, Santa Fe.

Nicholson, Henry B. 1971. Religion in pre-Hispanic central Mexico. In Handbook of Middle American Indians, 10, pp. 395-446. University of Texas Press, Austin.

Norton, Arthur P. and J. Gall Inglis 1959. Norton's star atlas and telescopic handbook. Sky Publishing Corporation, Cambridge, Massachusetts.

Nuttall, Zelia 1901. The fundamental principles of Old and New World civilizations. Archaeological and Ethnological Papers of the Peabody Museum, Harvard University, 2, pp. 1-602. Cambridge.

———— 1906. The astronomical methods of the ancient Mexicans. Boas Anniversary Volume, edited by B. Laufer, pp. 290-298. Stechert, New York.

Oort, Jan H. 1957. The Crab Nebula. Scientific American, 196, no. 3, pp. 53-60.

Oppolzer, Theodor R. 1962. Canon of eclipses. Translated by Owen Gingerich. Dover Publications, New York.

Ortiz, Alfonso 1969. The Tewa World. University of Chicago Press, Chicago.

Owen, Nancy K. 1972. Astronomical events on the dates of the Dresden codex. Cal. San Dieguito Pressman Association, Encinitas.

_____ 1973. Mayan calendrical correlation numbers based on the Dresden codex and eclipses visible in Chiapas. Phto-offset circulated Sept. 1973.

Paddock, J. 1966. Oaxaca in ancient Mesoamerica. In *Ancient Oaxaca*, edited by J. Paddock, Stanford University Press, Stanford, p. 126.

Parker, Richard A. 1972. The calendars and chronology. In *The Legacy of Egypt*. Edited by J. Harris. Oxford University Press, Oxford.

Parsons, Elsie Clews 1925. A Pueblo indian journal, 1920-1921. *Memoirs, American Anthropological Association*, no. 32.

_____ 1933. Hopi and Zuni ceremonialism. *Memoirs, American Anthropological Association*, no. 39.

_____ 1936. *Hopi journal of Alexander A. Stephen*. Columbia University Press, New York.

_____ 1939. *Pueblo Indian religion*. 2 vols. University of Chicago Press, Chicago.

Parsons, R. 1962. Indian mounds of northeast Iowa as soil genesis benchmarks. *Journal of the Iowa Archaeological Society*, 12, pp. 1-70.

Perez, Dictionary 1866-77. Diccionario de la lengua Maya. Compiled by J. Pio Perez, completed by C.H. Berendt. Merida.

Pogo, Alexander 1937. Maya astronomy. *Carnegie Institution of Washington, Year Book*, 36, pp. 2435. Washington.

Pollock, H.E.D. 1936. Round structures of aboriginal Middle America. *Carnegie Institution of Washington, Publication* 471. Washington.

Proskouriakoff, Tatiana 1946. An album of Maya architecture. *Carnegie Institution of Washington, Publication* 558. Washington.

Redfield, Robert, and Alfonzo Villa Rojas 1967. <u>Chan Kom, a Maya village</u>. University of Chicago Press, Chicago.

Reiche, M. 1949. Mystery on the desert. Lima, Peru.

_____ 1968. Mystery on the desert. Nazca, Peru.

Reyman, Jonathan E. and Frank C. Sanders n.d. Archaeoastronomy: theory, method, and techniques, (manuscript in preparation).

_____ 1971. <u>Mexican influence on southwestern ceremonialism</u>. Ph.D. dissertation, Southern Illinois University, University Microfilms, Ann Arbor.

_____ 1972. Review of <u>Megalithic lunar observatories</u>, by A. Thom. <u>American Anthropologist</u>, 74, pp. 945-946.

_____ 1973. Sources for the formulation of archaeoastronomical hypotheses. Paper presented at the Southern Anthropology Association Meeting, Wrightsville Beach, North Carolina.

Richards, Francis Shakespeare 1932. Notes on the age of the Great Temple of Ammon at Karnak. <u>Survey of Egypt</u>, no. 38. Cairo Gov't. Press, Cairo.

Ricketson, Olver G., Jr. 1928. Astronomical observatories in the Maya area. The Geographical Review, 18, pp. 215-225.

_____ and E.B. Ricketson 1937. Uaxactun, Guatemala, Group E, 1926-1931. <u>Carnegie Inst. Wash., Pub</u>. 477. Washington.

Riley, Carroll L., J. Charles Kelley, Campbell W. Pennington, and Robert L. Rands 1971. <u>Man across the sea: problems of Pre-Columbian contacts</u>. University of Texas Press, Austin and London.

Robertson, D. 1963. <u>Pre-Columbian architecture</u>. George Braziller, New York, Fig. 96.

Rowe, J.H. and D. Menzel 1967. *Peruvian archaeology*. Palo Alto, California.

Roys, Ralph L. 1933. The book of Chilam Balam of Chumayel. *Carnegie Institution of Washington, Publication* 438. Washington.

Ruppert, K. 1935. The Caracol at Chichen Itza, Yucatan, Mexico. *Carnegie Inst. of Washington*, Washington.

———— 1940. A special assemblage of Maya structures. In *The Maya and their neighbors*, pp. 222-231. D. Appleton-Century Company Inc., London.

Sahagun, Fray Bernadino de 1905. Edicion parcial en facsimile de los Codices Matritenses. Fototipia de Hauser y Menet, Madrid.

———— 1938. Historia general de las cosas de la Nueva Espana. 5 vols. Ed. P. Robredo. Mexico.

———— 1953. Florentine Codex, Book 7. Translated by Charles E. Dibble and Arthur J. O. Anderson. School of Am. Research and the Univ. of Utah, Santa Fe.

———— 1956. *Historia general de las cosas de Nueva Espana*, Tomo 8-11. Editorial Porrua, Mexico City, S.A.

———— 1957. Forentine Codex, Books 4 and 5. Translated by Charles E. Dibble and Arthur J. O. Anderson. School of Am. Research and the Univ. of Utah, Santa Fe.

Satterthwaite, Linton 1961. Inscriptions and other dating controls. Appendix to *Tikal Report Number 6: The Carved Wooden Lintels of Tikal*, by William Coe, Edwin Shook, and Linton Satterthwaite. University Museum, Philadelphia, Pennsylvania.

Scargle, J.D. 1967. The Crab Nebula--913 years after its outburst. *Astron. Soc. Pacific Leaflet*, no. 457.

Schaafsma, Polly 1962. Rock art of the Navajo reservoir. El Palacio, 69, no. 4, pp. 193-212. Santa Fe.

―――― 1963. Rock art in the Navajo reservoir district. Museum of New Mexico Pap. in Anth., no. 7. Museum of New Mexico Press, Santa Fe.

―――― 1965. Southwest indian pictographs and petroglyphs. Mus. of New Mexico, Vergara Printing Co., Santa Fe.

―――― 1966. Early Navajo rock paintings and carvings. Museum of Navajo Ceremonial Art, Santa Fe.

―――― 1972. Rock art in New Mexico. State Planning Office, Santa Fe.

Schellhas, Paul 1904. Representation of deities of the Maya manuscripts. Papers of the Peabody Museum, Harvard University, 4, no. 1. Cambridge.

Schram, Robert 1908. Kalendarigraphische und chronologische tafeln. J.C. Hinrichs, Leipzig.

Scott, O. 1942. The stars in myth and fact. Caxton, Caldwell.

Seler, Eduard 1904. Venus Period in the picture writing of the Borgian Codex group. Bureau of American Ethnology, Bulletin 28, pp. 355-91. Washington.

―――― 1910. Die Ruinen von Chichen Itza in Yucatan. 16th International Congress of Americanists, pp. 150-239. Vienna.

―――― 1960-1967. Gesammelte abhandlungen zur amerikanischen Sprach-und altertumskunde (new edition). Akademische Druck-u. Verlagsanstalt, Graz.

―――― 1963. Comentarios al Codice Borgia. 3 vols. Fondo de Cultura Economica, Mexico (translation of 1904 edition).

Shetrone, H. 1930. The mound builders. Kennikat Press, Port Washington.

Shklovsky, I.S. 1968. Supernovae. Wiley Press, London.

Silverberg, R. 1968. Mound builders of ancient America. New York Graphic Society Ltd., Greenwich.

Simmons, Leo 1942. Sun Chief: The autobiography of a Hopi Indian. Yale University Press, New Haven.

Smiley, Charles 1960. A new correlation of the Maya and Christian calendars. Nature, 188, pp. 215-216.

_____ 1965. Orientation by sextant and sun. Astronomical Society of the Pacific, Publications 77, pp. 241-245.

_____ 1973. The Thix and the Fox, Mayan solar eclipse intervals. Jour. Roy. Astr. Soc. Can., 67, pp. 175-182.

Smith, L. 1950. Uaxactun, Guatemala: excavations of 1931-1937. Carnegie Inst. Wash. Pub. 477. Washington.

Soustelle, Jacques 1940. La pensee cosmologique des anciens Mexicains. Hermann et Cie, Paris.

Spence, Lewis 1912. Civilization of ancient Mexico. Putnam, Cambridge.

Spinden, Ellen 1933. The place of Tajin in Totonac archaeology. American Anthropolosits, 35, no. 2.

Spinden, Herbert J. 1916. The question of the zodiac in America. American Anthropologist, n.s., 18, pp. 53-80.

_____ 1948. Mexican calendars and the solar year. Smithsonian Institution, Annual Report, pp. 393-405.

Squire, E.G. and F.M. Davis 1948. Ancient monuments of the Mississippi Valley. Smithsonian Contributions to Knowledge, no. 1.

References

Stahlman, William D. and Owen Gingerich 1963. *Solar and planetary longitudes -2500 to +2000*. University of Wisconsin Press, Madison, Wisconsin.

Star-charts 1972. Monthly star-charts. *Sky and Telescope*. Sky Publishing Corporation, Cambridge, Massachusetts.

Stephen, Alexander M. 1936. Hopi journal. Edited by Elsie Clews Parsons. *Columbia University Contributions to Anthropology*, 23.

Stephens, J. 1843. *Incidents of travel in Yucatan*, 1. Harper & Bros., New York, p. 221.

Stevenson, Matilda Coxe 1904. The Zuni indians. *Bureau of American Ethnology, Annual Report*, 23, pp. 1-634. Washington.

Steward, J.H. 1927. Petroglyphs of California and adjoining states. *University of California Pubs. Am. Arch. and Ethn.*, 24, p. 47.

Stirling, M.W. 1940. An initial series from Tres Zapotes, Vera Cruz, Mexico. *National Geographic Society Contributed Technical Papers, Mexican Archaeological Series*, no. 1, Washington.

Stubbs, Stanley A. 1950. *Bird's Eye View of the Pueblos*. University of Oklahoma Press, Norman.

Swartz, B.K., Jr. 1963. Klamath basin petroglyphs. *Archives of Archaeology*, no. 21, University of Wisconsin Press.

_____ 1964. Archaeological investigations at Lava Beds National Monument, California. Doctoral thesis, Dept. of Anthropology, University of Arizona.

Teeple, John D. 1930. Maya astronomy. *Carnegie Institution of Washington, Publication* 403, Contribution 2. Washington.

_____ 1937. Astronomia Maya. Trans. Cesar Lizardi Ramos, Mexico.

Tezozomoc, Hernando Alvarado 1944. Cronica mexicana. Editorial Leyenda, Mexico.

Thom, Alexander 1967. Megalithic sites in Britain. Clarendon Press, Oxford.

_____ 1971. Megalithic lunar observatories. Clarendon Press, Oxford.

Thompson, J.E.S. 1931. Archaeological investigations in the southern Cayo District, British Honduras. Field Museum of Natural History, Anthropological Series, 17, no. 2. Chicago.

_____ 1938. Sixteenth and seventeenth century reports on the Chol Mayas. American Anthropologist, 40, pp. 584-604.

_____ 1944. Variant methods of date recordings in the Jatate drainage, Chiapas. Carnegie Institution of Washington, Division of Historical Research, Notes on Middle American Archaeology and Ethnology, no. 45. Cambridge.

_____ 1950. Maya hieroglyphic writing: an introduction. Carnegie Institution of Washington, Publication 589.

_____ 1958. Symbols, glyphs and divinatory almanacs for diseases in the Maya Dresden and Madrid codices. American Antiquity, 23, pp. 297-308.

_____ 1960. Maya hieroglyphic writing. Introduction. University of Oklahoma Press, Norman.

_____ 1962. A catalog of Maya hieroglyphs. University of Oklahoma Press, Norman.

_____ 1965-70. Preliminary decipherments of Maya glyphs, nos. 1-5. Ashdon, Saffron Walden, England.

_____ 1970. *Maya history and religion*. University of Oklahoma Press, Norman.

_____ 1972a. A commentary on the Dresden codex, a Maya hieroglyphic book. *Memoirs of the American Philosophical Society*, 93. Philadelphia.

_____ 1972b. *Maya Hieroglyphs without tears*. Trustees of the British Museum, London.

Thompson, S. 1929. *Tales of the North American Indian*. Indiana University Press, Bloomington.

Titiev, Mischa 1944. Old Oraibi, a study of the Hopi Indians of third mesa. *Papers of the Peabody Museum of American Archaeology and Ethnology, Harvard University*, 22, no. 1. Cambridge.

_____ 1972. *The Hopi Indians of Old Oraibi, change and continuity*. University of Michigan Press, Ann Arbor.

Tozzer, Alfred M. 1907. A comparative study of the Mayas and the Lacandones. Peabody Museum. Cambridge Massachusetts.

_____ 1910. Animal figures in the Maya codices. *Papers of the Peabody Museum of Archaeology and Ethnology, Harvard University*, 4, no. 3. Cambridge.

_____ 1941. Landa's relacion de las cosas de Yucatan. *Papers of the Peabody Museum of Archaeology and Ethnology*, Harvard University, 18. Cambridge.

Trimble, V. 1968. Motions and the structure of the filamentary envelope of the Crab Nebula. *Astron. J.*, 73, pp. 535-547.

_____ 1971. Dynamics of the Crab Nebula. In *The Crab Nebula*, R.D. Davies and F.G. Smith, eds. Reidel, Holland, pp. 12-21.

Tuckerman, B. 1962. <u>Planetary, lunar and solar positions 1601 B.C. to A.D. 1</u>. Amer. Phil. Soc., Philadelphia.

―――― 1964. <u>Planetary, lunar and solar positions A.D. 2 to A.D. 1649 at five-day and ten-day intervals</u>. Am. Phil. Soc., Philadelphia.

Vaillant, George, C. 1944. <u>Aztecs of Mexico</u>. Doubleday, Garden City, New York.

Velazquez, Primo Feliciano, Trans. 1945. <u>Codice Chimalpopoca</u>. (Anales de Cuauhtitlan y Leyenda de los Soles). Imprenta Universitaria, Mexico.

Villacorta, J. Antonio and Carlos A. Villacorta 1930. <u>Codices Mayas</u>. Reproducidoes y Desarrollados, Guatemala, C.A.

Vivian, Gordon and Paul Reiter 1960. The great kivas of Chaco Canyon and their relationships. <u>The School of American Research and The Museum of New Mexico</u>, Monograph 22. Santa Fe.

Watson, Editha L. 1964. Navajo sacred places. <u>Navajoland Publications, Series</u> 5. Navajo Tribal Museum, Window Rock, Arizona.

Wheelwright, Mary C. 1940. Myth of Sontso Hatral (big star chant) and the myth of Ma-Ih Hatral (Coyote Chant). <u>Museum of Navajo Ceremonial Art, Bull</u>. no. 2. Santa Fe.

White, Leslie A. 1942. The pueblo of Santa Ana, New Mexico. <u>American Anthropological Association, Memoir</u> 60.

Willoughby, C. 1922. The Turner group of earthworks, Hamilton Co., Ohio. <u>Papers of the Peabody Museum of American Archaeology and Ethnology</u>, 8, no. 3.

Willson, R.W. 1924. Astronomical notes on the Maya codices. <u>Papers of the Peabody Museum, Harvard University</u>, 6, no. 3. Cambridge.

Wisdom, Charles 1940. *The Chorti Indians of Guatemala*.
University Publications in Anthropology, Ethical
Series. University of Chicago Press, Chicago.

Wyman, Leland C. 1970. *Blessingway*. University of
Arizona Press, Tucson.

Xi, Ze-zong and Shu-jen Po 1966. Ancient oriental
records of novae and supernovae. *Science*, 154,
pp. 597-603.

Yazzie, Ethelou 1971. *Navajo History*, 1. Navajo
Community College Press, Many Farms, Arizona.

Zimmermann, G. 1956. Die hieroglyphen der Maya-handschriften. *University of Hamburg. Abhandlungen aus dem Gebiet der Auslandkunde*, 62. Hamburg.

Index

Abu Simbel, temple of, 142, 147-151
Acanceh, Stucco Palace at, 165
Acoma, 72, 84, 85
Adams County, Ohio, 221
Adena Indians, 218, 222, 227
Aldebaran, 85
alignment: azimuths of, 163, 165; of buildings at Chichen Itza, 181, 182; of buildings in Egypt, 136-151; of buildings in Mesoamerica, 191-204, 382; of buildings at Monte Alban, 165, 173, 174, 177, 214; and cardinal directions, 42, 86, 200; of desert markings, 151-162; discernment of, 139-141, 193; of horizon markers, 34; importance of, to archaeoastronomy, 212; of Konsu temple, 145; measuring techniques of, 35-37; of mounds, 227-233, 382; of sun towers, 213; template, 218, 219, 227-233, 383; two-point, 217. See also azimuths, alignments to; moon, alignments to; orientations; Pleiades, alignments to; solstices, alignments to; sun, alignments to; Venus, alignments to
Algonquin Indians, 222, 234-235
Amatlan, Morelos, 277
American Indians. See Indians (American)
Amon (Egyptian deity), 138, 141, 145, 147. See also Amon-Ra
Amon-Ra (Egyptian temple complex), 138-147. See also Amon; Ra
Anasazi, 89-90, 92, 95, 96; plateau, 97; region, 212
Andes, lines at foothills of, 151
Antares, 122
Aquarius, 119, 120
Aquila, 120
archaeoastronomy, 131-132, 158, 379-380, 385, 386; methods and procedures of, 35-37, 42, 109-119, 163, 206, 209, 210, 213-215; problems of, 4, 5, 23, 90, 93, 97, 99-100, 154, 162, 164, 205-211, 380. See also archaeoastronomy, tools of; astroarchaeology
archaeoastronomy, tools of: alidades, 173; books, 210-211; camera, 36; compass, 36, 37, 178; computer, 163-192; maps, 162, 210; photogrammetric survey, 138, 139, 151, 155; plane table, 173; sextant, 206; theodolite, 138, 139, 178; transit, 35-36, 163, 166, 172, 173, 175, 178, 187
Archaic Period, 219
Argyllshire, 208
Astra Motif Index, 189
astroarchaeology, 157-158; methods of, 132-134, 151-154; methods applied, example of, 134-162. See also archaeoastronomy
astronomer-priests, 244, 256, 340, 350
astronomers, pre-Columbian, 360
astronomical orientation, 197, 199, 203. See also alignment; orientations
astronomy, prehistoric knowledge of: evidence of, 136, 192; in reckoning of Tajin Ballcourt panels, 273-274; reflected in pre-Columbian buildings, 191
Atum (Egyptian deity), 143
Augural Bands, 302-303, 304-305, 315-316, 321, 328, 331
Auriga, 116, 119
axial distribution of Mesoamerican cities, 166-189
azimuths, 137, 139, 163, 164, 165; alignments to, 141, 147, 151, 163, 175, 178, 181, 184, 206
Aztec Indians, 22, 23, 24, 85; calendar of, 9, 11-14, 31, 32, 213; and cardinal direction colors, 8-9; deity of, 11 (see also deities, Mexican); priests

of, 4, 14, 172; Winter Solstice Ceremony of, 76

Bacabs, 72
Badger Clan, 72
Badger Man, 81
Ballcourt panels: at Monte Alban, 165; at Piedras Negras, 193, 200; at Tajin, 263-274, 277, 278, 383; at Tula, 168; of the Underworld, 279; at Uxmal, 184, 188; at Yagul, 165. See also ballcourts
ballcourts, 197, 268. See also Ballcourt panels
Ball Game, ceremonial, 263; in the Underworld, 268, 270
Bear Mounds, 232
Beast Gods, 78
Big Dipper, 221, 305, 319, 380. See also Great Dipper
big dipper people, 104
Big First One, the, 101, 102
Big Horn Medicine Wheel, 384
Big Star (Morning Star), 62, 73
Big Star Chant, 97, 104
Big Star-Sky god, 82
Binding of the Years, 24
Bird Effigies, 227, 230, 234
Bird Mounds, 232-233
Black God, 100, 259
Black Star People, 104
Blessingway ceremony, 100,
Bodley Codex, 196
bolide, 119
Bootes, 120, 124
Borgia Codex. See Codex, Borgia
Bow Priest, 77, 78
British Honduras: Altar Q at Copan in, 256; Caracol Stela 3 in, 257-262; people of, 290
Broad Star, the, 85
Bronze Age, 135

Caballito Blanco, 172, 173-177
cacique, 71, 73, 78, 82. See also Town Chief
Cacique's Society, 83
Cahokia culture, 354
calendar, 13-14, 18, 19-22, 30, 84, 141, 189; agricultural, 24; American Indian, 374; astronomical, 151, 283-340; Aztec, 9, 11-14, 31, 32, 213; Chaldean-Babylonian, 374; Chinese, 374; Christian, 253, 255; eclipse, 31; Egyptian, 12, 374-376; Greek, 372; Gregorian, 327, 337; Hopi, 65, 67, 70; Inca, 213; lunar, 64, 207; Julian, 12, 237-246, 247, 250-252, 253, 316, 327; in Madrid Codex, 283-340; Maya, 11-14, 15-18, 30, 65, 213, 237-246, 254, 255, 258, 275, 289, 314, 343-347; Mexican, 11, 65; in Peruvian desert, 151; priests, 86, 315, 324; Pueblo, 59, 63, 65-70; ritual and, 14; Roman, 374; round, 247, 255, 261; serpent, 290, 295, 296; solar, 11, 13, 17, 207; star, 289; stick, 61; Venus, 63, 75, 180, 281. See also ceremonial year; Chamula calendar board; lesser year; lunisolar calendar; sidereal year; solar year; 260-day count; Vague Year; year
Calendar Round, 11, 19, 24, 247, 255, 261
Calendar Stone (Aztec), 31, 32
Camelopardis, 119
Cancer, 116
Canopus, 101
Canyon de Chelly National Monument, 89-90, 95-99; glyphs in, 99-107; planetaria in, 99-107; significance of, to Navajo, 90-92, 107; sites in, 92-95
Canyon del Muerte, 81, 91, 92, 93
Capella, 173, 176, 177, 189
Capricorn, 119
Caracol: at Chichen Itza, 172, 178-182, 193, 197, 382; British Honduras, Stela 3 at, 257-262
cardinal directions, 80, 290; building alignments and orientations to, 42, 86, 200; Chicchan serpent associated with, 335, 336; color associations of, 8-9, 10, 302, 305; stars of, 85. See also cardinal points
cardinal points, 13, 78; alignments and orientations to, 178, 182. See also cardinal directions
Carson, Kit, 91
Casa del Gobernador, 184. See also Governor's Palace
Casa Grande, 77
Casa Rinconda, 35, 39, 40; axis of symmetry, 36; solstice marker at, 38; orientation of, 41

Index

Cassiopeia, 101, 104, 116
Castor, 226
cave paintings (Navajo), 57.
 See also glyphs; pictographs;
 star ceilings
Ce Acatl Topiltzin, Prince, 278, 280
celestial bands (Maya), 20, 28
celestial observations, 84. See
 also sky watching; sun watching
Cemetery Group plaza, 187, 188
Cepheus, 116
ceremonial Ball Game, 263
Ceremonial Cave (Navajo), 97
ceremonial centers; alignments
 of, 200; Maya, 187, 195, 197,
 208; as observatories, 198
ceremonial cycle, 83
ceremonial dances, 268
ceremonial year, 72; Hopi, 70
ceremonies, 74-87 passim; solstice,
 33, 66, 68, 69, 76, 79, 81, 83,
 211-212 sun, 84. See also
 ceremony
ceremony: Aztec, 76; Blessingway,
 100; Maya, 264; Pamurti, 67;
 Powamu, 66-67, 72, 83; prayer-
 stick, 61; solstice, 3, 66, 68,
 69, 76, 81, 83. See also ceremonies; feasts; festivals;
 Shalako celebration
Cerro Colorado (Teotihuacan), 168, 169
Cerro Gordo, 192
Cetus, 122
cervids, 120
Chac, 301; as God B (Madrid Codex),
 298, 302, 321, 328, 331, 333;
 in Madrid Codex, 286, 298, 299,
 319, 324, 333; as rain god, 286, 321
Chaco Canyon, 34, 53-54, 75, 86;
 observations at, 40, 41, 42;
 orientation at, 86; petroglyphs
 at, 37, 54, 59, 60, 67, 74, 86-
 87; pictographs at, 56; sol-
 stice observations at, 38, 40;
 symbols at, 62, 76. See also
 Casa Rinconda; Chetro Ketl I
 Great Kiva; Penasco Blanco;
 Pueblo Bonito
Chakewa (Hopi), 67
Chama, 87
Chamula calendar board, 343-347, 371, 372
Chetro Ketl I Great Kiva, 35, 36, 39, 41, 42, 43, 208
Chiapas, Mexico, 13; Chamula cal-
endar board at, 343-347, 371,
 372; eclipses at, 238, 239,
 245; Initial Series at, 325-
 327; Santa Elena Poco Uinic,
 Stela 3, at, 325-327
Chicchans, 335-338
Chicchan serpent, 285, 335, 336
Chichen Itza, 28, 29, 210, 382;
 azimuth alignments at, 165;
 Caracol at, 172, 178-182,
 193, 197, 382; orientation of
 buildings at, 166, 197; sight-
 ing markers at, 193
Chichimecs, 20
Chic Kaban festival, 336-338
Chilam Balam (Maya), 19, 281
Chimalma (Mexican deity), 279
Chimalman (Maya deity), 277-278
Cholula, 166
Chorti Maya, 13, 25; calendar
 year of, 336; Chicchan serpent
 among, 335
Christian calendar, 253, 255.
 See also calendar
Chumash, 57
Cihuacoatl, 278, 279. See also
 Quillaxtli
circle glyph, 298, 302, 316, 319,
 328, 384; interpreted, 299, 338-339
Citlalco, 8
Citlalcolotl (constellation), 25, 26
Citlallicue (deity), 8, 27
citlaltlachtli, 23, 27
Ciudadela, the (Teotihuacan), 168
Codex, Bodley, 196
Codex, Borgia, 21, 276, 279;
 Venus cycle in, 275, 276, 277
Codex, Cospi, 11
Codex, Dresden, 14, 17-18, 20,
 21, 180, 211, 276, 278-281
 passim, 284, 315, 382; cal-
 endar correlations in, 237-
 246; solar eclipse warning
 table in, 247-256; Venus Cycle
 in, 275, 276, 277
Codex, Fejervary-Mayer, 211
Codex, Florentine, 31, 211
Codex, Grolier, 20
Codex, Madrid, 2, 9, 284-285,
 300, 304, 313, 319, 322, 339,
 380, 381; astronomical calendar
 in, 283-340; circle symbol in,
 337; crossed bands in, 302;
 Draco in, 339; glyphs in, 285-
 289, 297-333; pages from, 301,
 306, 307, 309, 311, 312, 317,

320, 323, 326, 329, 332; serpent in, 290, 295, 296, 304, 334. See also Chac; gods in Madrid Codex
Codex, Maya Dresden. See Codex, Dresden
Codex, Paris, 28, 29
Codex, Tro-Cortesianus. See Codex, Madrid
Codex, Vaticanus A, 6, 7, 8
codices, 4, 191, 193, 268, 273; Maya, 19; Mexican, 14, 19, 20. See also Codex, Bodley; Codex, Borgia; Codex, Cospi; Codex, Dresden; Codex, Fejervary-Mayer; Codex, Florentine; Codex, Grolier; Codex, Madrid; Codex, Paris; Codex, Vaticanus A
colors: of cardinal directions, 8-9, 10, 302, 305; of four offspring, 10; of layered heavens, 8; of Navajo star paintings, 95, 104; of six directions, 70. See also cardinal directions
Colossi of Memnon, 142, 147
colotlixayac, 23
colures, 110, 113, 116, 120, 300
Coma Berenices, 120
comets, 8, 22, 57, 119
constellations, 22-28, 31, 74, 99-107, 220-222, 310, 383, 384; Aztec, 22, 32; Citlalcolotl, 25, 26; Citlaltlachtli, 23, 27; Colotlixayac, 23; creation of (Navajo), 100-106; in effigy mounds, 218, 233; the Fire Drill, 8, 25; frog, 324; in Great Basin petroglyphs, 109-130; in Madrid Codex, 324; in Navajo paintings, 57, 97, 98; Sahagun drawings of, 22; She of the Starry Skirt, 8; turtle, 324. See also individual constellations by name
Copan, 9, 183, 256
Corona Borealis, 109, 114, 116, 120
Corvus, 101; in Great Basin petroglyphs, 114, 122
cosmology, Maya, 187
cosmos (Mesoamerican), conception of, 10. See also heavens, conception of; world, conception of; universe, conception of
Cospi Codex, 11
Cottonwood Canyon, 93
Coyote (Navajo), 100
Crab, in Fern Cave pictographs, 52
Crab Nebula, 46, 48, 60, 383; as Guest Star (Chinese), 57-58; petroglyph of, 56. See also supernova
Crater, 120, 122
Crier Chief, 66
Cuba, New Mexico: star painting near, 95, 96
Cuicuilco, 166
Cuzco: Temple of the Sun near, 384
Cygni, Gamma, 302, 380; Right Ascension of, 292, 313, 314, 315, 324-325, 327
Cygnus, 232, 234, 302, 311, 315, 325, 327; in Great Basin petroglyphs, 119; in effigy mounds, 226

day: count, 285-286, 297, 298-299; divisions of, 14; glyph for, 285-286, 308; symbol for, 263
days without names, 11
Death God, 308, 310, 318, 321, 322, 330, 333
De-coo-dah, 219
deities, 4, 10, 20, 28, 277-278; Bacabs as, 10; Black Star People as, 104; dual nature of, 8, 10, 15, 23; Fire Man (Navajo), 100; Maya, 4, 10, 277-278; (see also Maya, gods of); Mexican, 4, 10; pulque, 15; quadripartite nature of, 10; sun, 10, 83, 211-212; in Tajin Ballcourt panel 1, 266. See also gods
Delaware Indians, 222
Delgadito Canyon, 95, 96
Dilyechi (constellation), 101
Dinetah phase in New Mexico, 90, 91
Diophantine equation, 255
directions: marked by ancient astronomers, 192; Maya, 187. See also cardinal directions
Distance Number, 327
Divine Twins, 271. See also Twin War Gods
divinity, dual, 10. See also deities
Draco, 124, 296, 335, 381; in Madrid Codex, 286, 289, 296, 300, 310, 311, 339; in petroglyphs, 116, 119; serpent representation of, 286, 336. See also Draconis

Index

Draconis: Beta and Gamma, 296, 310, 319, 324; Eta, 289, 290, 304, 305, 310, 314, 315, 319, 331, 381; Eta, right ascension of, 290, 292, 295; Eta, stability of, 338, 339; Eta, transit of, 293, 295, 296, 297, 337
Dresden Codex. See Codex, Dresden
dual creator god. See deities, dual nature of
Dulce (site), 95

Eagle Mound, 226, 227
Eagle Pueblo, 81
earth: on Ballcourt panel, 263; conception and representation of, 6-9; Venus comes to, 266, 267
Earth Mother, 84. See also Mother Earth
eclipse, 85, 238, 379, 382; calendar, 31, 213; in Central America, 238, in Dresden Codex, 247-256; glyph for, 238, 325-327; lunar, 17-18; in Maya calendar, 237-246; 247; Maya Eclipse Cycle (MEC), 238; solar, 325-327, 250-252; Solar Eclipse Warning Table, 247-256; table, 28, 237, 239. See also ecliptic
ecliptic: cycle, 264, 273; of Eta Draconis, 289
effigy mounds, 218-234, 383; cult of, 218; National Monument, 230, 231. See also Marching Bear
Egypt, 141; alignments in, 136-151; astronomers in, 380; awareness of azimuthal swing in, 150; calendar in, 12, 374, 376; Sky gods in, 136-138; temples in, 136-151. See also gods, Egyptian
Ehecatl, 86
ephemeris, 237-238; Maya, 245, 246, 286
equinox, 12, 13, 75, 256, 313, 314, 316, 380; alignment to, 180, 181; glyph for, 314; in Madrid Codex, 289, 314, 325; orientation to 172; precession of, 170, 290
Eskimo Indians, 222
ethnoastronomy, 5
ethnoscience, 22
Europe, 218, 380
Evening Star, 73, 74, 83, 87, 264, 266, 271-273, 275, 276. See also Big Star; Morning Star; Venus

Fasting House, 282
Father Sky, 97, 98, 101. See also sky gods
Father Sun, 63, 74
feasts, 24, 76. See also ceremonies; festivals
Feathered Serpent (Maya, 20, 263, 336. See also Kukulcan; serpents
Fejervary-Mayer Codex, 211
Fern Cave, 50-51, 52, 53
festivals, 13, 172, 173, 336, 337. See also ceremonies, feasts
Fire Drill (constellation), 8, 25
Fire Man (Navajo), 100
Fire Serpents, 8. See also serpents
First Man (Navajo), 100
First Mesa village (Hopi), 67, 71-72, 73
First Woman (Navajo), 100, 101
584-day Venus Calendar, 63, 75, 180, 281
Florentine Codex, 31, 211
Flute ceremony, 68
Flute Society, 72
Fone, 255
Four Feet, Venus as, 281
Four Steps, Venus as, 281
Fox Indians, 221; mounds of, 226 232
Franco-Egyptian Research Center, 138
frog constellation, 324

Galactic Equator, 120
Galaxy, 85. See also Milky Way
Gallup, New Mexico: star paintings near, 104
Gemini, 26, 27; in Great Basin petroglyphs, 120; in Madrid Codex, 324, 339; in Maya, 303
glyphs, 17, 20, 62; of constellations, 31; for day, 285-286, 308; of eclipses, 238, 325-327; in Madrid Codex, 285-289, 297-335; at Monte Alban, 196; of moon, 46, 73, 276; of Morning Star, 73, 74; at Naranjo, 257; Fornight, 286, 314; of Orion, 73, 74; of Pleiades, 73, 74; of solar-sidereal year, 314; spiral, 330; star, 109-130, 257; of sun, 62, 73, 286, 289, 314; of sun symbol, 75; of supernova, 40, 46, 50, 52, 53, 54, 57; on

Tajin Ballcourt panel 1, 276; writing, Maya, 62, 238, 369-371; writing, Mixtec, 370, 371. See also cave paintings; celestial bands; circle glyphs; Kan glyphs; katcinas; kin glyphs; mounds; petroglyphs; pictographs; reliefs, rock art; sandpaintings; solstices, markers of; star ceilings; stars, paintings of; stellar bands

Gobernador phase, 91

godesses, 268, 278, 279

gods, 8, 23, 66, 86, 268, 274; "arrival of the Gods," 76; Big Star-Sky, 82; Christian, 62; four s eps of, 274; old black (Maya), 259; sun, 143, 266, 272; of "two horizons" (Egyptian), 141-142; Wind God, 86. See also deities; godesses; gods, Egyptian; gods in Madrid Codex

gods, Egyptian: Amon, 138, 141, 145, 147; Atum, 143; Horakhety (see Ra-Horakhety); Horus, 143, 145; Khonspekherod, 139; Khonsu, 138-146 passim; Mut, 138-141, 142; Opet, 138; Ptah, 147; Ra, 138, 139, 143, 145; Ra-Horakhety, 138, 141-143, 144, 147, 149, 150, 152; Sechat, 141. See also Amon-Ra

gods in Madrid Codex: Chac, 286, 298, 299, 301, 302, 310, 319, 321, 324, 328, 331, 333; Corn God E, 279; Death God, 308, 310, 317, 321, 322, 330, 333; God B, 303, 321; God C, 303, 310, 314, 315, 316, 322, 333; God D, 308, 310, 321, 322; God K, 305, 322, 328; God M, 311, 316; Maize God, 308, 310

Governor's Palace at Uxmal, 165, 172, 182-187, 188; alignment of, 193

Great Basin region, 53, 113; petroglyphs in, 109, 110, 113, 114, 130

Great Dipper, 74, 85. See also Big Dipper

Great Kivas. See kivas

Great Mountain, 78

Great Star, 67, 75, 87; as Morning Star, 75; petroglyph, 59; as Venus, 8, 19

Great Star Evil Chasing Chant (Navajo), 96

Great White Horse (effigy), 234

Green County, Ohio: mound in, 224

Gregorian calendar, 327, 337

Grolier Codex, 20

Guest Star (Chinese), 58

Gulf Coast, 275, 280

Hall of Festivals (Egyptian), 138, 142; inscription in, 143; murals in, 141; in temple of Ra-Horakhety, 141-143

Hamilton County, Ohio: mound in, 223, 227, 230

hand design, 54, 59, 76, 87, 221, 222

Hano, 73. See also Hopi-Tewa

heavens: conception of, 6-10; as home of gods, 28; layers of, 6-8. See also cosmos, conception of; universe, conception of; world, conception of

Hercules: in Great Basin petroglyphs, 113, 114, 116; sky maps of, 114, 115, 116

Hero Twins, 20. See also Quetzalcoatl-Kukulcan

Herrera Chronicles, 290, 293

High Room of the Sun, 138, 141, 142, 144, 145

Histoire du Mechique, 12

"home of the sun" (Puerblo), 75

Hopewell Indians, 218, 222

Hopi Indians, 61, 62, 66, 72, 82, 96, 97, 209; calendar of, 65, 67, 70; and celestial objects, 57; ceremonial year of, 72; ceremonies of, 64, 67; curing ritual of, 85; and Star Chief, 73; sunwatching by, 64, 72; War Priest of, 72; Winter solstice ceremony of, 62. See also Hopi-Tewa; Pueblo

Hopi (site), 71; Hopi-Tewa at, 73; Oraibi (Third Mesa village) at, 72; orientations of, 85; solstice observation sites at, 72

Hopi-Tewa, 70, 73. See also Tewa

Horakhety. See gods, Egyptian

horizon markers, 33-34. See also markers; observatories

Horus (Egyptian deity), 143, 145

"hours," 13-14

House of Tepozteco. See Tepozteco, House of

House of the Dwarf, 184. See also House of the Magician

House of the Magician, 182, 184, 187, 188

Index

House of the Old Woman, 187, 188
Huave Indian, 263
Huicholobos, 172
Huitzilopochtli, Temple of: orientation of, 172
Huiznahua: in Maya legend, 277
Hyades, 109; Navajo name for, 101; pattern of, in glyphs, 22; on war shield, 97. See also Lower Regions; Underworld
Hydra: in Great Basin petroglyphs, 114, 116, 120, 122, 129

Ikil (site), 165
Inca Indians, 213
Indians (American), 5, 222, 276, 381; alignments of, 218, 233; lore of, 220-223; mounds of, 218-235; symbols of, 218-235; treatment of stars by, 222. See also individual Indian tribes and cultures by name
Initiates' moon (Pueblo), 70
intercalary calendar system, 314, 373-376
Iowa: site in, 230, 231
Ipet Sout (Egyptian), 138
Isleta (site), 84
Itzamna (Maya god), 23
Ixil (Maya), 14
Izamal (site), 165

Jacaltec (Maya), 14
jaguar: glyphs of, 298, 318; tradition, 380; Ursa Major associated with, 305, 319; warfare associated with, 318, 319
Jemez (Indians), 72
Julian calendar. See calendar, Julian
Jupiter, 22, 258-259; at Caracol Stela 3, 257-262; at Naranjo, 257

Kan glyphs, 286, 289, 298, 302, 309, 311, 317, 318, 329
Karnak: Egyptian temples at, 138-146; alignments at, 143; archaeoastronomy methods at, 162
Katcina clan chief, 72. See also katcinas
katcinas, 57, 60, 72, 78; calendar, 70; in ceremony, 62, 65, 66, 68; in Hopi, 62, 72; of Shalako complex, 64; of sun, 67; as symbol, 61
Kayenta: building orientation of, 86
kehtahn, 96, 98
Keys of St. Peter (constellation), 23, 25
Khonspekherod (Egyptian), 139
Khonsu (Egyptian), 138-146 passim
kin glyphs, 286, 289; in Madrid Codex, 299, 300, 314, 324, 325, 328
Kinich Kakma (building), 165
Kinsey, Alfred Charles, 278
kivas, 35, 36, 39, 41, 73; alignments of, 36, 42; hand design in, 76; murals on, 57; orientations of, 85-86; Orion on, 73; petroglyphs at, 56; Rio Grande, 86. See also Pueblo Bonito A; Village of the Great Kivas
Klamath Basin region, 53
Koshare religious society, 73, 81
Kukulcan: Maya festival of, 336-338; temple of, 336. See also Feathered Serpent

Labna (site), 165
Lacandon Indians, 25, 27
Lake Nasser, 147
Lamat (Maya): glyphs for, 309, 317, 323, 329
Las Bocas mosaic pendant (Olmec), 341-377; analysis of, 356-363; anomalies in pattern of, 363-367; cultural bases of, 367-371; description of, 347-356; implication of, for archaeoastronomy, 372; intercalary month in, 376; main face of, 342
Lava Beds National Monument, 50
La Venta: ceremonial complex at, 380
leap day: in Dresden Codex, 253, 255; in Madrid Codex, 313, 316
legends: Mexican, in Tajin Ballcourt panels, 266-273, 276-278, 279-280, 282; of Venus, 266, 268, 270-280, 383. See also mythology
Leo, 116, 120
"lesser year," 70
lightning markers (Zuni), 78
Long Count Calendar, 11

"Long Walk, The," 91, 95
Lord of the Night, 26
Los Tuxtlas (artifact), 370
Louisiana: Poverty Point Bird effigies in, 227
Lower Regions: represented on Ballcourt panels, 263. See also Hyades; Underworld
lub: or starting date, 248, 253; of Solar Eclipse Warning Table, 255
Lunar Table: of Dresden Codex, 248
lunisolar calendar, 16, 17, 373, 380. See also moon
Luxor (Egyptian), 138
Lyra, 113-116 passim

macaw: as Maya symbol, 10
McGregor, Iowa (site), 230
Madrid Codex. See Codex, Madrid
Maize God, 308, 310, 316, 318, 322
Mamalhuaztli (constellation), 25, 26
Mani, provinces of, 336
"Man With Feet Ajar" (constellation), 101
Marching Bear: effigies, 234; Group, 230, 231; National Monument, 230
Maricopa Indians, 221, 222, 226
markers, 87; petroglyph, 110; pillar at garden of Zuni as, 77-78; of solstices, 37, 38, 40, 42, 176; sun symbol as, 75. See also alignments; observatories; orientations
markings: on Egyptian temples, 136-151; in Peruvian desert, 151-157, 159-161, 383; on pillar at Matsakya, 62; Teotihuacan circles as, 168, 192; types of, 194; Uaxactun circles as, 192. See also glyphs
Mars, 21-22; "4 Mars beasts," 20; synodic period of, 21; in Tajin Ballcourt panels, 273
Massacre Cave (site), 91
Mat'saka. See Matsakya
Matsaki. See Matsakya
Matsakya, 76-77; glyphs at, 62; pillar at, 78; prayer sticks at, 79; sun shrine at, 79, 211; little "tower" at, 80, 86
Maya, 4, 277, 297, 299, 304; alignments in, 200; astronomical system of, 191, 238, 260; calendar of, 11-14, 15-18, 30, 65, 213, 237-246, 255, 254, 258, 275, 289, 314, 343-347; ceremonial centers of, 187, 195, 197, 208; Chiapas, 13, 343-347; Chorti, 13, 15; Classic, 13, 284; codices of, 19; cosmology of, 187; dates of, 258; dating system of, 237-246; day counts of, 238, 276, 283, 315, 339; deities of, 4, 10, 277-278; in Dresden Codex (see Codex, Dresden); glyph writing of, 62, 238, 369-371; gods of, 259, 278-279; ephemeris of, 245, 246, 286; festivals of, 66, 336; folklore of, 180; Lacancon, 23; leap days of, 316; leap year of, 377; Maya Eclipse Cycle (MEC), 238, 239, 244, 245, 246; observatory of, 200; orientations of, 164; priest-astronomer of, 244; Quiche, 27; sighting markers of, 178; sky watching of, 84; Solar Eclipse Warning Method of, 254; Solar Eclipse Warning Table, 255; symbols of, 10, 263, 281, 321; texts of, 283, 325; Underworld conception of, 8; universe conception of, 187; Venus in, 180 281; Yucatec, 15-18, 300, 303
Mayapan, 336
medicine man (Navajo), 92, 98, 100, 101, 102, 104, 107. See also astronomer-priests; cacique; priests; shamans
Megalithic period, 208, 217, 218
Memnon, Colossi of, 142, 147
menhir, 217. See also markers; observatories
Mercury, 22, 262; cycle, 264; on Tajin Ballcourt panel, 271, 273
Mesa Verde: cave sites at, 214; Far View House of, 208, 210; kivas at, 86; orientations at, 86; sun towers at, 213
Mesoamerica, 273, 289, 293, 341, 381; alignments in, 204; archaeoastronomy techniques applies in, 192, 283; architecture in, 187, 189, 200, 204; astronomical knowledge in, 191, 390; axial distribution of cities in, 164, 166-189; cardinal directions in, 290; earth representations in, 6-9; jaguar in, 319; observatories in, 191;

Index

orientations in, 163; symbols in, 286, 302; Ursa Major in, 321
meteors: in petroglyphs, 57
Metonic cycle, 373-374
Mexico, 13, 72, 75, 86, 209, 263, 284, 325-327; alignments in, 189; calendar in, 11, 65; deities in, 4, 10; eclipses in, 238; Las Bocas pendant in, 341-373; legends of, 266, 275; orientations in, 164, 166; Quetzalcoatl in, 72 (see also Quetzalcoatl); sky watching in, 86; symbols in, 81, 220; Valley of, 166
Mexico City, 166
Milky Way, 27, 383; in Hopi art, 57; in mounds, 226, 227; in Navajo sandpaintings, 98, 100-101; and Pomo Indian, 22; Pueblo symbol of, 85
Mimixcona (stars), 279
Mishongnovi (Hopi), 85
Mixcoatl (Maya deity), 28, 273-278
Mixcoatl-Camaxtli, 20. See also Mixcoatl; Quetzalcoatl-Kukulcan
Mixtec, 196; glyph writing system of, 370, 371; star symbol of, 334
Mixtec-Cholula (Mexico), 279
Moctezuma Xocoyotzin, 22
Monte Alban: alignments at, 165, 173, 174, 177, 214; Building J at, 172-173, 176, 196, 197, 201; glyphs at, 196; Monte Alban II at, 173; Mound J at, 196, 197, 201, 214; Mound P at, 193, 201; zenith passage of sun at, 173
Montezuma, 172
Monument Canyon (site), 89
moon, 47, 62, 70, 81, 84, 85, 139, 209, 217, 380, 383; alignments to, 139, 145, 154, 178, 180, 181, 187, 188; on Ballcourt panels, 263-264, 268, 273-274; in calendar, 68 (see also calendar, lunar); in ceremony, 62, 64; in conjunction with supernova, 46-49 passim, 52, 53, 58; in conjunction with Venus, 54; crescent, 46, 52-53, 57, 58, 59, 62, 67, 75, 97, 130, 138; cycles, 264-266, 275, 276, 281; as deity, 14, 16, 63, 81, 83; in Dresden Codex, 253; in Fern Cave pictograph, 52, 53; glyph of, 46, 73, 276; in Great Basin petroglyphs, 130; in Hopi, 57, 62, 67; on kivas, 73; in Madrid Codex, 308; in Maya, 266; in mythology, 229; in Navajo, 97, 98, 101; petroglyph of, 59, 130; in pictograph, 52, 53; prayer stick for, 78; in rock art, 46, 52-53; Sahagun drawings of, 22; in sandpainting, 98, 101; in Shalake ceremony, 64; watching, 70-74 (see also sun watching); in Zuni, 64, 78
Moon Mother, 78
Morning Star, 63, 74, 84, 275, 383; as deity, 73, 77, 83, 86, 282; to Evening Star, 271-273; in Hopi, 57, 62; on kivas, 73; observation of, 84, 85; as symbol, 63, 72, 75, 87; Venus as, 23, 75, 264, 266, 268, 276, 281; watched, 74. See also Big Star; Evening Star; Venus
mosaic pendant. See Las Bocas mosaic pendant
Mother Earth, 63, 71, 74, 78, 81, 82, 84
Mound Builders, 219, 222, 233
mounds, 183; alignments of, 227-233, 382; building of, 384; effigy, 218-234, 383; Indian lore represented on, 220-223; types of, 223-227. See also Serpent Mound; serpents
murals (Egyptian), 141, 142, 143
Mut (Egyptian deity), 138-141, 142
Muu (constellation), 220
mythology: American Indian, 220-223; Maya, 180, 266-273, 276-278, 282, 336; Navajo, 100; Pueblo, 81, 83, 85; Venus, 280. See also legends

Nacxit (deity), 274, 281
Nahua. See Aztec
Nambe, 74
Naranjo (site), 257
Narbona, Lieutenant Antonio de, 91
Nasca, Peru: desert markings at, 151-161 passim, 383
Navajo Indians, 87, 89-107; astronomy of, 5, 6; cave paintings of, 57; lore of, 63; pictographs of, 91, 93, 102; planetaria of, 92-95, 99-107; symbols of, 57
Navajo Canyon, 87
Navajo Reservoir District, 95
Nazca, 234. See also Nasca
Neolithic period, 134-135, 136, 218

Newark Mound Group, 227, 228
New Mexico, 71, 212
Nezahualpilli, 3
night: glyph for, 286, 314
Night People, 85
Night Sun, 273
Nile, 136, 138, 141, 147
Niwan katcina ceremony, 68
Nine Lords of the Night, 14
Nohpat (site), 183-184, 186, 188, 193
north, glyph for, 315. See also cardinal directions
North Star, 85, 97, 101, 104, 384. See also Polaris
numerology, 30, 280-282
Nunnery Quadrangle, 187, 188

Oaxaca, 164, 172
observatories, 142, 172; lunar, 208; solar, 34, 42, 176; types of, 191-204, 218. See also markers; markings
Ohio: Adams County in, 221; earthworks of, 218; Green County in, 224; Hamilton County in, 223, 227, 230; Newark Mound Group in, 227, 228; Ross County in, 222, 225; Serpent Mound in, 220-233 passim; Valley, 220, 233
Old Fire God, 10, 23
Olmec culture, 341, 343-, 370-371, 380, 381
Olmec mosaic pendant. See Las Bocas mosaic pendant
Opet (Egyptian deity), 138
Ophiuchus (constellation), 116, 119
Oraibi (Hopi), 72
orientations, 35, 36, 41, 42, 85, 163-189 passim, 191-204, 214, 382. See also alignment
Orion, 13, 26, 27, 73, 74, 85; Belt of, 33, 74, 220, 226; as deity, 83; in Hopi, 57, 73; in mounds, 225; in Navajo, 57, 97, 101, 102; in Paviotso, 220; in petroglyph, 122; in Pueblo, 63; in Yavapai, 220
Outside Chief (Puerblo), 71, 72

Palace of the Governor. See Governor's Palace
Palacio tower, 208
Palenque (site), 208
Pamurti ceremony, 67

Papagos Indians, 61
Paris Codex, 28, 29
Patiabu, 81
Paviotso Indians, 220
paw-prints, 222, 223, 226, 227
Pegasus, 122, 124
Pekwin (Zuni), 72, 77, 78, 79, 83, 211. See also priests
Penasco Blanco, 40, 59, 60
Perseus, 116, 120
Peru, 209, 384; Nasca, 151, 234
Peten District, 166
petroglyphs: at Chaco Canyon, 37, 54, 59, 60, 67, 74, 86-87; in Great Basin region, 109, 110, 113, 114, 130. See also glyphs; pictographs; star paintings
pictographs, 54, 57, 76, 87; at Chaco Canyon, 56; in Fern Cave, 50-53; Navajo, 91, 93, 102; Serpent Mound as, 230; at Symbol Bridge, 53, 55. See also cave paintings; glyphs, petroglyphs; sandpaintings
Pieuris (culture), 209
Pidgeon's De-coo-dah, 219
Piedras Negras, 193, 200, 203
Pima Indians, 61
Pinching Stars (constellation), 101
Pisces, 119, 122
planetaria (Navajo), 92-95, 99-107
planets, 19-22, 257, 259. See also individual planets by name
Plateau Anasazi, 97
Plataforma Adosada, 150
Pleiades, 13, 23, 24, 25, 33, 63, 83, 85, 169-170, 171, 300, 335, 380, 382; alignments to, 182, 189; in Hopi, 57, 73, 74; in Madrid Codex, 304; 319, 322, 339; in mounds, 220, 221, 225, 226; in Nasca, 154, 155; in Navajo, 57, 96, 97, 101, 102; at Teotihuacan, 168-169
Polaris, 101, 173, 220, 228-233 passim. See also pole star
pole star, 220, 229, 232. See also North Star; Polaris
Pollux, 226
Pomo Indians, 221
Popol Vuh (deity), 271
Poverty Point, Louisiana, 227
Powamu ceremony (Hopi), 66-67, 72, 83
prayer sticks, 61, 64, 65, 78, 211
precession (of celestial objects), 37, 110, 112, 158, 169, 214, 290
Priestess of Fecundity, 78

Index

priests, 4, 14, 64, 71-74, 82, 214; astronomer-, 244, 256, 340, 350; Aztec, 4, 14, 172; calendar, 86, 315, 324; Chorti, 13; Maya, 244; Pueblo, 75, 206; Zuni, 207. See also medicine man; shamans
Ptah (Egyptian deity), 147
Puebla, Mexico: Las Bocas pendant in, 341-373
Pueblo Indians, 33, 60-64 passim, 71, 84, 85, 209, 212, 213; alignments by, 34; calendar of, 59, 63, 65-70; ceremonies of, 63, 82; marker of, 87; orientations of, 85; priests, 75, 206; sun conception of, 80-81; symbols of, 59, 74, 75, 76. See also Chaco Canyon; Zuni
Pueblo Bonito, 43, 75; A, 35, 37, 39, 41. See also Chaco Canyon
Puerto Rico, Mexico, 182, 183, 193
Pyramid of the Sun, 168, 381. See also Teotihuacan

Quetzalcoatl, 10, 28, 72, 266, 271, 279, 280, 282; Kukulcan, 20
Quillatzin, 280
Quillaxtli, 278, 279
Qintana Roo, Mexico (site), 284
Quito, 245

Ra (Egyptian deity), 138, 139, 143, 145. See also Amon-Ra; Ra-Horakhety
"Rabbit Tracks" (constellation), 102
Ra-Horakhety (Egyptian deity), 138, 141-142, 143, 147, 149, 150, 152. See also gods, Egyptian
Rameses II, 138, 147, 150
Rameses III, 138, 139
Regulus, 156
reliefs, 92, 193. See also glyphs; rock art
"Revolving Female," 101, 104
"Revolving Male," 101
Right Ascensions (in Madrid Codex), 287, 289, 290, 292, 295
Ring Nebula, 113
Rio Grande: designs in, 76; kivas in, 86; observations on, 80; Pueblo (people of), 71, 73, 87
ritual calendar. See calendar
rock art, 40-41, 46, 48, 52, 53, 57, 58, 95, 383. See also glyphs

Ross County, Ohio: Stoneworks in, 222, 225

Sacred Almanac of 260 days, 281
sacred places (Navajo), 92, 97, 107. See also markers
Sacred Round, 286, 308, 328
Sagittarius, 120
Sahagun (Spanish historian), 4, 9, 18, 22, 23, 24; constellation designated by, 25-26; Historia General de las Cosas de Yucatan (The Florentine Codex) by, 211
St. James, cross of (constellation), 23
Sandia pueblo, 85
sandpaintings, 97, 98, 101, 102. See also glyphs
San Juan's Day, 82
Santa Ana (site), 71
Santa Elena Poco Uinic, 325-327
Sapawe (site), 73, 74
Saturn, 257-262
Scorpio (constellation), 23, 26, 101, 102, 109, 120, 122, 130, 380
Sechat (Egyptian deity), 141
Serpent Mound, 220, 221, 227, 228, 230, 231, 232, 233
serpents, 61, 229, 276-277, 290, 300, 310, 335-336, 381; calendar, 270, 295, 296; Chicchan, 285, 335, 337-338; on deer bone, 334, 335; deity, 337; glyph of, 122; in Madrid Codex, 285, 295, 300, 304, 319, 322, 324, 328, 331; in mounds, 228-229, 232; pictures of, 286-289; in Tajin Ballcourt panels, 263, 266; as Venus, 263. See also Feathered Serpent; Fire Serpents; snakes
Serpent Throne, 276-277
Seti II: temple of, 138
seven: in Maya, 259; as Orion, 73; in Venus numerology, 280-282
Shabik'eschee: village of, 74
Shalake celebration (Zuni), 64, 66, 76, 80, 83
shamans, 113, 279, 343. See also priests
Sherente (tribe), 169
sidereal periods, 258-262
sidereal year, 286, 289, 290, 313, 314, 337
singers, Navajo, 5. See also priests

Sioux, glyphs of, 57
Sirius, 176, 177, 189, 375, 376
sky gods, 63, 72, 74, 82, 84, 136; representations of, 97, 263, 265, 308. See also Star God
sky watching, 84, 213. See also sun watching
Slim First One (constellation), 101
Snake-Antelope ceremony, 68
snakes: on mounds, 222; rattles of, 266, 271, 300, 319, 322. See also serpents
Snake society, 68
so'ahots'i'i (constellation), 101
solar calendar, 11, 13, 17, 207
Solar Eclipse Warning Table, 247-256
solar observations, 13, 86. See also sun watching
solar system: creation of, 10
solar year, 286, 289, 293, 311, 314
solstices, 13, 33, 34, 40, 64, 72, 73-74, 78, 82, 83, 84-85, 142, 193, 211, 324, 380; alignments to, 176, 200, 213; in calendars, 70, 77, 141, 207; ceremonies, 33, 66, 68, 69, 76, 79, 81, 83, 211-212; as four directions, 13; markers of, 37, 38, 40, 42, 176; sunrise and sunset values for, 39; symbols of, 62, 116-117, 289, 300, 324, 328. See also sun
Sothis (Egyptian deity), 375
Sound of Jura, 208
Southern Cross, 13, 27
Spider Rock, 91, 95, 97
Spring Canyon, 89, 93
star ceilings, 92-95, 98, 102. See also stars
Star Chief, 73
Star God, 63, 72, 73, 82. See also sky gods
stars, 73, 293, 299; alignments to, 34; in calendar, 22, 289, of cardinal directions, 85; clusters, 220; color of, 104; gazing, 5, 22, 34; godesses, 279; in glyphs, 109-130, 257; in mounds, 226; in mythology, 22, 96; in Navajo, 5, 92-95, 98; paintings of, 93-99, 102, 104; in Pueblo, 73, 75; symbols of, 54, 61, 62, 63, 73, 75, 93-104 passim, 334, 335; treatment of, 220, 222-223, 244. See also star ceilings
Star-watcher, 82. See also sun watching
Stela 3: of Caracol, British Honduras, 257-262; of Santa Elena Pocouinic, Mexico, 325-327
stellar bands, 28. See also stars, paintings of
Stonehenge, 131, 132, 134-136, 207, 208
Street of the Dead, 168, 169. See also Teotihauacan
Stucco Palace (at Acanceh), 165
sun, 14, 22, 63, 72, 74, 75, 78, 84, 217, 264, 268, 273, 274, 275, 383; alignments to, 139, 154, 180, 181, 188; azimuths, 34, 42; ceremonies, 84; cult of, 81; cycles of, 33, 264-266; deities of, 10, 83, 211-212; Father Sun, 63, 74; glyphs of, 62, 73, 286, 289, 314; in Hopi, 57, 64, 266; in Maya, 10, 62, 281; in mythology, 14, 211-212; in Navajo, 98, 101; observations of 84; observatory, 42, 86; in petroglyphs, 54; priests, 77; in Pueblo, 59, 63, 71, 80, 81, 213; shrine, 79, 80, 81; symbols of, 10, 40, 41, 54, 59, 61, 73, 74, 75, 80, 81, 87, 116, 263, 264, 268-273; 281; watcher, 33, 65, 68, 71-74, 75, 81, 82, 213; worship, 137; zenith passage of, 13, 170, 173, 176, 177; in Zuni, 62, 78, 86, 211-232
Sun Chief, 72, 77, 81
Sun Father, 78
Sun Man, 81
sun watching, 33, 61, 62, 64, 65, 67, 68, 71-74, 75, 81, 82, 87, 213
Sun Youth, 64, 66, 67
supernova, 47, 58, 59, 63-64; in glyphs, 40, 46, 50, 52, 53, 54, 57
Symbol Bridge Site, 53, 55

Taharqa (Egyptian), 174
Tajin (Mexico), 1, 275-281 passim; Ballcourt panels at, 263-274, 277, 278, 383
Takelet II (pharaoh), 143
Tancah (site), 284

Index 435

Tanoan pueblo, 80
Taurus, 48, 109, 154, 169, 300
Temple Wood, 218
Tenayuca, 166, 168, 170, 171, 381
Tenochtitlan (site), 172
Teotihuacan, 10, 14, 40, 158, 166-171 passim, 192, 209, 214, 382
Teotihuacan Mapping Project, 166-168, 210
Tepozteco, House of, 168, 170, 171, 382
Tepoztlan, 168
Tewa, 33, 71, 73, 74. See also Hopi-Tewa
Tezcatlipocas (deities), 10, 28
Tezozomoc, 23, 25, 27
Thebes, 138-151, 162
Thom, Alexander, 384
Thunder Mountain, 77, 78, 79
Thutmosis III (pharaoh), 138, 141
Tiahuanaco (site), 384
Tikal (site), 210, 259, 260
Tizimin (site), 281
Tlacaxipeualistli, festival of, 172
Tonactecuhtli (goddess), 279
Tonatiuh (Aztec), 10
Topiltzin, Prince Ce Acatl, 278, 280
Torquemada, 3
Tosodio Canyon, 96
Town Chief, 71, 73, 78, 82, 83
Toxiuhmolpilia feast, 24
Tremper Effigy, 226, 227, 229
Tres Zapotes, 370
Tula, 166-171 passim, 279, 280
Tule, 282
Turner mounds, 227, 230
turtle glyphs, 303, 324, 328, 330, 331
Twin War Gods, 63, 72-73, 74
260-day count, 9, 10, 14, 19, 20, 255, 281, 305, 308, 319, 324, 328, 340, 366, 380; in Borgia Codex, 276; in Madrid Codex, 285, 286, 300, 316, 321, 333
Tzeltal (Indians), 27

Uaxactun, 176, 192, 193, 208
Underworld, 268, 270, 271, 273, 279, 281; conception of, 7, 8, 28. See also Hyades; Lower Regions
UNESCO engineers, 147
Universe, conception of, 10, 69, 187
Ursa Major, 231, 232, 284, 305, 380; in Madrid Codex, 296, 321, 328; in mounds, 226, 229, 233, 234; in Navajo, 101, 104; in petroglyphs, 116, 119, 120, 124
Ursa Minor, 220, 228-229, 233; in Madrid Codex, 296; in Navajo, 101; in petroglyphs, 114, 116, 117
Uxmal, 165, 172-187, 188, 193

Vague Year, 11-13, 19, 20, 24-25, 30, 66, 259
Valley of Mexico, 166
Vaticanus A Codex, 6, 7, 8
Vega, 113
veintena, 11, 12, 15, 30
Velarde region, 212
Vela supernova, 48
Venus, 8, 19-20, 22, 54, 245-247, 257, 260-262, 382; alignments to, 180, 181, 184, 185, 189; in Ballcourt panels, 263-275 passim; calendar, 63, 75, 180, 281; cycle of, 274, 275, 276, 277; in glyphs, 257; gods of, 21, 277, 279; legend of, 266, 268, 277-280, 383; in Maya, 30; numerology, 280-282; Revolution (VR), 280-281; symbols of, 266; table, 30, 237, 247, 256. See also Evening Star; Morning Star
Viking Group (Teotihuacan), 168
Village of the Great Kivas, 56, 86
Virgo, 120

Walum Olum, 222
War Captain, 82
War Chief, 87. See also War Priest
War God(s), 73, 77, 78, 79
War Priest, 71, 72, 73, 82. See also Outside Chief; War Chief
War Star God, 72. See also Star God
White Mesa (site), 87
Wild Cherry Canyon, 89
Wind God, 86
Wisconsin: effigy mounds of, 218, 227, 231
Woodland period, 218, 219
world, conception of, 69. See also heavens, coneption of; Universe, conception of
Wuwuchim, 66

Xochipilli (deity), 268
Xochiquetzal (goddess), 263, 266
Xolotl (god), 271
Xonecuilli (constellation), 23, 26, 27

Yacatecuhtli (deity), 28
Yagul (site), 165
Yavapai Indians, 220, 222, 226
Yaxchilan (site), 200
year: Pueblo, 78; sidereal, 286, 289, 290, 313, 314, 337; solar, 286, 287, 293, 311, 314; tropical, 258, 260, 262; Zuni, 70. <u>See also</u> "lesser year"; 260-day count; Vague Year
yeis (Navajo), 93, 95, 96, 99
Yucatan, 66, 164, 178, 238, 290, 306, 307, 310, 312, 315, 331, 332, 336
Yucatec Maya, 12, 27, 300, 322

Zapotec, 14
zenith passage, 9, 13, 291, 386; alignment to, 150; of Pleiades, 24; of sun 13, 170, 173, 176, 177
Zia (Pueblo), 80, 85
Zocalo (site), 166
Zodiac, 28-30
Zuni Indians, 70, 76, 77, 80, 84, 87, 209; calendar system of, 70, 207, 212; mythology of, 211-212; orientations by, 85; petroglyph at, 54, 59; priests, 207; Shalake celebration of, 64, 66, 76, 80, 83; sun tower of, 86, 212; sun watching by, 62; symbols of, 85; Valley, 76; year at, 70. <u>See also</u> Pekwin